Cytopathology

Edited by

Behdad Shambayati
Ashford and St Peter's Hospitals Foundation Trust

OXFORD
UNIVERSITY PRESS

OXFORD
UNIVERSITY PRESS

Great Clarendon Street, Oxford OX2 6DP

Oxford University Press is a department of the University of Oxford.
It furthers the University's objective of excellence in research, scholarship,
and education by publishing worldwide in

Oxford New York

Auckland Cape Town Dar es Salaam Hong Kong Karachi
Kuala Lumpur Madrid Melbourne Mexico City Nairobi
New Delhi Shanghai Taipei Toronto

With offices in

Argentina Austria Brazil Chile Czech Republic France Greece
Guatemala Hungary Italy Japan Poland Portugal Singapore
South Korea Switzerland Thailand Turkey Ukraine Vietnam

Oxford is a registered trade mark of Oxford University Press
in the UK and in certain other countries

Published in the United States
by Oxford University Press Inc., New York

British Library Cataloguing in Publication Data

Data available

Library of Congress Cataloging in Publication Data

Data available

Typeset by MPS Limited, a Macmillan Company
Printed in Italy on acid-free paper by
L.E.G.O. S.p.A.

ISBN 978-0-19-953392-3

1 3 5 7 9 10 8 6 4 2

Cytopathology

TRANSFUSION &
TRANSPLANTATION SCIENCE

edited by Robin Knight

BIOMEDICAL SCIENCE
PRACTICE

EXPERIMENTAL & PROFESSIONAL SKILLS

edited by Hedley Glencross, Nessar Ahmed,
Chris Smith & Gwyn Wong

CYTOPATHOLOGY

edited by Behdad Shambayati

CLINICAL BIOCHEMISTRY

edited by Nessar Ahmed

IMMUNOLOGY

edited by Angela Hall & Christine Yates

HAEMATOLOGY

Andrew Blann, Gavin Knight,
& Gary Moore

MEDICAL
MICROBIOLOGY

edited by Michael Ford

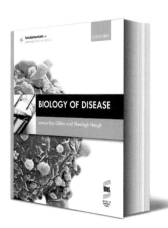

BIOLOGY OF DISEASE

edited by Roz Gibbs and Sheelagh Heugh

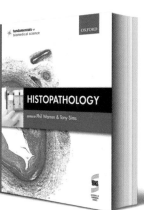

HISTOPATHOLOGY

edited by Phil Warren & Tony Sims

fundamentals OF
biomedical science

'From cradle to grave, strive in search of knowledge'
Saadi (1184–1283) Persian poet from Shiraz

Contents

An introduction to the Fundamentals of Biomedical Science series

Biomedical Scientists form the foundation of modern healthcare, from cancer screening to diagnosing HIV, from blood transfusion for surgery to food poisoning and infection control. Without Biomedical Scientists, the diagnosis of disease, the evaluation of the effectiveness of treatment, and research into the causes and cures of disease would not be possible.

However, the path to becoming a Biomedical Scientist is a challenging one: trainees must not only assimilate knowledge from a range of disciplines, but must understand—and demonstrate—how to apply this knowledge in a practical, hands-on environment.

The *Fundamentals of Biomedical Science* series is written to reflect the challenges of biomedical science education and training today. It blends essential basic science with insights into laboratory practice to show how an understanding of the biology of disease is coupled to the analytical approaches that lead to diagnosis.

The series provides coverage of the full range of disciplines to which a Biomedical Scientist may be exposed – from microbiology to cytopathology to transfusion science. Alongside volumes exploring specific biomedical themes and related laboratory diagnosis, an overarching Biomedical Science Practice volume provides a grounding in the general professional and experimental skills with which every Biomedical Scientist should be equipped.

Produced in collaboration with the Institute of Biomedical Science, the series

- Understands the complex roles of Biomedical Scientists in the modern practice of medicine.

- Understands the development needs of employers and the Profession.

- Places the theoretical aspects of biomedical science in their practical context.

Learning from this series

The *Fundamentals of Biomedical Science* series draws on a range of learning features to help readers master both biomedical science theory, and biomedical science practice.

CASE STUDY 10.2

A 66-year-old man who had recently emigrated from Vietnam to the UK presented with fever, shortness of breath, and **haemoptysis** (coughing up blood) to the A&E department. He was admitted for observation. When giving his medical history he also mentioned that he had suffered from diarrhoea for a long time. During his stay in hospi-

humans. It is a rare infection is prevalent in societies whe faecal matter.

The larvae exist in two form larvae, and the free-living rh which lives in the soil. The

Case studies illustrate how the biomedical science theory and practice presented throughout the series relates to situations and experiences that are likely to be encountered routinely in the biomedical science laboratory.

Additional information to augment the main text appears in **boxes**

BOX 5.3 Alternative cytological terminology for cervical precancer

The World Health Organization (WHO) advises the use of the terms **dysplasia** and **carcinoma *in situ*** to describe the cytological features of cervical precancer, and this system is in use in several countries. In the United States the Bethesda terminology is employed. This is a two-tier grading system using the terms low-grade squamous intraepithelial

Method boxes walk through the key protocols that the reader is likely to encounter in the laboratory on a regular basis.

METHOD Epidemiological methods

One of the purposes of epidemiological studies is to establish disease causation by comparing disease frequency with the prevalence of suspected risk factors in the environment. There are several ways of doing this.

Geographical comparisons look for an association between the distribution of a disease and the presence of suspected

environmental causes of di... migrants in USA have lower... than native Japanese. This s... mental cause of the disease.

Longitudinal studies are diffe... that they are prospective, tha...

Health & Safety boxes raise awareness of key health and safety issues related to topics featured in the series, with which the reader should be familiar.

HEALTH & SAFETY

Categories of hazard in the cytology laboratory

Microbiological hazards

Every clinical specimen submitted to the cytology laboratory must be regarded as potentially infectious and should be handled with great care by well-trained staff. Drivers, porters, and nurses delivering specimens to the laboratory, as well as scientific and clinical staff receiving them, should be made aware of the potential dangers inherent in handling these specimens.

Chemical hazards

Chemicals can be corrosive, irritant, poisonous, carcinogenic, teratogenic, narcotic, flammable,

Key points reinforce the key concepts that the reader should master from having read the material presented, while **Summary** points act as an end-of-chapter checklists for readers to verify that they have remembered correctly the principal themes and ideas presented within each chapter.

glandular abnormalities that can be detected cytologically, albeit with less accuracy than for squamous lesions.

Key Points

The cervical screening test is not ideal for detecting endocervical glandular lesions. In fact, the main intention of cervical screening programmes is to reduce the burden of squamous cell carcinoma.

Key terms provide on-the-page explanations of terms with which the reader may not be familiar; in addition, each title in the series features a **glossary**, in which the key terms featured in that title are collated.

...as type I and type II pneumocytes. Type I pneumo-...ss a large surface and cover the areas where gaseous...ytes produce **surfactant**, which has a dual function of...breathing cycle and protecting the lungs from inju-...bodies and pathogens. Type II pneumocytes also act as...an differentiate into type I pneumocytes when they

Surfactant
A lipoprotein secreted by alveolar cells, its function is to reduce the surface tension of fluids in the lung.

...s histiocytes) are macrophages found in the pulmo-...eir role is engulfing, digesting, and removal of inhaled...ulate lymphocytes and other immune cells to respond

Self-check questions throughout each chapter and extended questions at the end of each chapter provide the reader with a ready means of checking that they have understood the material they have just encountered. Answers to these questions are provided in the book's Online Resource Centre; visit www.oxfordtextbooks.co.uk/orc/shambayati/

SELF-CHECK 10.7

Why is staging of lung cancer important?

10.8 **Cytology of lung cancer**

Squamous cell carcinoma

Cross references help the reader to see biomedical science as a unified discipline, making connections between topics presented within each volume, and across all volumes in the series.

...sinophils. They are seen in patients with allergic disor-...umonia (Figure 10.10 and Case Study 10.1).

...n) or **asbestos bodies** are formed when filamentous...iron. This is often due to inhalation of asbestos (see...estos) where small asbestos fibres are enveloped by...s bodies are often called asbestos bodies.

...ments of plant tissue and meat fibres are sometimes...tissue has a characteristic rectangular shape (Figure...be recognized by the presence of cytoplasmic cross

Cross reference
Case Study 10.1.

Cross reference
Chapter 9, Section 9.5.

Online learning materials

online resource centre

The *Fundamentals of Biomedical Science* series doesn't end with the printed books. Each title in the series is supported by an Online Resource Centre, which features additional materials for students, trainees, and lecturers.

www.oxfordtextbooks.co.uk/orc/fbs

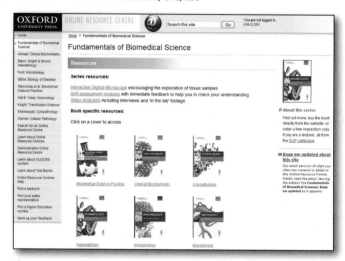

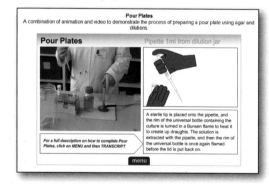

Guides to key experimental skills and methods

Multimedia walk-throughs of key experimental skills—including both animations and video—to help you master the essential skills that are the foundation of Biomedical Science practice.

Biomedical science in practice

Interviews with practising Biomedical Scientists working in a range of disciplines, to give you valuable insights into the reality of work in a Biomedical Science laboratory.

Digital Microscope

A library of microscopic images for you to investigate using this powerful online microscope, to help you gain a deeper appreciation of cell and tissue morphology.

The Digital Microscope is used under licence from the Open University.

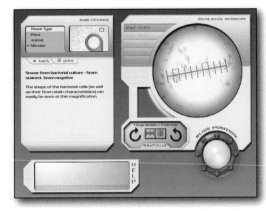

'Check your understanding' learning modules

A mix of interactive tasks and questions, which address a variety of topics explored throughout the series. Complete these modules to help you check that you have fully mastered all the key concepts and key ideas that are central to becoming a proficient Biomedical Scientist.

We extend our grateful thanks to colleagues in the School of Health Science at London Metropolitan University for their invaluable help in developing these online learning materials.

Answers to self-check questions and tips for responding to discussion questions

Answers to questions posed in the book are provided to aid self-assessment.

Lecturer support materials

The Online Resource Centre for each title in the series also features figures from the book in electronic format, for registered adopters to download for use in lecture presentations, and other educational resources.

To register as an adopter visit **www.oxfordtextbooks.co.uk/orc/shambayati/** and follow the on-screen instructions.

Any comments?

We welcome comments and feedback about any aspect of this series.
Just visit **www.oxfortextbooks.co.uk/orc/feedback/** and share your views.

Acknowledgements

I am always thankful to my mum and dad who have supported me from far away.

This book could not have been written without the help of many colleagues. My sincere thanks go to the chapter authors for the patience and commitment they have shown throughout this project.

I would like to thank the following colleagues at Ashford and St Peter's Hospital for critique of the text: Dr Salih Ibrahim, Consultant Pathologist, and Mr Krishnar Patil, Consultant Urologist, for reviewing Chapter 8; Dr Michael Wood, Consultant Respiratory Physician, for his valuable suggestions on Chapter 9; Dr Paul Murray, Consultant Respiratory Physician, for valuable remarks on Chapter 10 and Mr Martin Thomas, Consultant Surgeon, for his detailed comments on clinical aspects in Chapter 11.

I would also like to express my gratitude to Dr Bridget Wilkins, Consultant Pathologist at Guy's and St Thomas' NHS Foundation Trust, for her critical comments on Chapter 13.

I would like to thank Public Health Wales for granting permission to use photo micrograph of cytological preparations for Chapters 4 and 5.

I am indebted to Dr Michael Oatey, a good friend, who gave us invaluable advice on writing style.

Many people have helped along the way: I would particularly like to thank Mrs Beverley Crossley and Mrs Janice Tew for proof reading many of the chapters.

I am also grateful to Jonathan Crowe, Editor in Chief, Natural, Health, and Clinical Sciences at Oxford University Press, for his patience and encouragement.

And last, I would like to express my appreciation to my family who put up with many lost weekends.

Behdad Shambayati
September 2010

Contributors

Viv Beavers
Consultant Biomedical Scientist
Cytology Dept
Victoria Hospital
Blackpool, Wyre, and Fylde NHS Trust
Blackpool

Stephen Blackman
Senior Biomedical Scientist
Cytopathology Dept
Ashford and St Peter's Hospitals Foundation Trust
Chertsey
Surrey

John Crossley
Advanced Biomedical Scientist Practitioner in
Cervical Cytology
Cytology Department
Royal Hallamshire Hospital
Sheffield

Andrew Evered
Consultant Biomedical Scientist Public Health Wales
and Principal Lecturer in Biomedical Science,
University of Wales Institute Cardiff (UWIC)

Margaret Morgan
Consultant Biomedical Scientist/Head of Service
Tissue Sciences
GSTS Pathology
St Thomas' Hospital
London

Behdad Shambayati
Consultant Clinical Cytologist
Cytopathology Dept
Ashford and St Peter's Hospitals Foundation Trust
Chertsey
Surrey

Allan Wilson
Consultant Biomedical Scientist
Monklands Hospital
Monkscourt Avenue
Airdrie

Online materials developed by

Behdad Shambayati
Consultant Clinical Cytologist
Cytopathology Dept
Ashford and St Peter's Hospitals Foundation Trust

Dr Ken Hudson
Lecturer in Biomedical Sciences
Faculty of Human Sciences
London Metropolitan University

Abbreviations

ABA	Association of Biomedical Andrologists
ABC	Avidin-Biotin complex
aCGH	Array comparative genomic hybridization
AIDS	Acquired immune deficiency syndrome
ALL	Acute lymphocytic leukaemia
BAC	Bronchoalveolar cell carcinoma
BAL	Bronchoalveolar lavage
BAS	British Andrology Society
BCG	Bacillus Calmette-Guerin
BDFP	BD FocalPoint GS Imaging
BSCC	British Society for Clinical Cytology
BSCCP	British Society for Colposcopy and Cervical Pathology
CGH	Comparative genomic hybridization
CGIN	Cervical glandular intraepithelial neoplasia
CIN	Cervical intraepithelial neoplasia
CIS	Carcinoma *in situ*
CISH	Chromogenic *in situ* hybridization
CLL	Chronic lymphocytic leukaemia
CMV	Cytomegalovirus
CPA	Clinical Pathology Accreditation (UK)
CT	Computerized tomography
DAB	3,3'-diaminobenzidene
DAPI	4',6-diamidino-2-phenylindole
DNA	Deoxyribonucleic acid
DTT	Dithiothreitol
DWP	Department for Work and Pensions
EA	Eosin azure
EBUS	Endobronchoscopic ultrasound
EBUS-TBNA	Endobronchial ultrasound guided transbronchial needle aspiration
EGFR	Epithelial growth factor receptor
EQA	External quality assessment/assurance
ER	Oestrogen receptor
EUS	Endoscopic ultrasound
EUS-FNA	Endoscopic ultrasound-guided fine needle aspiration

FCA	Flow cytometry analysis
FDA	United States Food and Drug Administration
FDG	Fluorodeoxyglucose
FISH	Fluorescent *in situ* hybridization
FNA	Fine needle aspiration
FNAB	Fine needle aspiration biopsy
FNAC	Fine needle aspiration cytology
FOV	Fields of view
FRET	Flurochrome resonance energy transfer
FSH	Follicle stimulating hormone
GI	Gastrointestinal
GIST	Gastrointestinal stromal tumours
GnRH	Gonadotropin releasing hormone
GSK	GlaxoSmithKlein
H&E	Haematoxylin and eosin
HBPC	Hospital based programme coordinator
HC2	Hybrid capture 2
hCG	Human chorionic gonadotropin
HEPA	High efficiency particulate air filter
HER2	Human epithelial receptor
HIV	Human immunodeficiency virus
HNPCC	Hereditary non-polyposis colorectal cancer
HPV	Human papillomavirus
hrHPV	High-risk HPV
HRT	Hormone replacement therapy
HSIL	High-grade squamous intraepithelial lesion
HSV	Herpes simplex virus
HTA	Health technology assessment
HTA	Human Tissue Act
IAC	International Association of Cytology
IAP	Inhibitor of apoptotic proteins
IBMS	Institute of Biomedical Science
ICC	Immunocytochemistry
ICS	Immotile cilia syndrome
IUCD	Intrauterine contraceptive device

IVU	Intravenous urogram		QA	Quality assurance
JCVI	Joint Committee on Vaccination and Immunization		QARC	Quality Assurance Reference Centre
KSF	Knowledge and skills framework		QC	Quality control
LBC	Liquid based cytology		QM	Quality management
LH	Luteinizing hormone		RCOG	Royal College of Obstetricians and Gynaecologists
LLETZ	Large loop excision of the transformation zone		RNA	Ribonucleic acid
LMD	Laser microdissection		RRN	Risk rating number
LOH	Loss of heterozygosity		RS	Reed-Sternberg
LSI	Locus specific indicator		SCJ	Squamocolumnar junction
LSIL	Low-grade squamous intraepithelial lesion		SCLC	Small cell lung cancer
MAR	Mixed antiglobulin reaction		SIL	Squamous intraepithelial lesion
MCM	Minichromosome maintenance		SLE	Systemic lupus erythematosus
MGG	May-Grünwald-Giemsa		SOB	Shortness of breath
MLA	Medical laboratory assistant		SOP	Standard operating procedure
NCR	Nucleocytoplasmic ratio		SV40	Simian virus 40
NFR	No further review		TB	*Mycobacterium tuberculosis*
NHAIS	National Health Application Infrastructure Services		TBNA	Transbronchial fine needle aspiration
NHL	Non-Hodgkin's lymphoma		TCC	Transitional cell carcinoma
NHS	National Health Service		TEM	Tuboendometrioid metaplasia
NHSCSP	National Health Service Cervical Screening Programme		TEQA	Technical external quality assurance
NPV	Negative predictive value		TFTs	Thyroid function tests
NSCLC	Non-small cell lung carcinoma		TK	Tyrosine kinase
NSE	Neuron specific enolase		TMA	Tissue microarray analysis
OG	Orange G		TNF	Tumor necrosis factor-alpha
OGD	Oesophagogastroduodenoscopy		TOP2A	Topoisomerase II alpha
PAP	Papanicolaou stain		TPIS	ThinPrep® imaging system
PAS	Periodic acid Schiff		TTF1	Thyroid transcription factor 1
PCR	Polymerase chain reaction		TURBT	Transurethral resection of bladder tumour
PCT	Primary care trusts		TZ	Transformation zone
PET	Positron emission tomography		UKNEQAS	United Kingdom National External Quality Assessment Service
PNL	Prior notification list		US	Ultrasound
PPV	Positive predictive value		VATS	Video-assisted thoracoscopic surgery
PR	Progesterone receptor		WHO	World Health Organization
PUNLUMP	Papillary urothelial lesions of low malignant potential		ZN	Ziehl-Neelson

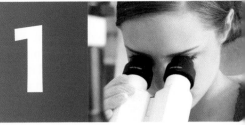

Introduction: a glance at the past, a glimpse of the future

Viv Beavers

Welcome to the fascinating world of cytology. In this chapter you will be taken on a journey starting from the early origins of cytology, when it was a mere curiosity, to the present day where it contributes greatly to the diagnosis of disease and the management of the patient. Cytologists prepare and examine a whole range of specimens. In any working week a cytologist might screen many cervical samples to detect the earliest signs of cancer, analyse semen specimens for infertility analysis, identify crystals in synovial fluids to help in the diagnosis of joint disease, and might even be called to a hospital ward to assist in the preparation of a needle aspirate at the bedside. The main tool of the cytologist's trade is the light microscope and their primary skill is the ability to visually interpret the **morphology** of cells accurately and consistently.

The ability of an experienced cytologist to make a reliable diagnosis without the help of complex analysers that are so prevalent in other pathology disciplines makes cytology uniquely attractive to those with good visual skills. The decision to embark on a career in cytology should not be taken lightly. It is only through rigorous training and years of experience that a cytologist can truly be called an 'expert'. For those who enjoy hands-on laboratory work there is plenty of stimulation in the specimen preparation laboratory. In addition to undertaking routine staining techniques that involve simple dyes and chemicals, cytologists are adept at complex immunocytochemical procedures and are increasingly becoming involved in advanced techniques that would be more familiar to a molecular biologist.

In this book we will explore all of the things that make cytology such an interesting discipline to work in and give you an introduction to what will be involved if you choose to join this profession. Chapter 2 includes a detailed description of the full range of preparation techniques that are used to extract diagnostic material from what are often quite technically challenging specimens.

Morphology
The study of the form and structure of a cell or organism.

Next, a series of chapters will cover cervical screening and the microscopic features of normal and abnormal cells found in cervical cytology samples, followed by the techniques and specimens of non-gynaecological cytology including semen analysis. We will finish with a chapter on advances in cytopathology, looking at what the future may offer for the profession and those working in it. By the end of this book you will have gained the knowledge that will equip you with the foundations needed for a successful career as a cytologist.

Learning objectives

After studying this chapter you should confidently be able to:

- Describe what a cytologist does.
- Describe how cytology developed as a profession.
- Have a basic understanding of the role of cytology as a screening and diagnostic tool.
- Be aware of recent developments which may impact on the future of cytology.

Screening tool
Testing apparently healthy people for presence of a disease or disorder.

Cervical intraepithelial neoplasia (CIN)
An abnormal growth of tissue in the cervix that is restricted to epithelium. This is a pre-cancerous lesion.

Diagnostic cytology
The discipline involved in examining cells from the symptomatic patient in an effort to diagnose the reason for illness.

Non-invasive technique
Procedures that do not involve cutting or entering a body cavity.

Minimally invasive technique
Medical or surgical procedures that cause least possible trauma to the patient.

Fine needle aspiration (FNA) cytology
The method of extracting small tissue or cell samples from a lesion by using a small gauge needle, with or without suction applied.

Cross reference
Chapters 7–11.

1.1 **What is cytology?**

Cytology is the study of the microscopic appearance of cells. Cytologists are trained to identify the type of cells that are normally present in different areas of the body and to use that knowledge to detect either changes seen in the morphology of the cells themselves or the presence of inappropriate cells that are associated with particular conditions.

Cytology may be used as a **screening tool** in populations of apparently healthy people who are at risk of a particular disease. Cervical cytology screening is performed in order to detect a condition known as **cervical intraepithelial neoplasia (CIN)**, a pre-cancerous lesion that does not produce symptoms, and is often invisible to the naked eye. When found, these affected tissues can usually be treated or removed before they become malignant and life threatening. At the present time, only cervical screening uses cytology as the primary screening tool.

Diagnostic cytology is used in the symptomatic patient, and uses either **non-invasive** or **minimally invasive techniques** in order to obtain cell samples which we can examine and use to make a diagnosis. The specimens are often easy to collect, like urine (Chapter 8) and sputum (Chapter 10), but also include body cavity fluids (Chapter 9) and endoscopy samples, for example bronchial specimens (Chapter 10). Cellular material can also be obtained by using a needle and syringe to extract cells from a variety of sites in the body, a technique known as **fine needle aspiration (FNA)** (see Chapter 11).

Key Points

Cytology screening aims to detect early changes in the cells in the symptomless patient, which may signal the possibility of risk of developing a malignant lesion at a later date.

Diagnostic cytology, on the other hand, is used in the symptomatic patient in an effort to diagnose a condition, already present, which is causing the patient's symptoms.

1.2 **The development of cytology**

The history and development of cytology as a diagnostic discipline cannot be discussed without reference to those individuals who developed and refined the microscope. The conception and evolution of this fundamental tool gave rise to cytology in its most basic form as early as the seventeenth century and led to the extensive, complex science that it is today.

Anton van Leeuwenhoek (1632–1723) developed his own microscope, and could be deemed to be the world's first cytologist. He was the first person to record the appearance of human cells, such as red blood cells and spermatozoa. He also described bacteria, yeasts, and diatoms. Robert Hooke (1635–1703), an English inventor, improved on the design of van Leeuwenhoek's microscope, and is credited with introducing the word 'cell' to the science of biology.

Very little was recorded regarding the morphology of human cells from van Leeuwenhoek's observations in the late seventeenth century until the early nineteenth century. Schwann and Schleiden published their cell theory in 1838.

Matthias Schleiden proposed that all plants were made of cells. Theodor Schwann proposed all life began as a single cell and all animal tissues were composed of cells.

Donne published his work on the cells found in the respiratory tract in 1845 and, almost simultaneously, Mueller published a book in which considerable attention was given to the microscopic appearances of **cancer** cells. Later in the nineteenth century the design, magnification, and clarity of the image was much improved by the American Charles Spencer (1838–1881). The Carl Zeiss corporation produced excellent microscopes around this time, and it was August Köhler (whilst working at Zeiss), who introduced the principle of **Köhler illumination** in 1893.

> **Köhler illumination**
> A form of bright field microscopy that results in two major advantages: even illumination across the field of view and the elimination of glare.

By the early twentieth century, the microscopic appearance of cancer cells from numerous sites of the body had been described. However, little clinical importance was attached to these findings by the medical profession, and cytology as a credible diagnostic tool was not accepted until much later.

George Papanicolaou

One of the most influential individuals in the acceptance and development of cytology as a clinical tool was George Papanicolaou (see Figure 1.1). His extensive work in cytology of the female genital tract in the twentieth century helped cytological diagnosis gain credibility. His work also helped form the basis of numerous successful cervical screening programmes in the developed world, greatly reducing the incidence and mortality of cervical cancer.

George Papanicolaou was born on the island of Evia in Greece in 1883. Following in his father's footsteps, he graduated in medicine at the University of Athens at the age of 21, and rather than going into general medical practice like his father, progressed his career by becoming a physiologist. He later gained a doctorate of philosophy at the University of Munich. At the age of 30, after serving in the Balkan War, and having recently married, he went on to obtain a position at the world famous Cornell University in New York. There he did extensive research into the **oestrous cycle** of animals. He later extended his research to study the human menstrual cycle. During this research he examined vaginal specimens from women who incidentally had **carcinoma** of the cervix, and learned to recognize the morphology of malignant cells from this site.

> **Oestrous cycle**
> Hormone dependent cycle in female mammals involving ovulation and periods of heat.
>
> **Carcinoma**
> Malignancy originating in epithelial tissues. Epithelial tissue is the cellular covering of internal and external body surfaces.
>
> **Fixation**
> This preserves cells or tissues in a lifelike form for microscopical examination and renders them resistant to putrefaction and bacterial growth. A fixative is an agent that is used for this purpose.

During this research, Papanicolaou realized the value of immediate **fixation** of cellular samples in retaining the fine cellular detail required for accurate morphological evaluation. He also devised a staining procedure using a mixture of dyes.

FIGURE 1.1
George Papanicolaou.
Reproduced with permission. Copyright
© 2010 UPS.

Cross reference

Fixation and staining are discussed in more detail in Chapter 2.

These dyes facilitate the visualization of cells during microscopy and help distinguish between various components of the cell itself, and in this aspect, the Papanicolaou was revolutionary. In addition to enhancing the appearance of the nucleus, the technique stains the cytoplasm of squamous cells different colours depending on their maturity, which in turn depends on the hormonal state of the patient at the time of sampling. Papanicolaou developed this staining technique to estimate the time of ovulation and hence aid gynaecologists in the investigation of a patient's fertility. Although proving useful in those early times, this method of estimation of ovulation has been superseded by more accurate biochemical tests.

Although Papanicolaou is acclaimed for describing the morphological appearances of malignant cells of the cervix, credit has to go to Babes, a Romanian pathologist, who almost simultaneously, in 1928, published his own findings on the subject. Babes' technique for the evaluation of cells did not, however, include the **wet fixation** method employed by Papanicolaou, and thus would not have lent itself to mass screening such as that for the cervical screening programme.

Key Points

The wet fixation method developed by Papanicolaou involved spreading the cells directly onto a glass slide and rapidly flooding them with alcohol ether, which evaporated leaving a dry slide.

In cervical screening programmes, specimens had to be transported to the laboratory, sometimes over long distances. This method of fixation was easy to perform at the bedside, and resulted in specimens that were easily transported and did not deteriorate over time.

Following Papanicolaou's example, a wet fixation technique for the detection of **neoplastic** cells in sputum samples was described by Dudgeon and Wrigley in 1935.

SELF-CHECK 1.1

Why is it useful to stain a cytology sample?

The publication of a book by Papanicolaou and Traut in 1943 contributed much to the acceptance of cytology as a useful diagnostic tool. Publications on new techniques, and others describing the morphological presentations of neoplastic cells from various other sites and differing tumour types, then increased. By the early 1950s diagnosis by cell sampling became an integral, albeit small, part of the diagnostic pathology service.

Papanicolaou published his atlas of exfoliative cytology in 1956. This is a comprehensive reference work containing detailed drawings of the morphological presentation of normal, **dyskaryotic** and malignant cells. The fine detail and accuracy of these illustrations can be appreciated even today, over half a century later.

Neoplastic
A term used to describe conditions where there is an uncontrolled proliferation or multiplication of cells. Neoplastic lesions may be benign, premalignant, or malignant. The word originates from the Greek: *neo* = new and *plassein* = to form.

Dyskaryosis
A term derived from the Greek term *dys* (meaning abnormal). and *karyon* (meaning nut or nucleus). So, dyskaryosis is a morphological abnormality of the nucleus of a cell.

Key Points

The main indictor of neoplasia lies in the morphological appearance of the nucleus. Dyskaryotic cells may come from neoplastic lesions.

Whilst the medical world came to accept the benefits of cytological diagnosis Papanicolaou continued his research unstintingly until his death in 1962 at the age of 79.

The expansion of cytology

As cytology became accepted as a valid diagnostic tool, many professional societies with a particular interest in the subject were established around the world and numerous publications became devoted to the subject. The first cytology society journal to be produced was *Acta Cytologica* by the International Association of Cytology (IAC) in 1957. This publication lists over 50 cytology societies around the world that contribute to its scientific content.

Leopold Koss and Grace Durfee published a groundbreaking book in 1961 which encompassed images and descriptive text of diagnostic cytological samples with the correlating histology of biopsies from the same site. This book has been updated and greatly expanded over the years and is currently composed of two volumes, including a CD-ROM. The British Society for Clinical Cytology (BSCC) was founded in 1961 and first published its official scientific journal, *Cytopathology*, in 1990.

Many factors have contributed to the development of cytology as a medical laboratory discipline; these include better sampling techniques, improvement in diagnostic imaging, and development of techniques that allow specific identification of cells.

Sampling

In the early days, cytological diagnosis relied exclusively on examination of cells that were either **exfoliated** (naturally shed), or forcibly removed from a surface using a sampling device, such as a wooden spatula or scalpel. Whether exfoliated or forcibly removed, the cells could only be sampled from relatively accessible sites within the body.

Endoscope
A lighted tube that is passed into the body and through which the observer can visualize the surfaces within. The word endoscope originates from the Greek: *endon* = within and *skopos* = target.

Cross reference
Chapter 10, Section 10.2.

Cross reference
Chapter 11.

Palpable lesion
One that can be felt by the clinician during manual examination of the patient.

The development of the **endoscope** and the incorporation of small brushes allowed surfaces deep within the body to be sampled. Bronchial brushings using the early rigid tube endoscope device were routinely examined for evidence of malignant cells from the 1950s. This procedure, however, being very uncomfortable for the patient, was mostly undertaken during general anaesthetic. The invention of fibre optics allowed the development of flexible endoscopes that facilitated the sampling of surfaces deep within the body. The flexibility of the instrument, coupled with its small circumference, allows it to be used in an outpatient setting, making an in-patient stay unnecessary in most cases.

The acceptance in the 1980s of **fine needle aspiration (FNA)** sampling of **palpable** lesions added a new concept in cytological diagnosis.

Fine needle aspiration yields cells and small fragments of tissue from deep within the lesion. The obtained sample closely resembles a histological biopsy, albeit on a much smaller scale. For this reason fine needle aspiration cytology (FNAC) specimens are often referred to as fine needle aspiration biopsies (FNAB). To perform fine needle aspiration of a palpable lesion, the clinician must be able to steady the lesion with one hand whilst inserting the fine needle into it with the other. Skill is required in order to manoeuvre the needle in the correct manner for appropriate sampling, prior to the extraction of the needle and the contained sample. This method of diagnostic sampling was widely accepted in other European countries, with the Karolinska Institute in Sweden leading the way, long before its acceptance in the UK.

Imaging

Developments in the field of imaging, such as the use of **ultrasound (US)**, **computerized tomography (CT)**, **mammography**, and **angiography** added an exciting new dimension to the field of diagnostic cytology. The imaging techniques and their inclusion in sampling processes have enabled image-guided fine needle aspiration of deep-seated, previously unreachable lesions to be performed.

Image guided FNA was extensively used in the breast screening programme, contributing much to its success. An imaging technique, for example ultrasound or mammography, enables the sample taker to observe both the lesion and the needle on the screen as it is inserted through the skin. Using the imaging screen as guidance, the operator can accurately position the needle prior to extracting the sample from the lesion. Samples taken in an outpatient setting from internal organs such as the liver are now commonplace. Fine needles can also be incorporated into endoscopic devices, and these, too, can be used to take samples from deep within the body, using image guidance if necessary.

Impalpable lesions
These cannot be felt by the clinician on manual examination of the patient.

Prior to the introduction of image guided FNA, the only way to sample deep-seated and **impalpable lesions** was by open surgery, with the accompanying risks to the patient and a prolonged hospital stay. Where diagnostic cytology is routinely performed, this is rarely the case nowadays.

SELF-CHECK 1.2

How does the fine needle sampling of palpable lesions differ from the sampling of impalpable lesions?

Special stains

During the latter half of the twentieth century cytology embraced developments in preparation and staining methodologies which revolutionized the diagnosis of disease at the tissue level. Special staining techniques that had been used in the histology laboratory for many years were successfully applied to cytology preparations. These techniques highlight specific cellular constituents such as **mucin** which can aid in the identification of tumour type and origin.

> ## Key Points
>
> Constituents of cells may identify the tissue type from which they originated. Mucin is produced by glandular epithelium. If cells from a malignant tumour contain mucin it can be assumed with a high degree of confidence that the tumour originated in glandular epithelium.

Immunohistochemical techniques applied to the cytology preparation add further, often quite specific information regarding the tumour type. A panel consisting of a number of these immunohistochemical markers can yield invaluable information that could never be gained from morphological assessment alone. By utilizing special techniques in addition to the routine stains, cytologists can not only give a diagnosis of malignancy but are able to give more information, such as the tissue of origin and the likely reaction of the tumour to certain treatment regimens.

1.3 Cytology in the late twentieth and early twenty-first century

Cytology has two broad applications in modern clinical practice: as a population screening tool for the early detection of neoplastic conditions of the cervix and as a diagnostic tool to assess patients in whom there is a clinical suspicion of malignant disease. Let us examine the current and future status of these two separate but related applications.

Cervical screening

The development of the national cervical screening programmes in the UK in 1988 was the culmination and coordination of many small, locally organized programmes that were run and funded on a fairly ad hoc basis. The national coordination of these programmes prompted a vast increase in workload in the cytology departments throughout the country. The programmes aim to reduce the incidence and mortality of cervical cancer. Eligible women are invited for initial screening, and then recalled at specific intervals which vary between three and five years. If pre-cancerous cell changes are found on their cervical smear, women are referred to a **colposcopy** department for assessment and treatment before the lesion has a chance to progress.

This increase in workload required the recruitment of many more staff for cytology departments. Mandatory training requirements were established, and examinations for cytology screeners and biomedical scientist staff employed within cytology were developed.

Until fairly recently a cytological diagnosis has relied solely on morphological changes in cells, but is now embracing technological and diagnostic advances. These advances help to refine the diagnosis and add information necessary to determine appropriate treatment options. Cervical screening is undergoing transformation at a rapid rate due to advances in preparatory techniques, greater understanding of the disease process itself, and automation.

By the very nature of screening (testing many to identify the few individuals at risk), cervical cytology necessitates screening numerous slides, the vast majority of which are negative, to detect relatively few positives. This is very labour intensive, and the quest to develop and implement automation in cytology screening began as early as the 1950s, when automated haematological analysers were developed. However, these systems relied upon evaluation of each individual cell. A **conventionally spread** smear contains small tissue fragments composed of numerous cells strongly adherent to each other. Also, the cells are often overlaying each other and obscured by other elements contained within the sample. Therefore these automated systems were not useful for evaluating cervical smear preparations.

Other systems were developed over time, and in the 1990s a computerized system was produced that displayed panels of images onto a monitor screen for review by the cytologist. Although it showed great promise, and by capturing electronic images was perhaps a glimpse into the future of training and standardized assessments, it was withdrawn from the market prior to implementation.

In 1998 in the United States, the **Food and Drug Administration** (**FDA**) gave approval for automated systems to be introduced for cervical screening. These automated systems, however, have proven less sensitive than manual screening within the cervical screening programmes in the UK.

It was recognized that a system of preparation that avoided cell overlap was required to optimize results of automated systems. As a by-product of the development of these automated screening devices **liquid based cytology** (**LBC**) preparatory techniques were developed. LBC methods use automated or semi-automated systems to process the samples contained in the liquid transport medium. The resultant slide is a **monolayer** preparation, facilitating the automated evaluation of the individual cells, but more importantly in the UK, making the microscopic evaluation much easier for the cytologist. In addition, these preparation methods considerably reduce the number of **leucocytes** (white blood cells) and **erythrocytes** (red blood cells) present on the slide. The removal of leucocytes and erythrocytes was seen as a great advantage in the cervical screening programme. Conventional cervical smears often had excessive numbers of these cells which hampered the visualization of the epithelial cells and rendered the sample unsuitable for screening. Consequently, numerous repeat tests had to be taken. In addition to the monetary costs involved in repeat sampling and repeat evaluation, repeat tests often caused unnecessary anxiety for many patients. The removal of excess blood and leucocytes, reducing the number of repeat tests, was the main reason that liquid based cytology was introduced to the screening programmes in the UK.

Key Points

Liquid based cytology involves immediately dispersing the cells in a liquid transport medium. In the laboratory setting with the aid of automated processors a proportion of the total cells present is deposited on the slide. The end result is a thin layer of cells in a prescribed circular area devoid of excess leucocytes and erythrocytes.

For the NHS Cervical Screening Programme, conversion to LBC, which began in 2004, was not an overnight process. All smear takers had to be trained in the new method, which involved a new sampling device and transferring the sample to a vial of liquid fixative rather than directly onto a glass slide. Because LBC gives subtle but often very important morphological differences to some cell types when compared to conventional cytology, all cytologists, from cytology screeners to pathologists, were required to attend courses on the evaluation of LBC samples, and become certified as competent. The last laboratory to convert to LBC was in late 2008, and according to information on the NHSCSP website (www.cancerscreening.nhs.uk), the conversion to LBC reduced the inadequate rate from 9% with conventional cytology to a national rate of 2.5% in 2009.

The monolayer preparation, being a slide with a thin layer of cells within a defined area, is quicker to screen than conventional smears, and this has led to an increase in productivity within laboratories. This, coupled with the reduction in workloads due to fewer repeat tests, has led to laboratory mergers and centralization.

Key Points

The introduction of liquid based cytology into the NHSCSP resulted in a 6.5% reduction (from 9% to 2.5%) in tests considered inadequate for reporting in 2009. This reduced the need for many repeat tests and resultant anxiety for the patient.

Liquid based preparation methods do not require all cells present in the original sample to be deposited on the slide. A random but fully representative portion of the cells present is processed onto the slide, leaving a residual sample which may be utilized for producing additional slides for teaching purposes, or for performing ancillary tests.

The recognition of the importance of high-risk **human papilloma virus (HPV)** in the development of cervical cancer has added another dimension to cervical screening. Combining cervical screening with HPV testing of the residual sample will allow a more scientific approach to the diagnosis of significant lesions, and help in the follow-up regime of those patients that have been treated for premalignant disease. The reduction in the requirement of repeat testing for those patients proving negative for high-risk HPV types will further reduce the workload of the cytology laboratories. Further mergers of small laboratories are expected in the future, as HPV testing continues to roll out across the whole of the screening programmes in the UK.

The **HPV vaccination programme**, introduced in 2008 for girls aged 13–18 in the UK, is expected have an effect on the cervical screening workload when those immunized at the age of 18 in 2008 enter the programme.

Cross reference
Chapter 13, Section 13.3.

The vaccine protects against HPV types 16 and 18, which are the most common types associated with cervical cancer. The vaccine is said to give up to 70% protection against developing cervical cancer, and a reduction in significant lesions requiring treatment or repeat tests is expected in the years to come.

SELF-CHECK 1.3

What changes have occurred since the introduction of LBC and what new developments are likely to have an impact on the cervical screening service in the near future?

Diagnostic cytology

Cervical screening is not the only area of cytology to take advantage of the advances in technology. The LBC preparation method has the same advantages in the diagnostic setting, and is routinely employed in many laboratories for the preparation of many non-cervical specimens suitable for monolayer preparation. Many non-cervical specimens contain blood and leucocytes. Using the liquid based approach eliminates a large proportion of these which, if present in large numbers, mask the microscopical appearance of the pertinent cells. Liquid based cytology specimens also facilitate the preparation of multiple identical slide preparations. These may be used for special stains and immunocytochemical techniques when refinement of the diagnosis is needed.

The development of sophisticated techniques in both the diagnosis of disease and the advances in treatments will inevitably have a profound effect on the cytology service of the future. **Molecular techniques** are used to investigate disease at the most fundamental level, and help in refining the diagnosis in particular tumour types. They are also useful in deciding the most appropriate treatment options in some cases.

Molecular techniques
These investigate the cell at the submicroscopic level. They are used to identify genetic mutations and chromosome abnormalities. They are helpful in refining the diagnosis in particular tumour types and evaluating the probable effectiveness of various treatment options. They are also used to detect genetic mutations which may predispose a person to a particular disease or condition later in life.

The increasing requirement for sophisticated ancillary testing and the drive for cost effectiveness may well signal the end of the small cervical cytology laboratory, but skilled cytologists will still be valued at the local level. In the diagnostic setting, rapid diagnosis and treatment is becoming increasingly important. Fine needle aspiration cytology is likely to remain the preferred method of sampling at sites in the body where core biopsies would subject the patient to risk of excessive bleeding or damage to adjacent tissues.

Enhanced training for non-medical staff may propel more individuals towards more involvement at the clinical level. The cervical screening service has already embraced the role of consultant biomedical scientists. These individuals are highly trained and report abnormal cervical samples and recommend patient management.

Biomedical scientists are also involved in the reporting of normal diagnostic cytology samples such as bronchial specimens, urines, and body cavity fluids, but a consultant equivalent role for the biomedical scientist in diagnostic cytology has yet to be developed.

Multidisciplinary working is becoming increasingly important as new technologies and methods, such as molecular biology, cross traditional discipline boundaries. Newer techniques will forever be on the horizon, to enhance or replace those which are at the moment in their infancy.

As cytologists we will welcome and embrace new developments, and adapt them to suit cytological samples, thus ensuring that the patient will continue to receive the best possible outcome from the least invasive procedure.

 SUMMARY

- It was almost a hundred years after Mueller's 1985 publication on the microscopic appearance of malignant cells, before cytology became accepted as a valuable diagnostic tool.

- George Papanicolaou's work contributed much to the field of cervical cytology and helped form the basis of successful cervical screening programmes around the world, reducing the incidence of cervical carcinoma considerably in screened populations.

- Advances in diagnostic procedures such as endoscopy and imaging, combined with developments in cytology sampling devices and methods such as FNA, enable a timely and cost effective diagnosis to be reached with minimal risk to the patient.

- The adoption of LBC methodology in all areas of cytology has enhanced cytological diagnosis, and reduced the need for repeated sampling in many cases.

- Special staining, immunocytochemistry, and molecular techniques have added refinement to the cytological diagnosis, and contribute significantly to the management of the patient.

 # FURTHER READING

The following chapters expand on and clarify aspects of different organ systems of the body and cytological diagnosis which are merely touched upon in this introductory chapter. You are encouraged to read on.

2

Preparation techniques

Allan Wilson and Andrew Evered

After reading the first chapter you will realize that the great strength of cytology lies in its simplicity of principle and approach. The essence of the cytologist's job is to search for, and as far as possible classify, neoplastic cells in clinical specimens. Unfortunately, the majority of specimens submitted for cytological assessment contain many more normal cells and irrelevant material than neoplastic cells. Cytologists must therefore find ways of concentrating the few neoplastic cells that might be present onto a relatively small area of a microscope slide. This chapter explores in detail the tried and tested specimen preparation techniques that cytologists have perfected over the past few decades. Many of the techniques are wonderfully simple and straightforward. Others are more technically challenging and demand a high degree of skill. Yet others rely on semi-automated modern technology to overcome some of the difficulties we have in handling the wide range of material we receive in the laboratory. Suffice to say that the full repertoire of cytological preparation techniques will challenge the hands and minds of even the most technically gifted cytologist.

Key Points

Cytology preparation is not rocket science. The aim is to transfer a sample of cells from the specimen to the glass slide and to present them to the cytologist's eye in a manner that will facilitate a diagnosis.

For cytologists to play their part in the provision of high quality patient care, the role of the clinician in procuring good quality specimens must not be underestimated. An improperly taken sample will yield useless results—'rubbish in, rubbish out', as the saying goes. It is therefore quite appropriate for us to spend some time at the beginning of this chapter explaining the techniques used by clinicians to obtain the best possible specimens for cytology.

By the end of this chapter you will therefore be quite familiar with collection and preparation techniques for a wide range of specimen types received in the cytology laboratory. You will also have a firm grasp of an array of cell demonstration techniques that cytologists use to reach their diagnosis. Beyond the scientific and technical aspects of cytology preparation is the crucially important area of health and safety. It is extremely important for you to be aware of the potential health risks associated with handling and disposing of clinical specimens. Students often claim this topic to be uninteresting and even dull. You might be tempted to skip the subject altogether, but be warned: *ignore the last section of this chapter at your peril!* Health and safety issues consume the everyday lives of practising biomedical scientists.

Learning objectives

To summarize, by the time you have finished reading this chapter you should be able to:

■ Describe the principles and techniques of specimen collection.

■ Discuss the purpose and mode of action of cytological fixatives.

■ Give examples of the common types of specimens received in the cytology laboratory.

■ Describe the processing techniques that are used to ensure high quality preparations.

■ Outline the staining and other demonstration techniques routinely used in cytology.

■ Describe the health and safety risks of transporting and processing cytology specimens and how these risks are minimized.

■ Discuss the principles of clinical and chemical waste disposal.

2.1 Principles and techniques of specimen collection

In order to describe specimen collection techniques it is useful to remember that the cytologist's main concern is whether the specimen contains **neoplastic** cells.

The term *neoplastic* provides us with the first hint of how best to collect specimens for cytology. One of the best known characteristics of neoplastic cells is their tendency to lose **cohesiveness**, that is, they become less 'sticky'. Cells that lose their stickiness tend to **exfoliate** readily so if a neoplasm involves the surface of a tissue or organ then, as you can imagine, it should be a fairly straightforward job to collect these cells.

> **Exfoliation**
> The shedding of loosely held cells from their parent tissue.

All that is needed is an appropriate device to 'sweep up' the cells. The sweeping up process is the basis of **exfoliative cytology**. On the other hand, if a neoplasm involves a body site that is out of direct contact with a tissue surface then it is unlikely to be accessible using any of the methods of exfoliative cytology. Instead, it is far more sensible to use a needle and syringe to suck the cells away from **suspicious lesions**. This is the basis of aspiration cytology or, more specifically, **fine needle aspiration cytology (FNAC)**.

> **Suspicious lesion**
> A tissue mass that the clinician thinks could be neoplastic.

Now that you understand the very straightforward principles of specimen collection we can look at the various types of collection techniques available to clinicians.

> **Fine needle aspiration**
> The use of a narrow gauge needle (22–27 gauge) and syringe to aspirate a sample of cells from suspicious lesions.

BOX 2.1 Two ways to collect cells

Collection techniques for cytology can be broadly classified into exfoliative and aspiration methods. Exfoliative cytology is the collection of cells that have spontaneously shed into a body cavity or from the surface of a tissue. Aspiration cytology involves the forcible removal of cells from lesions or masses using a needle and syringe. The latter technique is called fine needle aspiration.

Exfoliative cytology

Collection techniques that rely on the harvesting of exfoliated cells may involve **scraping**, **brushing, or washing** body surfaces or cavities, but perhaps the simplest of all techniques is the collection of bodily fluids into which cells have spontaneously **exfoliated**. These fluid products of metabolism are simply collected into a suitable container and sent directly to the laboratory for processing. Two examples are **voided urine** cytology and **sputum** cytology, both of which are probably the earliest examples of the application of cytology in the diagnosis of human disease.

Now we will take a look at the various ways in which cells can be scraped, brushed, or washed from tissue surfaces.

Perhaps the best known application of the scraping technique is the cervical sample, important because of its use in cervical screening.

The devices used to collect cervical samples have changed a number of times over the years but all are designed to collect a representative sample of cells from the **transformation zone** of the cervix, as it is from this area that most cervical neoplasms originate.

Figure 2.1 shows a variety of sampling devices that have historically been used to collect cells from the cervix. These devices include wooden spatulas, plastic spatulas, and an assortment of broom-like instruments.

You will notice that all these devices have one thing in common—*their shape is intended to match the anatomical profile of the cervix*. The main disadvantage of the spatulas is their inflexible shape—they simply cannot accommodate the natural variability in the shape and size of the cervix in the female population. For this reason the wooden and plastic spatulas have gradually given way to the **cervical broom**, the 'bristles' of which can flex as the device is swept over the cervix. Unlike the typical fine bristles of the various brushes described next, the individual bristles of the cervical broom are in fact flexible paddle-shaped scraping devices, as can be seen in Figure 2.2.

The cross section of the 'bristles' on a cervical broom is such that cells are effectively scraped from the cervix as the broom is rotated in a clockwise motion.

Cross reference
Chapters 8 and 10.

Cross reference
Chapters 3, 4, and 5.

Transformation zone
An area where the transformation from columnar epithelium to squamous epithelium occurs.

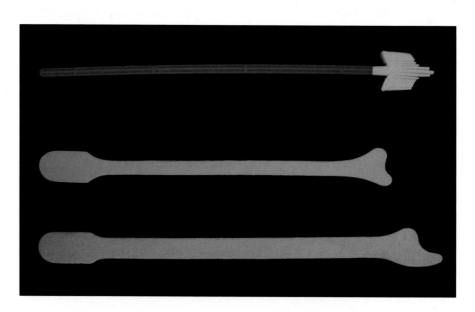

FIGURE 2.1
Cervical broom and spatula.

As cervical sampling devices have evolved there have been huge improvements in the adequacy of samples in terms of collecting *the right types of cells in sufficient quantity*.

Now let's look at an example of a brushing technique. **Bronchial brushing** is a tried and tested technique for collecting cells from the epithelium of the respiratory tract.

The technique is described in detail in Chapter 10; for this section it is enough to state that the sampling device consists of fine nylon bristles that are gently rubbed over the surface of suspicious lesions found during a **bronchoscopy** examination. A bronchial brush is shown in Figure 2.3.

Cross reference
Chapter 10.

Bronchoscopy
Allows the direct visualization of bronchial epithelium using a miniature camera fixed to the end of a flexible tube known as a bronchoscope. The technique also allows the collection of cell and tissue samples.

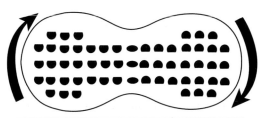

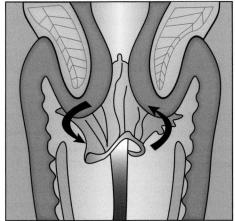

FIGURE 2.2
The cross section of the 'bristles' on a cervical broom is such that cells are effectively scraped from the cervix as the broom is rotated in a clockwise motion.
Reproduced with kind permission from Rovers Medical Devices, Oss, the Netherlands.

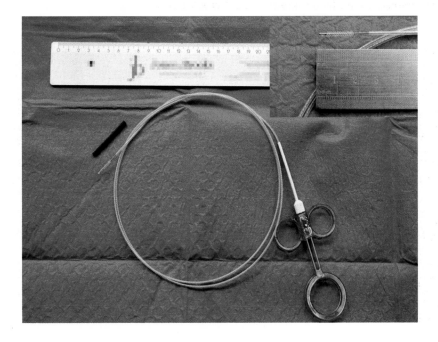

FIGURE 2.3
A bronchial brush.

Finally, there is the washing technique, in which a harmless fluid (usually **physiological saline**) is used to rinse a body surface or cavity in order to dislodge cells that are only loosely bound to their parent tissue.

Following the washing process the rinse fluid is collected into a watertight container and sent to the cytology laboratory for processing. For example, a very useful clinical procedure is the **peritoneal washing**, which is performed under general anaesthetic when a patient is suspected of **cancer** that has spread to the abdominal cavity.

In addition to helping in the initial diagnosis of cancer in the abdominal organs, the presence or absence of cancer cells in a peritoneal washing can also be useful for **staging** cancer.

We will now turn our attention to aspiration cytology, which uses quite a different approach to specimen collection.

Aspiration cytology

In the 1930s a small number of enthusiastic clinicians began to experiment with a technique for sampling suspicious lesions that could not be easily accessed by any of the exfoliative methods described above. The technique involves the use of a fine needle to puncture a suspicious site, followed by several movements of the needle in and out of the lesion. The movement of the needle is sufficient to dislodge cells, which therefore accumulate in the barrel of the needle. Curious as it may seem, suction may not be required to aid this process, but many clinicians prefer to apply negative pressure by attaching a syringe to the needle during the procedure. This technique has therefore been appropriately named fine needle aspiration (FNA). The collected cells are expelled from the needle using positive pressure from a syringe, either directly onto a glass slide for subsequent staining and microscopy, or into **collection fluid** for laboratory processing later. **Fine needle aspiration cytology (FNAC)** is covered in detail in Chapter 11.

FNA is a more invasive technique than the exfoliative methods described above, so under what circumstances should FNA be used? The simple answer is that FNA is useful whenever a specific lesion or mass is 'seen' by the clinician, but is outside the direct reach of the exfoliative sampling techniques described in the previous section. In the early days of FNA, the only lesions that could be sampled were those that could be directly visualized by the clinician (superficial breast lumps, for instance). Nowadays, clinicians have at their disposal an array of **imaging** techniques and specially designed long needles, making aspiration of virtually any body site possible, even bone. Figure 2.4 shows a selection of fine needles.

This book is not the place to go into any of the imaging procedures in great detail. Suffice to say that modern imaging techniques enable real-time visualization and sampling of small deep-seated masses with a clarity that could scarcely be believed just ten years ago. Two imaging modalities used for guiding needles to the correct body site are **ultrasound** (US) and **computed tomography** (CT).

Sampling accuracy and sampling error

The previous discussion of exfoliative and aspiration cytology would be incomplete without a mention of the importance of *adequate sampling* and the concept of **sampling error**.

Peritoneal washing
A procedure where physiological saline is introduced into the peritoneal cavity and then removed by suction. The aspirated fluid is examined for tumour cells.

Cancer
The common term used to describe a neoplasm that grows into the surrounding tissue and spreads to distant parts of the body. The stage of a cancer is the extent to which it has spread.

Cross reference
Chapter 11.

Sampling error
In terms of specimen collection for cytology, sampling error can be defined as the failure to collect cells that are representative of the site being sampled, or to collect them in sufficient quantity to make a reliable diagnosis.

Key Points
The 'in/on' rule: If the cells are not *in* the sample they won't be *on* the slide.

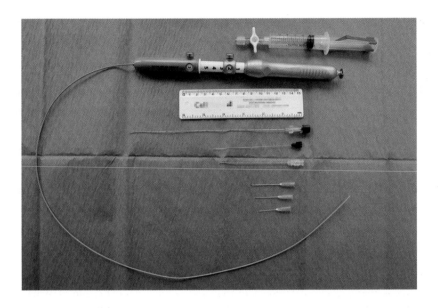

FIGURE 2.4
A variety of needles used for FNA.

What is an **adequate specimen**? Once again, let us remind ourselves that cytologists are generally concerned only with looking for neoplastic cells. A purist would say that the only time we can truly claim that a specimen has been adequately collected is when we find neoplastic cells in it! But this is not very helpful for specimens collected from patients with no underlying neoplastic disease. So how can we define an adequate specimen more pragmatically? Put simply, an adequate specimen is one that contains cells that are **representative** of the site sampled. If the site is a tumour, then tumour cells are the only truly representative cells. If the site happens to be a normal healthy cervix then an adequate cervical sample should contain cells that are representative of the area that is most susceptible to neoplastic change, that is, the transformation zone.

Now we turn to sampling error. Having defined an adequate sample as one that contains cells that are representative of the site being sampled, we can state that sampling error is the failure to collect such cells, or failure to collect them *in sufficient quantity*. So, when neoplastic tissue is not sampled adequately there is a *significant chance* that the diagnosis will be missed, leading to a **false negative** report.

> **Adequate specimen**
> In cytological terms, an adequate specimen is one that contains a sufficient quantity of cells that are representative of the site being sampled.

Key Points

A false negative cytology report erroneously reassures the clinician and patient that tumour cells are not present in a specimen, when in fact the patient has a neoplasm. The problem might be related to sampling error (tumour present but not sampled) or cytologist error (tumour cells present in the specimen but are not detected or are incorrectly interpreted).

By now you will appreciate that excellent specimen collection technique is absolutely crucial for the success of cytology. Importantly, *clinicians and cytologists must work together* to procure the best possible samples for cytological assessment. Cytologists must be in regular communication with clinicians, both informally (e.g. telephone discussions) and in the formal setting (e.g. **multidisciplinary team** meetings). This is a theme that will be repeated time and time again throughout this book, but perhaps it will be most noticeable in Chapters 10 and 11. The value of the cytologist's assistance and advice during an FNA or endoscopic procedure should not be underestimated.

> **A multidisciplinary team**
> is a group of healthcare professionals from a variety of disciplines who work together to provide an integrated approach to patient care.

Key Points

Cytologists must work closely with clinicians to ensure exemplary specimen collection technique. Failure to observe this golden rule is the single biggest cause of erroneous reporting and can be detrimental to patient care.

Specimen containers, transport media, and transportation to the laboratory

Following on directly from the previous discussion, the cytologist is of central importance in deciding on the most appropriate specimen containers to use, how to achieve optimum preservation of specimens (a process known as **fixation**), and how best to transport specimens to the laboratory.

Specimen containers come in many shapes and sizes and may or may not be pre-filled with chemicals such as **anticoagulants** (substances used to prevent the formation of clots which might otherwise interfere with cytological processing) or fixatives (the chemicals used to preserve cells and prevent the overgrowth of any microorganisms). In terms of specimen containers, common sense rules the day. The chosen container must be *fit for purpose*, which means that it must be of *adequate capacity* to contain the entire volume of the sample (there's nothing more annoying than receiving 100 ml of voided urine in five separate 20 ml containers). The container must also be sufficiently *robust* and properly *sealed* to prevent leakage of the specimen, even if it is accidently dropped. Surprising as it may seem, the specimen container need not be sterile; the presence of a small number of contaminant microorganisms in a cytological sample will generally not affect the diagnosis. The most common containers used for cytology specimens are 25 ml and 60 ml transparent plastic containers. These two types of container are shown in Figure 2.5. The benefit of the 25 ml container is that its base is cone-shaped, thus facilitating immediate **centrifugation** upon receipt in the laboratory.

Centrifugation
The use of centrifugal force (g-force) to isolate cells or other particles from the medium in which they are suspended. The suspension is rotated at a speed which forces cells away from the axis of rotation.

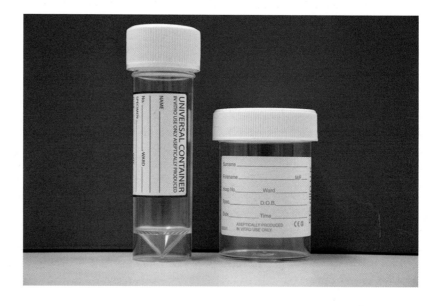

FIGURE 2.5
25 ml and 60 ml containers.

It is imperative that all unfixed specimens are transported to the laboratory immediately following collection. Any delay, particularly if specimens are stored at room temperature or above, could result in cellular decomposition and overgrowth of microbial contaminants and will inevitably adversely affect the cytological diagnosis. If a delay in transportation is anticipated then a number of options are available. Perhaps the simplest solution is to store the specimen at 4°C whilst transportation is being arranged. The cool temperature helps to slow down the process of cell degeneration. Alternatively, mixing the specimen with an appropriate **transport medium** is popular in some centres. Transport media may contain cell nutrients (thus helping to keep the cells alive) or a fixative (which inevitably kills the cells but preserves them in a life-like state). Commercially available **cell culture media** may be used if there is a need to keep cells alive. This is important if the intention is to subsequently produce air-dried slides for **Romanowsky staining** (see Sections 2.2 and 2.4). Common commercially available transport media that contain fixatives are CytoRich Red® (Tripath Imaging, Inc., Burlington, NC) and CytoLyt® (Hologic, Marlborough, MA). These agents have the added advantage that they promote **haemolysis**. The addition of **glacial acetic acid** to these fluids further helps to destroy red blood cells that could otherwise hamper the subsequent diagnosis. We now turn to the topic of cell fixation.

Haemolysis
The breakdown of red blood cells.

SELF-CHECK 2.1

Describe how sampling error might affect the reliability of a cytology report for the following types of specimen: (a) bronchial washings, (b) voided urine, and (c) FNA of a suspicious mass.

2.2 The mode of action and effects of cytological fixatives

Cells begin to decay very soon after they are removed from their natural environment. This must be prevented as far as possible if the cytologist is to provide the clinician with reliable and consistent reports. In cytology, the term 'fixation' refers to the preservation of cytoplasmic and nuclear morphology. If subsequent processing is to include the demonstration of **antigens** or other chemical components of cells then fixation must aim to preserve these cellular elements as well.

The aims of fixation can be summarized as follows:

1. Prevention of **autolysis**.
2. Prevention of microbial attack.
3. Leaves cells in a condition that allows subsequent staining techniques.
4. Cells should remain as close to their living state as possible.

Antigen
A molecule that is recognized by the immune system and stimulates the production of an antibody.

Autolysis
The disintegration of cells caused by the release of damaging intracellular enzymes.

Fixation achieves these aims by stabilizing cell proteins and inactivating enzymes. By rendering proteins insoluble, the whole cell structure is given mechanical strength and autolytic enzymes are inactivated. In trying to understand the mechanism of action of fixatives it is worthwhile reflecting on the term 'fixation' itself. Fixation refers to the chemical bonding of soluble protein to structural proteins, thus rendering the whole milieu insoluble and stable. Fixation also acts by disrupting the chemical bonds which give many proteins their three-dimensional structure. Of particular importance to cytologists is nuclear fixation; nuclear morphology is the foundation of diagnostic cytology. Stabilization of **nucleic acids** and nuclear **chromatin** is paramount for successful staining and microscopic presentation.

Key Points

The nucleic acids are the fundamental molecules of life, controlling all aspects of cell structure and function. They comprise deoxyribonucleic acid (DNA) and ribonucleic acid (RNA). Chromatin is the combination of nucleic acid and protein within the nucleus of the cell. When optimally fixed, chromatin gives the cytologist important morphological clues that relate to the biological behaviour of cells. Rapid and appropriate cell fixation is a key tenet in cytology.

The most commonly used cytological fixative is ethyl alcohol or **ethanol** (chemical formula C_2H_5OH) or one of its derivatives. Ethanol is a member of the group of fixatives known as **precipitating agents**, also known as *denaturing* or *coagulating* fixatives.

Precipitating fixatives
So-called because their action results in a product that is insoluble in water and therefore resistant to further chemical and structural change.

Ethanol provides excellent preservation of nuclear and cytoplasmic detail without causing excessive cell shrinkage or distortion. Moreover, fixation in ethanol is rapid: typical fixation times are in the range of 10–15 minutes. Ethanol is unpopular as a histological fixative because of its poor penetration of tissue. Ethanol is often combined with a water-soluble waxy substance called **Carbowax®** (polyethylene glycol), which coats the cells and acts as a barrier to help prevent cell damage. The ethanol/Carbowax mixture is available in dropper bottles or in spray cans, both of which are convenient ways of fixing samples once they have been deposited on a glass slide. For reasons that will become clear in the next section, fixation that is achieved by immersing or spraying cell preparations with a fixative solution is known as **wet fixation**.

Fixation by air-drying

Cross reference
See Chapter 10, Section 10.7 for a discussion of small cell carcinoma.

In contrast to wet fixation, air-drying achieves its fixation effect by removing water from cells. For best results slide preparations must be rapidly air-dried, preferably within ten seconds of preparing the slides. This can be achieved by ensuring thinly spread preparations. Air-drying has the effect of flattening cells to the glass slide, thus creating the visual impression of cell enlargement when viewed under the microscope. This method of preparation is ideal for demonstrating small cells in diagnostic cytology preparations, such as lymphocytes or the cells of **small cell carcinoma**.

Air-dried slide preparations are only acceptable for Romanowsky staining, and are an ideal complement to wet-fixed smears stained using the **Papanicolaou** technique (see Section 2.4 for an explanation of the Papanicolaou and Romanowsky stains).

Key Points

Fixation should be viewed as the introduction of controlled artefacts into cells and tissues. The key objective is to stabilize the protein framework of cells to achieve consistency and reliability in subsequent demonstration techniques.

Now that we have adequately dealt with the principles of specimen collection, transportation, and fixation we can turn our attention to the ways in which specimens are prepared prior to microscopical analysis.

SELF-CHECK 2.2
Why is fixation of cytology specimens necessary?

2.3 Preparation techniques

For ease of discussion we can divide specimen processing into eight categories:

1. The 'direct smear'
2. Large-volume centrifugation
3. Small-volume centrifugation ('cytocentrifugation')
4. Density gradient centrifugation
5. Gravity sedimentation
6. Membrane filtration
7. Cell blocks
8. Combinations of techniques

We will deal with each category in turn, mentioning examples as we go along. By 'processing' we mean the techniques and protocols employed to ensure that a sample of cells is deposited on a glass slide in readiness for staining or other demonstration technique.

Key Points

The selected preparation technique must make the job of detecting and classifying neoplastic cells as *easy* and *quick* as possible.

The direct smear

Over the years a large number of imaginative techniques have been employed to prepare glass slide preparations directly from a collected specimen. Some of the more sensible approaches are illustrated in Figure 2.6. The method is generally referred to as the **smear** technique.

Examples of the use of direct smears include those prepared from **mucoid** specimens such as **sputum** and those prepared from fine needle aspirates.

Mucoid specimens

Mucoid specimens, as the name suggests, contain **mucus**. Mucus consists of **mucin** (a **glycoprotein**) and inorganic salts suspended in water. It may also trap cells which can be extracted for cytological assessment. It is produced by some types of **epithelial** surfaces and serves many different functions in the body.

The most common mucoid specimens sent for cytological processing come from the respiratory tract and include **sputum** and other bronchial specimens. The consistency of mucus is variable but is often highly viscous.

The laboratory processing of mucoid specimens poses a unique challenge. An early technique, which is still used to this day in many cytology laboratories, is perhaps the simplest processing method of all. Sometimes referred to as the '*pick and smear*' method, the specimen is poured

Smear
A cytological preparation made by spreading the specimen (or a part of it) directly onto a glass slide.

Sputum
A thick, mucoid, secretory product of the lower respiratory tract. The production of large quantities of sputum usually indicates respiratory disease.

Epithelium
One or more layers of cells lining a tissue or organ.

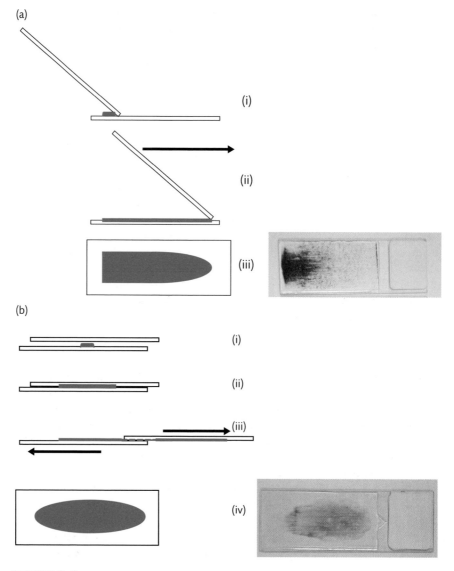

FIGURE 2.6

Techniques for producing thin and evenly spread smears.

(a) The 'blood smear' technique.

(i) Place a small drop of cell deposit near one end of a glass slide. Touch a second slide to the front edge of the drop. (ii) Push the second slide across the surface of the first slide, maintaining a constant angle with the horizontal. (iii) The final preparation is thin and uniform. This technique is very useful for producing thin rapidly air-dried smears for Romanowsky staining.

(b) The 'squash' technique.

(i) Place a drop of the cell deposit (red) near the centre of a glass slide. (ii) Squash a second slide against the first slide, ensuring a slight overlap. (iii) Pull the second slide away from the first slide, ensuring the two slides maintain contact and remain parallel. (iv) The final preparation is fairly thin and uniform, but should not be so thin as to rapidly air-dry. This allows sufficient time for the application of fixative prior to Papanicolaou staining.

into a **Petri dish** and examined against a black background. Any bloody or solid particles can be easily picked out against this background and spread directly onto a glass slide. In the absence of suspicious particles a small random portion of the specimen is selected. Immediate fixation with 95% ethyl alcohol or its equivalent will ensure optimum staining and presentation for microscopy.

Despite the simplicity of the 'pick and smear' technique it has been criticized for its relatively poor diagnostic **sensitivity**.

Part of the reason for the disappointing results of sputum cytology is that the selection of appropriate parts of the specimen for processing is operator dependent and difficult to standardize. A standard approach, and one which maximizes detection rates, is to break up the mucus before smear preparation. There are a number of ways of achieving such **mucolysis**. Mechanical disruption methods, such as the use of a blender (**Saccomanno technique**) or ultrasonic disintegration, permit the homogenization of samples. Subsequent processing by centrifugation, followed by spreading the deposit as a thin uniform film on a glass slide, is relatively straightforward. Chemical methods achieve the liquefaction of mucus to release the admixed cells, which can then be concentrated by centrifugation prior to the preparation of glass slides. Numerous chemicals have been used over the years to reduce the viscosity of mucus, including enzymes, peptides, and dilute mineral acids. **Dithiothreitol (DTT)** is a commonly used substance which is economical, safe, and easy to use.

Sensitivity
The probability of getting a correct diagnosis of disease rather than a false negative.

Fine needle aspiration smears

For the vast majority of sputum samples it is the laboratory staff who are responsible for preparing direct smears. For FNA samples, however, it is acceptable for clinicians to prepare smears, provided they have the correct training and employ due care and attention in their technique. However, the tradition in the UK of rotating junior medical staff has resulted in patchy training in FNA technique and the quality of samples received in the laboratory can be poor. Consequently, it is common practice for clinicians to request the assistance of the laboratory staff when planning to prepare FNA samples as this delivers consistency in approach and practice. A highly trained biomedical scientist, with wide knowledge of cell preparation and fixation techniques, is almost indispensable in the modern setting. Although a visit to a hospital ward can be time consuming for busy laboratory staff, it adds a new dimension to the work of the biomedical scientist, whose expertise is greatly appreciated by clinical staff. The ultimate beneficiaries of good preparation technique are, of course, the patients themselves.

A final word of warning regarding direct smears, *whoever* is responsible for preparing them. In all cases the smear *must* be of optimal thickness (not too thick and not too thin) and it *must* be immediately fixed. These two rules are all too easily broken and lack of consistency of approach can be a difficult problem to overcome. For this reason direct smearing of specimens is less popular than *cell concentration* techniques, to which we will now turn.

SELF-CHECK 2.3
State the benefits and possible limitations of the direct smear technique.

Large-volume centrifugation

Specimens of all types usually require some method of concentration to facilitate the deposition of the suspended cells onto the glass slide. Concentration of fluid specimens has been traditionally achieved by **large-volume centrifugation**. Large volume centrifuges typically handle centrifuge tubes with a capacity of 25 ml. Following the centrifugation of fluid specimens the objective is to select cells from the **buffy layer** for smear preparation. The buffy layer (so-called because it usually has a buff colouration), is the fraction of the centrifuged fluid sample that contains most of the white blood cells and any tumour cells that may be present. The buffy layer sits in the centrifuge tube as a thin layer between the fluid above and the red blood cells below. Figure 2.7 illustrates the typical appearance of a fluid specimen following centrifugation.

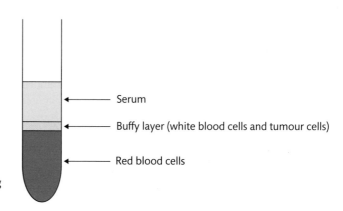

Serum

Buffy layer (white blood cells and tumour cells)

Red blood cells

FIGURE 2.7
Diagrammatic representation of the buffy layer following centrifugation of a fluid specimen.

Small-volume centrifugation (cytocentrifugation)

Low-volume fluids (generally less than 0.5 ml), or fluids that fail to yield a sufficient buffy layer for a smear, can be processed by **cytocentrifugation**, the principle of which is illustrated in Figure 2.8.

A cytocentrifuge is a device that uses centrifugal force to 'spin' cells in a fluid suspension directly onto a glass slide. This avoids the need to make smears from a deposit. Several different

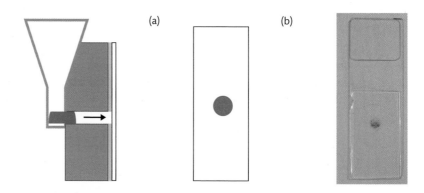

(a)   (b)

FIGURE 2.8
The principle of cytocentrifugation.
(a) A cytocentrifuge funnel (blue), absorbent paper (yellow), and glass slide (clear) are clamped together. A small volume of cell suspension (red) is added to the funnel and the whole apparatus is centrifuged. This forces the cell suspension in the direction shown by the arrow. (b) The final preparation is a small circular deposit of cells on the glass slide.

varieties of cytocentrifuge apparatus are available, as shown in Figure 2.9, including the single Cytofunnel® and Megafunnel®. The latter permits the cytocentrifugation of a larger volume of fluid (up to 6 ml) than is possible with Cytofunnels® (up to 0.5 ml) and is popular for processing urine specimens (see Chapter 8).

Cross reference
Chapter 8.

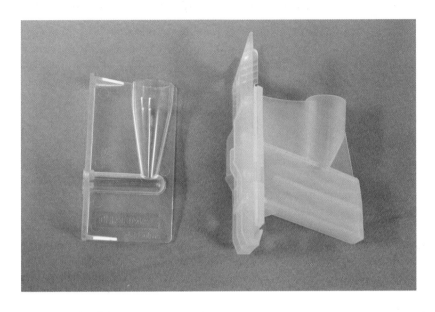

FIGURE 2.9
**From left to right:
single Cytofunnel® and
Megafunnel®.**

Density gradient centrifugation

In this technique a measured aliquot of the sample is carefully dispensed on top of a **density gradient fluid** in a centrifuge tube (see Figure 2.10). The density gradient fluid is a

(a)   (b)

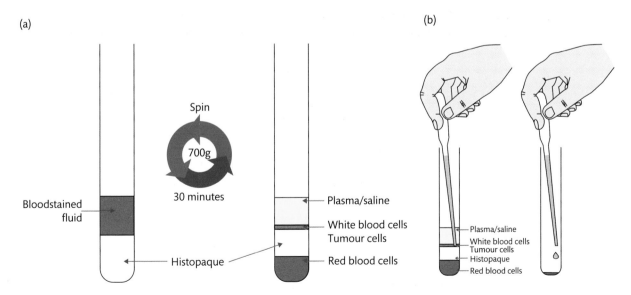

FIGURE 2.10
(a) Density gradient centrifugation.
(b) Isolation of the buffy layer using a careful pipetting technique.
Reproduced with kind permission from Victoria Burden.

BOX 2.2 Specific gravity

Specific gravity is the density (mass per unit volume) of a substance relative to water. So, a substance with a specific gravity of 1.0 has a density equal to that of water. Density gradient fluids for cytological use have higher values. For instance, Histopaque® has a specific gravity of 1.077, meaning that it is more dense, or 'heavier', than water. A cell suspension will generally have a lower specific gravity than this and when carefully added to Histopaque® in a tube it forms a layer that 'sits' neatly on top of the density gradient fluid.

concentrated solution of sugars or other substances formulated so that when a cell suspension is carefully added as a top layer of fluid and the whole mixture is centrifuged, different cell types are separated into layers according to their **specific gravity**.

The important point to remember is that the cells in which the cytologist is interested—epithelial cells—generally have a different specific gravity to non-epithelial cells and tend to concentrate in a layer by themselves. In reality, the separation of different cell types is imperfect. The best that can be achieved is a layer of cells that is *enriched* with neoplastic cells but which will also contain material with a similar specific gravity.

Gravity sedimentation

As you can see in Figure 2.11 the principle of gravity sedimentation is straightforward: cells are simply allowed to drift under their own weight to the base of the container in which the fluid is held. The usual procedure is to decant the fluid specimen (or an aliquot of it) into a **settling chamber** to allow suspended cells to settle onto a glass slide.

A common problem with gravity sedimentation (and other slide preparation techniques, for that matter) is that cells generally do not adhere very well to glass slides. For fluids that are naturally proteinaceous and 'sticky', such as serous fluids, this may not be a problem—the stickiness

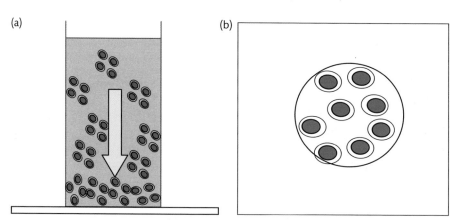

(a)

(b)

FIGURE 2.11

The principle of gravity sedimentation (e.g. the Surepath® technique).
(a) A settling chamber is clamped to a glass slide and the cell suspension is added.
Sufficient time is allowed for the cells to settle on the slide under gravity (arrow). Slides may be electrically charged or coated with a suitable material to facilitate cell adhesion.
(b) With Surepath® processing the final preparation consists of a 13 mm circular deposit of cells.

> **BOX 2.3** *The principle of density gradient centrifugation*
>
> When a cell suspension is carefully added as a layer above a density gradient solution and the whole system is centrifuged, cells will migrate according to their specific gravity. In the case of Histopaque® for instance, red blood cells will be forced through the density gradient to the bottom of the tube (see Figure 2.10a). Less dense cells, such as white blood cells and epithelial cells, reach equilibrium above the layer of density gradient fluid. The cells of interest (epithelial cells and any tumour cells) are therefore neatly separated from unwanted red blood cells. It is then a simple matter of aspirating the layer of interest for subsequent slide preparation.

of the fluid helps the cells to adhere to the slide. The lack of protein in other fluids, such as urine and needle washings from FNAs, means that the cells in these specimens will not readily adhere to the slide. For this reason slides can be coated with an adhesive, of which there are various types. **Albumin** is a naturally occurring cell adhesive which is commercially available. A tiny drop is spread thinly and evenly on the glass slide, which is then kept in a dust-free container until it is ready to use. **Poly-L-lysine** is another commonly used and potent cell adhesive.

Membrane filtration

Membrane filtration makes use of specially manufactured circular filters containing microscopic pores through which fluid samples can be drawn. The process is illustrated in Figure 2.12.

(a)   (b)   (c)

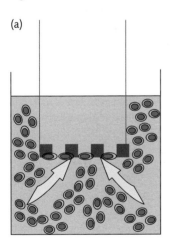
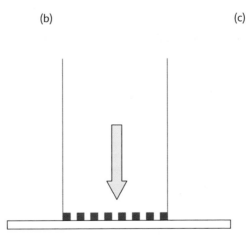
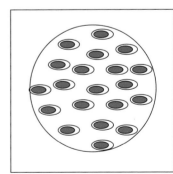

FIGURE 2.12
The principle of membrane filtration (e.g. ThinPrep® technique).
(a) A filter chamber (blue) is inserted into vial containing the cell suspension (grey). After vortexing the sample to ensure dispersal of cells the suspension is drawn through a membrane filter (arrows). Fluid passes through the filter whilst cells adhere to the membrane.
(b) The filter chamber is removed from the vial and pressed against a glass slide. A puff of positive pressure helps to transfer cells from the filter to the slide. Slides may be electrically charged or coated with a suitable material to facilitate cell adhesion. (c) With ThinPrep® processing the final preparation consists of a 20 mm circular deposit of cells.

The sample first undergoes a dispersion step to break up large groups of cells and unwanted debris, and is then thoroughly mixed. A negative pressure then draws the sample through the filter. The average size of the pores is such that fluid, small cells (such as erythrocytes and white blood cells), and unwanted debris pass readily through the filter. Larger cells, including tumour cells, if present, collect on the filter in a thin, even layer. The layer of cells is then transferred to a glass microscope slide in readiness for staining. An example of an automated membrane filtration technology is ThinPrep® (Hologic, Marlborough, MA), which can be used for processing a wide variety of fluid specimens, including cervical samples.

Cell blocks

A **cell block** is a way of processing a cytology sample just as if it were a solid tissue specimen. Although several methods are available the principle is to immobilize suspended cells in a solid or semi-solid medium which is then processed **histologically**. The end result is one or more slide preparations containing sections of thinly sliced 'tissue'. Cell blocks are particularly useful when **immunocytochemical** procedures are being planned (see Section 2.4). This is because immunocytochemical reactions work best on uniformly thin sections produced from **formalin**-fixed material. An additional advantage of the cell block method is that multiple similar slides can be produced, which is an absolute requirement when testing samples with panels of immunocytochemical markers.

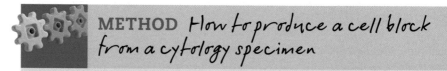

METHOD How to produce a cell block from a cytology specimen

Perhaps the best known method is the thrombin cell block. The fluid sample is first centrifuged to produce a cell pellet. The supernatant is discarded and a small volume of blood plasma and thrombin is added to the pellet. The mixture is gently agitated and within a few minutes a fibrin clot forms, trapping the suspended cells. The reaction exactly mimics the blood clotting process that occurs in the body when a blood vessel is damaged. The clot is then added to a pot of formalin fixative and processed as if it were a histology specimen.

A convenient alternative to thrombin and plasma is heated molten agar. This is added to the centrifuged pellet of cells and allowed to cool and solidify. The agar block with entrapped cells is then ready for histological processing.

Combinations of techniques

Although it is convenient to talk of categories of preparatory techniques it should come as no surprise to you that these divisions are not mutually exclusive. The use of multiple techniques *for the same specimen* is actually quite common practice. We will look at two examples.

Density gradient centrifugation followed by gravity sedimentation

This combination of techniques is the basis of SurePath® (Tripath Imaging, Inc., Burlington, NC) technology. This semi-automated method is designed primarily for batch processing cervical samples. The process is summarized in Figure 2.13.

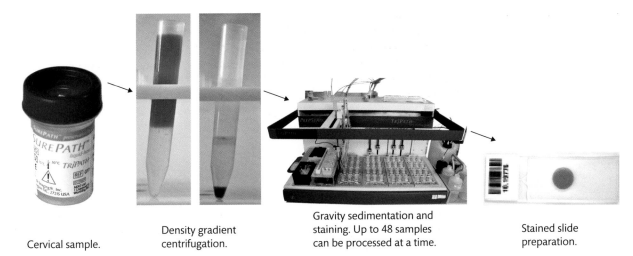

Cervical sample.

Density gradient centrifugation.

Gravity sedimentation and staining. Up to 48 samples can be processed at a time.

Stained slide preparation.

FIGURE 2.13
Density gradient centrifugation and gravity sedimentation (the SurePath® technique).

As with membrane filtration, the specimen is first mixed to break up large groups of cells. A measured aliquot of the sample is then carefully dispensed on top of a specially formulated density gradient fluid in a centrifuge tube. When the whole mixture is centrifuged at a relatively low speed, different cell types are separated roughly into two layers. The cells in which the cytologist is interested—cervical epthelial cells—have a fairly high specific gravity and are driven to the lower part of the centrifuge tube. The remaining unimportant cells—neutrophils, macrophages, lymphocytes, and the like—are lighter and are trapped in the upper zone of the centrifuge tube. The separation of cell types through this initial low-speed centrifugation is not perfect, so, after carefully discarding the top layer of fluid, the remaining fluid is centrifuged again at a higher speed to produce a firm deposit at the base of the tube. Following the high speed centrifugation the **supernatant** is discarded and the remaining contents of the tube are thoroughly mixed by **vortexing**. This double centrifugation technique is, in effect, a **cell enrichment** process; by reducing the number of unwanted cells and non-cellular debris the concentration of neoplastic cells (if present) in the final preparation is increased.

The final step in the SurePath® procedure is to decant the enriched sample into a settling chamber to allow suspended cells to settle onto a glass slide by gravity sedimentation.

Cell enrichment
The process by which the proportion of neoplastic cells in a specimen is increased, usually by depleting the sample of unwanted cells.

Large-volume centrifugation followed by density gradient centrifugation

This combination is popular for processing bloody serous fluids (see Chapter 9), which can be particularly troublesome to prepare. In highly bloodstained fluids it is often difficult to achieve a firmly packed deposit of red blood cells by simple centrifugation. As a result, the loose red cell layer is easily disrupted whilst attempting to sample the all-important buffy layer. Consequently, the relatively scanty tumour cells have only a poor chance of being sampled. Re-centrifugation of the buffy layer with density gradient solution is a useful technique for dealing with blood. By using a density gradient solution that is exactly dense enough to sit between the nucleated cells (including tumour cells) and the red cells after centrifugation, sampling the buffy layer becomes straightforward. A popular density gradient fluid for processing bloodstained serous fluids is Histopaque®. The procedure is illustrated in Figure 2.14.

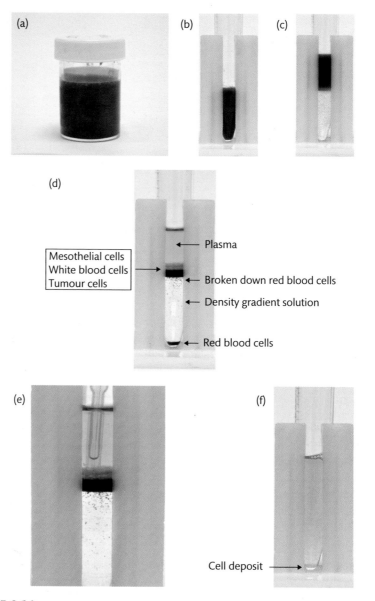

FIGURE 2.14

Large volume centrifugation followed by density gradient technique for bloodstained serous fluids.

(a) Heavily bloodstained fluid received in a 60 ml specimen container. (b) A portion of the fluid has been transferred to a centrifuge tube. (c) The sample has been carefully layered onto a volume of density gradient solution. (d) Following centrifugation the sample is separated into various levels. The white blood cells, meosthelial, and tumour cells appear as creamy buffy layer (arrow). (e) The creamy buffy layer is carefully pipetted off and placed in a solution of culture media and re-spun. This 'wash' removes any traces of density gradient that may have adhered to the cells. (f) The deposit after centrifugation is now ready for making the slide preparation.

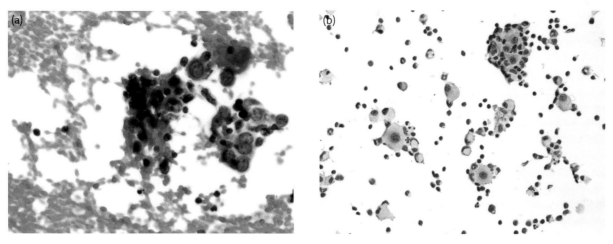

FIGURE 2.15
A bloodstained serous fluid preparation before (left) and after (right) density gradient centrifugation.

The main disadvantage of this elaborate protocol is that it can be time-consuming. The end results, however, can be quite stunning. Figure 2.15 shows slide preparations from a serous fluid before and after density gradient centrifugation.

Key Points

Cell *concentration* and *enrichment* techniques are the key processes involved in specimen preparation.

A final word about specimen processing techniques before we move on. One of the great strengths of cytology is the *simplicity* and *cost-effectiveness* of its preparation methods. It is all too easy to become swept up in the excitement and enthusiasm of new processing technologies. We must bear in mind that simple tried-and-tested procedures will produce consistently excellent results in skilful and well-trained hands. An over-reliance on expensive high-technology approaches to specimen preparation potentially weakens the discipline.

SELF-CHECK 2.4

Make a list of the various ways in which diagnostically important cells in cytology specimens can be concentrated onto a relatively small area of a glass slide.

BOX 2.4 Liquid based cytology (LBC)

The term 'liquid based cytology' refers to any cytology processing technique that relies on the immediate transfer of cells into a liquid preservative as soon as the specimen is collected. This ensures rapid cell fixation. Subsequent laboratory processing might include membrane filtration (such as ThinPrep® technology) or density gradient centrifugation followed by cell sedimentation (the basis of the SurePath® procedure).

2.4 **Demonstration techniques**

The choice of title for this section is quite deliberate. *Demonstration* techniques encompass a whole lot more than simple **tinctorial staining** methods.

In this section we will discuss the ways in which cells and their constituents can be made visible to the human eye through the microscope. There is no intention to turn this section into a list of staining protocols; we will concentrate only on the principles involved.

Tinctorial staining
The application of dyes to cell preparations in order to produce differentially coloured cell constituents.

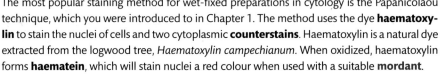

Key Points

Cells are virtually invisible until we apply the right chemicals to detect them.

We will discuss these methods in the following order:

1. Stains for wet-fixed preparations
2. Stains for air-dried preparations
3. Special stains
4. Immunocytochemistry
5. Molecular techniques

Stains for wet-fixed preparations

The most popular staining method for wet-fixed preparations in cytology is the Papanicolaou technique, which you were introduced to in Chapter 1. The method uses the dye **haematoxylin** to stain the nuclei of cells and two cytoplasmic **counterstains**. Haematoxylin is a natural dye extracted from the logwood tree, *Haematoxylin campechianum*. When oxidized, haematoxylin forms **haematein**, which will stain nuclei a red colour when used with a suitable **mordant**.

In the Papanicolaou method, a stained example of which is shown in Figure 2.16, haematoxylin is used as a **regressive stain**. This means that haematoxylin is added in excess and then the

Mordant
A substance that assists staining by forming a link between a dye molecule and a cell component. Commonly used mordants include the salts of aluminium, copper, iron, and potassium.

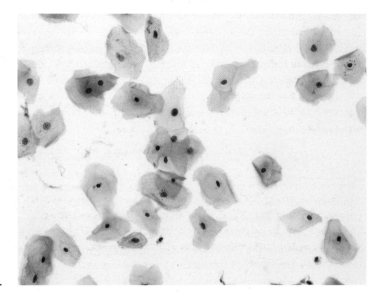

FIGURE 2.16
A Papanicolaou-stained cervical sample.

unwanted stain is removed using an acid solution referred to as a **differentiator**. **Differentiation** is allowed to proceed until the desired degree of stain intensity has been achieved.

Following differentiation the haematoxylin-stained preparation is *blued* using an alkaline solution. As the name suggests, **blueing** is the process of converting the red colouration in the nuclei to dark blue.

The first counterstain in the Papanicolaou method uses the synthetic dye **orange G** (OG). OG is used to stain the protein **keratin**, which is a normal component of the squamous epithelium of the skin but can also be found in the squamous epithelium of the cervix (and other sites) in certain conditions.

The second counterstain makes use of a dye combination known as **eosin azure** (EA), which is made up of eosin and light green. EA formulations are suffixed by a number (e.g. EA-36, EA-50, EA-65) which denotes the proportions of the constituent dyes. The most commonly used EA formulation is EA-50. EA-50 stains the cytoplasm of cells various shades of pink or green, depending on conditions inside the cell (such as pH).

In practice, the staining of slide preparations involves several washes in various solvents. Most laboratories make use of automated staining machines that are programmed to immerse the preparations in the correct solutions in the correct order for the correct duration to achieve the desired staining effect.

SELF-CHECK 2.5

Describe the purpose of the dyes used in the Papanicolaou technique.

An alternative routine staining technique for wet-fixed preparations is the **haematoxylin and eosin** (H&E) procedure. This is the most popular method for staining tissue sections in histology but there is no real reason why it could not be applied to cytological preparations. Its advantage over the Papanicolaou technique is the simplicity and speed of the procedure. However, H&E staining does not produce the range of cytoplasmic colouration that cytologists seem to prefer.

As with all other staining protocols in cytology the procedure ends with **dehydration**, **clearing**, and **mounting**.

Dehydration
The removal of water from a stained preparation.

Clearing
The process of rendering stained preparations optically clear.

Mounting
The application of a thin sheet of glass (a coverslip) or plastic tape to the stained preparation, thus protecting the preparation and helping to make it optically flat in readiness for microscopy.

Although these three stages play no part in the staining process they are nevertheless crucial for effective microscopy. Dehydration is achieved by several washes of the slide preparations in **absolute alcohol**. This step is necessary because if there is any residue of water left on the slide at the end of the staining procedure, the subsequent clearing step will be ineffective. The purpose of clearing is to replace the alcohol with a solvent (usually **xylene**) that has the same **refractive index** as the stained material within the sample. This ensures that the final preparation is transparent. Finally, the slide preparations are mounted. This involves adherence of a glass coverslip or a strip of optically clear plastic tape over the entire stained preparation using a **mounting medium**. Modern busy laboratories make use of automated coverslipping machines for this process.

Key Points

The refractive index (RI) of a substance is a measure of the extent to which light is slowed down or bent as it travels through a medium. It is important for the RI values of the clearing agent, the stained material, and the glass from which the slide and coverslip are made, to match closely. This ensures that light rays pass through the slide preparation with minimal distortion.

Stains for air-dried preparations

You will recall from Section 2.2 that air-dried smear preparations are a popular alternative to wet-fixed slides in cytology. Rapid drying introduces a controlled artefact in the preparation—cells tend to flatten against the slide, thus appearing larger than they do in the wet-fixed state. This can be very useful when there is a high probability that the specimen contains a population of small cells that will require interpretation, such as lymphocytes. Examples of such specimens include serous fluids and fine needle aspirates of lymph nodes. Unfortunately, air-dried preparations do not stain well with the Papanicolaou technique and alternative stains are required. Fortunately, a variety of haematological stains are available that will do the job very nicely. As a group these are referred to as the Romanowsky stains. Individual formulations include Giemsa, May-Grünwald, Jenner, Wright, Leishman, and others. The choice of method is largely made on personal preference, as each formulation yields a slightly different colour balance in the stained slides. One of the most popular methods, yielding good quality consistent results, is the **May-Grünwald-Giemsa (MGG)** method. Whichever technique is chosen the aim is to achieve the staining result known as the **Romanowsky effect**, defined as purple nuclei, blue cytoplasm, and pink red blood cells, as seen in Figure 2.17. The Romanowsky effect is achieved by the careful control of the acidity of the staining solutions, which should be between pH 6.8 and 7.2.

Cross reference
Chapter 11.

An additional benefit of the Romanowsky methods is that they produce very good results when applied as a rapid technique—particularly useful for the quick assessment of specimen adequacy at FNA clinics for instance (see Chapter 11).

Rapid Romanowsky staining kits are widely available from commercial companies and are sold under brand names such as **Rapi-Diff®** or **Diff-Quik®**, but most stain suppliers will have their own version of a rapid kit.

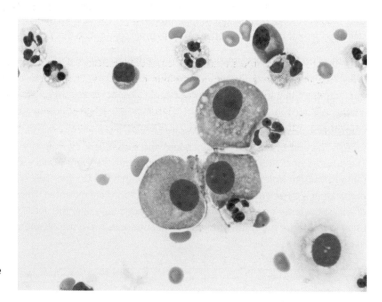

FIGURE 2.17
The Romanowsky effect: purple nuclei, blue cytoplasm, and pink red blood cells.

Special stains

Special stains to identify specific elements in cytological preparations are occasionally required after a careful assessment of the routine Papanicolaou and Romanowsky stained slides.

Examples of the application of special stains include the **periodic acid Schiff (PAS)** method for carbohydrate and **Grocott's methenamine silver method** for *Pneumocystis jiroveci*.

In practice, special stains can be variable in performance and may deliver non-specific background staining in cytology preparations. They have largely been replaced by immunocytochemistry, which generally provides more accurate and sensitive results.

Immunocytochemistry

Immunocytochemistry (ICC) can be defined as the cytological detection of specific cell constituents based upon their **antigenic** structure. The procedure uses **antibodies** and relies on the precise nature of antigen-antibody interactions and the availability of methods to visualize these interactions through the light microscope.

Key Points

The binding together of an antibody and antigen is specific—each antibody binds to a specific antigen.

ICC has numerous cytological applications, particularly in serous fluid and FNA diagnosis. These applications will be described in the relevant sections of this book. Here we will discuss the principles only.

Immunocytochemical techniques can be broadly divided into the *direct* and *indirect* methods. The direct methods rely on the direct visualization of antigen-antibody reactions using **primary antibodies** that have been **conjugated** with a visible marker, variously called a *probe*, *tag*, or *label*. Fluorescent dyes (**fluorophores**) were at one time quite popular as tags but these have generally been supplanted by methods that do not require an expensive fluorescence microscope. The principle of direct ICC is shown diagrammatically in Figure 2.18.

The direct method is simple and quick to perform but can suffer from poor sensitivity. The problem is that antigenic components within the preparation might be present in such low concentrations that they might be undetectable by direct methods. For this reason the indirect methods, which rely on **signal amplification**, are more popular. As you can see in Figure 2.19, a **secondary antibody** is applied after the primary antigen-antibody reaction takes place, and it is the secondary antibody that is tagged, not the primary antibody.

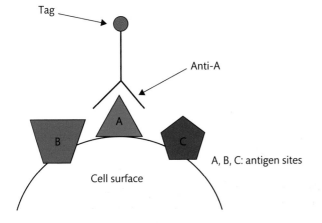

A, B, C: antigen sites

FIGURE 2.18
Immunocytochemistry: the direct method.

Cross reference
Chapter 10, Section 10.6.

Periodic acid Schiff technique
A method for demonstrating carbohydrate. The periodic acid oxides glucose residues forming aldehydes that react with Schiff reagent forming a magenta colour. The technique can be used for demonstrating pathogenic fungi and a variety of other constituents of cell samples.

Grocott's methenamine silver method
Uses a solution of methenamine silver that reduces to a black precipitate on contact with *Pneumocystis jiroveci* and other pathogenic fungi.

Antibody
A molecule that is produced by the immune system and binds to an antigen.

To summarize, the reaction product in the indirect method is a complex molecular interaction between the antigen to be demonstrated, the primary antibody, the secondary antibody, and the tag. This might seem unnecessarily complicated but the great advantage of the technique is that several secondary antibodies can bind to a single primary antibody, thus amplifying the signal. There are a number of alternatives to fluorescent dyes as tags, and some are quite innovative. In many cases the tag is an **enzyme**, which in itself is invisible. Only when a chromogenic substrate (or simply **chromogen**) is applied does the reaction become visible. A coloured product therefore indicates a positive **immunoreaction** between antibody and the antigen of interest. A popular choice of tag is the enzyme **horseradish peroxidase**. Although this enzyme cannot be seen directly, the addition of **hydrogen peroxide** and the colourless substance **3,3′-diaminobenzidene** (DAB) produces a brown reaction product which is easily recognized microscopically. An example is shown in Figure 2.20.

Another commonly used enzyme-substrate mixture is **alkaline phosphatase** and its chromogenic substrate New Fuchsin, which results in an appealing red colour at the immunoreaction site.

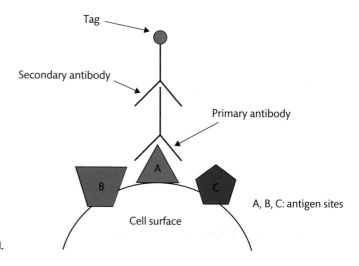

FIGURE 2.19
Immunocytochemistry: the indirect method.

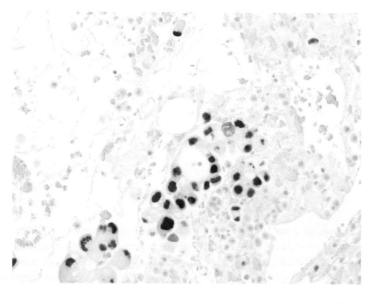

FIGURE 2.20
The brown immunocytochemical reaction product indicates the site of the antigen-antibody interaction.

This section would not be complete without mentioning the exquisite **avidin-biotin complex** method, known simply as the ABC technique. This widely used technique is illustrated in Figure 2.21 and is an adaptation of the indirect method. The reaction proceeds in several stages. The first stage involves the application of unlabelled primary antibody. Next comes the secondary antibody conjugated with several molecules of the low molecular weight vitamin **biotin**. The third step involves the addition of an excess of **avidin** molecules, each one bound to several tag molecules such as horseradish peroxidase. The addition of hydrogen peroxide and DAB completes the procedure and yields a brown reaction product. The clever part of the ABC method is the powerful signal amplification: each biotinylated secondary antibody is capable of binding to several avidin-peroxidase complexes, resulting in a very strong reaction signal at the site of the antigen-antibody reaction.

Biotin

A low molecular weight vitamin that can be easily conjugated to primary antibodies as a component of commercially available ICC kits.

Avidin

A large molecular weight glycoprotein with a high binding affinity for biotin. It can be readily combined with chromogenic tags such as horseradish peroxidase and, just like biotinylated antibodies, is also provided in commercially available ICC kits.

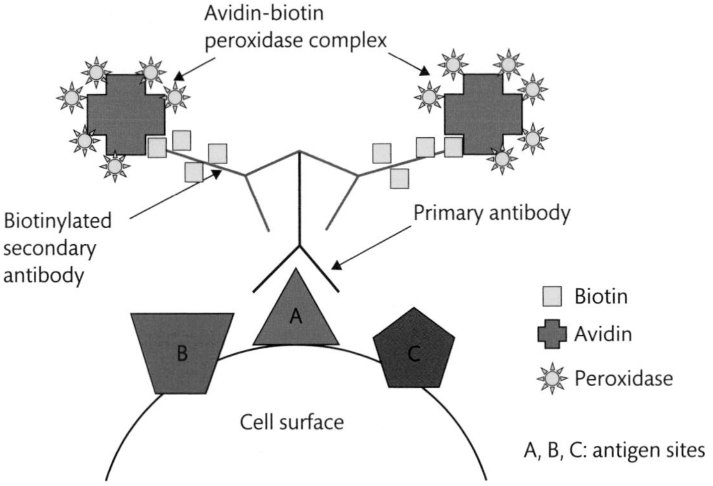

FIGURE 2.21
The avidin-biotin complex (ABC) method.

SELF-CHECK 2.6

List the similarities and differences between direct and indirect immunocytochemical methods.

One of the main uses of ICC is to determine the origin of tumour cells found in a serous fluid or in a lymph node aspirate. The use of ICC to 'type' tumours will often allow the clinician to determine the primary site of a tumour that has spread to the serous cavities or a lymph node. ICC markers are usually used in *panels* to differentiate between specific conditions or to confirm a preliminary diagnosis made on the routine stains.

ICC is mainly carried out on cell block sections rather than cytology preparations for the following reasons:

- Most ICC markers are designed for use on formalin fixed paraffin processed tissue such as cell blocks.
- Multiple sections can be easily prepared from cell blocks for ICC panels.
- The cell block method produces a permanent tissue block which can be used for retrospective studies.

It should be clear by now that immunocytochemistry is a technique that enables the identification and localization of cellular proteins and related products, which are of course the products of **DNA translation**. Immunocytochemical techniques have been in use for several decades and although they have many valuable applications their sensitivity and specificity can in some cases be lacking. In the next and final section dealing with demonstration methods, we will go upstream of the products of DNA translation and briefly explore the potential of **genetic markers** of disease.

Molecular techniques

Molecular pathology
Involves the manipulation of nucleic acids in patient specimens to identify alterations associated with disease.

Molecular biology
Involves the use of a range of techniques for studying genes and chromosomes and the molecules from which they are made—the nucleic acids (DNA and RNA).

Molecular cytology
The science of visualizing genomic events within cells.

Cross reference
Chapter 13.

Molecular pathology has existed as a discrete speciality within pathology since the early 1990s. It can be defined as the application of the methods of **molecular biology** to patient specimens in order to make specific predictions about disease processes. Molecular pathology has a broad range of clinical applications, including the diagnosis and **prognosis** of disease, predicting response to treatment, and monitoring patients for residual disease following treatment. **Molecular cytology** is still in its infancy and its full scope of application is yet to be defined.

The most commonly used techniques in molecular pathology are the **polymerase chain reaction** and its variants, **Southern blot** and **Northern blot** analysis, **cytogenetics**, and various **hybridization** techniques. The relevant techniques and their cytological applications will be described in detail in Chapter 13.

Health and safety in the cytology laboratory

To finish this chapter we must discuss a topic that is of central importance to all pathology disciplines—health and safety. Clinical laboratories can be the safest places in which to work but they can also be the most dangerous. As well as the microbiological **hazards** posed by clinical specimens there is a plethora of chemical and physical hazards that demand the highest attention of laboratory staff.

HEALTH & SAFETY

Categories of hazard in the cytology laboratory

Microbiological hazards

Every clinical specimen submitted to the cytology laboratory must be regarded as potentially infectious and should be handled with great care by well-trained staff. Drivers, porters, and nurses delivering specimens to the laboratory, as well as scientific and clinical staff receiving them, should be made aware of the potential dangers inherent in handling these specimens.

Chemical hazards

Chemicals can be corrosive, irritant, poisonous, carcinogenic, teratogenic, narcotic, flammable, etc. Examples of chemicals used in cytology include alcohol (flammable), xylene (irritant, flammable), hydrochloric acid (corrosive), and ammonia (irritant). For this reason the number and volume of chemicals stored in the cytology laboratory, as well as the level and duration of exposure to them, should always be kept to an absolute minimum.

Physical hazards

Physical hazards are everywhere. They include sharps (e.g. broken glass, needles, and scalpel blades, etc.), items that are heavy or awkward to lift (e.g. microscopes), spillages (slipping hazard), electrical appliances (risk of electric shock), and even the design of microscope workstations. There are many others. As long as common sense is applied the risk from physical hazards should be minimal. There is no room for complacency, however; risk assessments for physical hazards must be taken as seriously as any other hazard.

The promotion of high standards of health and safety is a central management priority in today's pathology laboratories. Hazard identification and risk management are the modern buzz phrases used by laboratory managers and safety representatives. Rather than provide an exhaustive account of health and safety management and legislation, the aim of this section is to describe the principles of safe working in the cytology laboratory.

First, a few definitions:

- A **hazard** is a situation with the potential to cause harm to a person's life, to their health, to property, or to the environment.
- **Risk** is the likelihood that harm from a particular hazard is realized.
- The extent of risk is the number of people who might be affected by a risk, that is, the number of individuals who may be exposed and the consequences for them.
- **Risk assessment** is a formal procedure for determining the level of risk from an identified hazard.
- **Risk control** refers to the practices, procedures, equipment, and training that are put in place to minimize or eliminate risk.

HEALTH & SAFETY

Hazard and risk explained

Imagine that you have been asked to prepare a solution of hydrochloric acid as the differentiating bath for the Papanicolaou staining procedure. A bottle of concentrated hydrochloric acid is stored on a low level shelf in the laboratory. The bottle of acid is clearly a hazard, but as long as it is securely sealed to prevent the release of fumes and the shelf on which it stands is well built there is little, if any, risk to the health and safety of laboratory staff. The moment you begin to handle the bottle and dispense the acid the risk increases. If you are well trained in the handling of laboratory chemicals and use the correct procedures, wearing appropriate **personal protective equipment**, and if the laboratory is well equipped with fume extraction hoods and pipetting equipment the risk will remain low. If on the other hand you are ill-prepared for the procedure and/or the laboratory is poorly equipped then you can expect the risk of a chemical spillage or other accident to increase dramatically, potentially harming yourself and others.

To summarize, a hazard is either present or it is absent. Risk, however, is measured on a sliding scale from near negligible to catastrophic, *and is under our control*.

Risk assessment and risk control are carefully considered logical processes that involve:

1. Identification of workplace hazards.
2. Elimination of the hazard if possible.
3. Assessing the magnitude of risk.
4. Introducing control measures if the risk is significant.
5. Monitoring the effectiveness of the control measures.
6. Reviewing all risk assessments periodically.

It is important that the person undertaking a risk assessment is systematic in identifying all the relevant hazards in the workplace. It is particularly important to identify groups or individuals who may be at greater risk than the general population (e.g. the young or inexperienced, pregnant staff, disabled staff, lone workers, etc.).

A risk assessment can take a variety of formats but the essential element in all types is a conclusion which states the *level of risk*. If the level of risk is judged to be low then no further action is required other than monitoring and periodic review of the activity. If risk is moderate then efforts should be made to reduce the risk, but the costs of prevention should be carefully measured and balanced against the benefits of risk reduction. In other words, we should do whatever is *reasonably practicable* to reduce health and safety risks. Any hazard that carries an unacceptably high risk means that any type of work involving the hazard must be prohibited until the risk has been reduced to acceptable levels.

HEALTH & SAFETY

One way of conducting a risk assessment

Risk rating number
A numerical measure of the degree of risk, calculated by multiplying the likelihood of an adverse event by its severity.

Perhaps the most objective method of performing a risk assessment for a work task or activity is to calculate the **risk rating number (RRN)**. This is done by multiplying the *likelihood* of an adverse event occurring by the *severity* of the adverse event. Both are measured on a scale of 1–5 as defined below.

Likelihood	Severity
1 = Remote	1 = Minor injury
2 = Possible	2 = Medical intervention required
3 = Probable	3 = Temporary major illness or injury
4 = Likely	4 = Permanent major illness or injury
5 = Certain	5 = Fatalities

Risk rating number = likelihood × severity

So, the risk rating number for a particular work activity can range from 1 (remote possibility of a minor injury) to 25 (fatalities are guaranteed).

The next step is to convert the RRN into a meaningful statement that describes the level of risk and the action that should be taken. One possible interpretation of the RRN is given below.

RRN	Risk level	Action and timescale
1–5	Low	No action required.
6–9	Moderate	Risk should be reduced but carefully balanced against the costs of doing so. Risk control should be implemented within a defined time period.
10–16	High	Work should not be started until the risk has been reduced. Considerable resources may have to be allocated to reduce the risk.
20–25	Unacceptable	Work is prohibited until the risk has been reduced.

CASE STUDY 2.1 A risk assessment exercise

Fresh (i.e. unfixed) body fluids frequently arrive in the cytology laboratory for analysis. Although there is a standard protocol for handling such specimens, you have been asked to perform a risk assessment for processing a serous fluid labelled as 'high risk'. The specimen was taken from a patient with a suspected but unknown tropical disease. Discuss the potential hazards and risks of handling the specimen and provide a report to the laboratory manager with your recommendations.

If the conclusion from a risk assessment is that there is an unacceptable risk to the health or safety of laboratory staff then action is clearly necessary. The choice will normally be to eliminate the hazard altogether (e.g. find an alternative chemical) or, more usually, to control the risk.

Controlling microbiological risks

Key Points

All clinical specimens are microbiological hazards.

In terms of the infectious risks from microbiological hazards clinical laboratories are built according to the **containment level** required. This in turn depends upon the *hazard group* to which the commonly handled microorganisms belong. The hazard groups are defined below and the corresponding containment levels are summarized in Table 2.1.

Group 1: Organisms unlikely to cause disease.

Group 2: Organisms that may cause disease in laboratory workers but are unlikely to spread in the community. Effective prophylaxis or treatment is usually available.

Group 3: Organisms that may cause severe human disease and might present a risk of spread in the community. There is usually effective prophylaxis or treatment available. Examples of group 3 organisms include *Mycobacterium tuberculosis*, human immunodeficiency virus, Hepatitis B virus, Creutzfeldt-Jacob disease agent, *Salmonella typhi* and *Legionella sp.*

Group 4: Organisms that cause severe human disease and pose a serious threat of spread in the community. There is usually no effective prophylaxis or treatment available. Examples include the haemorrhagic fever viruses, Ebola virus, Lassa fever virus, and Rabies virus.

Containment level

A numerical system (1–4) describing the standard to which a laboratory is designed and built for the purposes of containing hazardons microorganisms.

TABLE 2.1 Summary of laboratory containment levels.

Containment requirements	Containment levels			
	1	2	3	4
Laboratory isolation	No	No	Partial	Yes
Laboratory sealable for fumigation	No	No	Yes	Yes
Inward airflow into laboratory/negative pressure	Optional	Optional	Yes	Yes
Airlock	No	No	Optional	Yes
Airlock with shower	No	No	No	Yes
Wash hand basin	Yes	Yes	Yes	Yes
Effluent treatment	No	No	No	Yes
Autoclave	No	Yes (in suite)	Yes (in suite)	Yes (in lab)
Microbiological safety cabinet	No	Optional	Yes	Yes
Class of cabinet	–	Class I	Class I/III	Class III

Key Points

Most cytology laboratories are built to containment level 2 standard.

Microbiological safety cabinet (exhaust protective cabinet)

A ventilated enclosure providing protection, to the user and the environment, from airborne microorganisms that may be generated when handling clinical specimens.

In terms of specimen processing, **microbiological safety cabinets (exhaust protective cabinets)** require a special mention. There are three types of microbiological safety cabinet:

Class I cabinets (open-fronted exhaust protective cabinets) offer adequate protection to the worker against the inhalation of potentially infectious airborne particles. They can be used for all except hazard group 4 pathogens and are the usual type of cabinet fitted in cytology laboratories. The cabinet must exhaust through a **high efficiency particulate air (HEPA) filter** to the outside air or to the laboratory air extraction system.

Class II cabinets (vertical laminar flow) re-circulate some filtered air, exhaust some to atmosphere, and take in replacement air through the open front. They offer protection from contamination to the material handled and some protection to the worker. They are mainly used in tissue culture work and must not be used in cytology laboratories.

Class III cabinets (totally enclosed exhaust protective cabinets) are gas-tight and fitted with glove ports. They are used for handling group 4 pathogens when complete isolation of work from worker is required.

All safety cabinets must be regularly maintained, including at least weekly monitoring of airflow with an **anemometer** to ensure adequate airflow. The cabinet must also be regularly **fumigated** to remove any harmful microorganisms.

Fumigation

A method of decontamination that completely fills an area with a gaseous disinfectant.

SELF-CHECK 2.7

List the similarities and differences between the three types of exhaust protective cabinet.

Controlling chemical risks

Working with hazardous chemicals is part of everyday life in the cytology laboratory, but exposure to undue risk should be virtually unheard of. Chemical risks are controlled at several levels. First, only those chemicals *that are absolutely essential* should be used or stored in the laboratory. On no account should staff handle chemicals that are unnecessary for the provision of the cytology service. Second, when handling chemicals, staff are required to wear personal protective equipment according to the level of risk. A laboratory coat is an absolute must on all occasions, but other items might include safety goggles and gloves made of the appropriate material. At the third level of protection is the laboratory equipment. No cytology laboratory should be without efficient **fume extraction hoods**, good ventilation, and surfaces that are resistant to a wide range of chemicals. Appropriate pipettes and other glassware are other essentials. Finally, staff must receive appropriate information, instruction, and training in the safe handling of chemicals, as well as action to be taken in the event of an emergency, such as spillages, splashes, and evacuation procedures.

Fume extraction hood

A device used for handling noxious chemicals that draws in air and expels it from a building via suitable ducting. It is designed to limit the user's exposure to chemical fumes.

Controlling physical risks

The risk reduction methods for physical hazards will clearly depend on the nature of the danger. In common with the microbiological and chemical hazards, appropriate staff training is vital and control measures should be proportional to the risk. Rather than provide an exhaustive account of the control measures available we will take a look at one, and probably the most prominent, 'physical' risk that cytologists are exposed to.

The lasting impression that any visitor to a cytology laboratory has is of staff busily examining slides through a microscope for five or six hours a day. This apparently innocuous task can cause or exacerbate a number of conditions, collectively called **work-related upper limb disorders** (also known as **repetitive strain injury**) and aches and pains in the hand, wrist, arm, and shoulder. These conditions should not be taken lightly as they can result in long periods of absence from work or, worse still, retirement through ill health. Avoidance of the hazard is clearly not an option in this case, so once again we must return to the principle of *risk reduction*. How can we reduce the risks inherent in daily microscope work? The answer is that a holistic approach is needed. Not only should staff be provided with **ergonomically designed** microscopes and workstations, but they must also receive adequate training and advice on the use of the microscope, correct seating posture (see Figure 2.22), and the importance of taking sufficient breaks from microscopy during the working day. This investment in time, effort, and money will pay dividends as it will almost guarantee a highly productive and motivated workforce.

Work-related upper limb disorders (repetitive strain injury)
A term used to describe a variety of musculoskeletal problems that may be related to repetitive work tasks. The disorder is characterized by numbness, pain, and weakening of muscles in the hand, wrist, arm and shoulder.

HEALTH & SAFETY

The ergonomically designed microscope workstation

Ergonomics can be defined as the science of designing tasks, equipment, and the workplace for the purposes of human comfort, efficiency, and safety. An **ergonomic hazard** can be defined as any object, system, or environment with the potential to cause human harm through poor design.

Features of the microscope workstation that should be considered as ergonomic hazards include:

- Benches (height, width, depth, edge, surface)
- Chairs (adjustability, material, width and depth, lumbar support)
- Footrest (adjustability, width, depth, angle)
- Microscope (eyepiece angle and adjustability, position of controls, arm rest, quality of optics)
- Environmental conditions (temperature, humidity, ventilation, lighting, noise, etc.)

Figure 2.22 shows an example of an ideal workstation in cytology.

FIGURE 2.22
An ergonomic microscope workstation.

Procedures for the safe disposal of clinical and chemical waste

The disposal of waste from a cytology laboratory needs a lot more thought than disposing waste from your household. There are two broad categories of laboratory waste to consider: clinical waste and chemical waste.

Clinical waste can be defined as any material left over from the processing of specimens that is not required for diagnosis. There are strict laws concerning this type of waste and failure to comply can mean a heavy fine or even a prison sentence for those responsible. One important principle is the maintenance of *patient confidentiality* throughout the disposal process. A golden rule is that the patient's name on waste specimen containers must be removed or made unidentifiable before disposal. Ultimately, all clinical waste must be disposed of by incineration. This achieves the triple objective of reducing the volume of waste to manageable levels, rendering it non-infectious, and making it unidentifiable as clinical waste. It is then perfectly acceptable to dispose of the waste by landfill, just like ordinary domestic waste.

Chemical waste is potentially more noxious than clinical waste. The improper disposal of chemical waste presents a risk to personal and public safety as well as to the environment. As with clinical waste strict laws are in place to control how chemical waste is handled. The best way to dispose of waste chemicals is to commission a **disposal contractor** who will recycle waste wherever possible or dispose of it safely. The old days of rinsing waste chemicals down the sink with plenty of running water are long behind us (dilution is certainly *not* the solution to pollution).

Disposal contractor

A company authorized to dispose of waste safely and legally.

 SUMMARY

In this chapter you should have picked up the following points.

- A properly collected specimen by a well-trained clinician is the first step towards a clinically useful cytological diagnosis.

- Clinicians and cytologists must work together to procure the best possible samples for cytological assessment.

- Techniques for specimen collection can be grouped into exfoliative methods or aspiration methods. Both methods take advantage of the tendency of neoplasms to shed cells, thereby aiding cell collection.

- Sampling error is the failure to collect a sufficient quantity of diagnostically relevant cells, thus leading to the problem of false negative reporting.

- Fixation is the process by which cells and their constituents are chemically changed to preserve their morphology and to facilitate subsequent demonstration techniques.

- Specimen processing aims to maximize the chances of making clinically useful cytological diagnoses. Almost all techniques rely on methods that concentrate a cell population onto a relatively small area of a glass slide.

- Slide preparations should contain cells that are *representative* of the site being examined or, better still, should be *enriched* with diagnostically relevant cellular material.

Membrane filtration and density gradient centrifugation are the most widely used enrichment techniques.

- Methods for the demonstration of cells or their constituents range from sublimely simple staining procedures to advanced molecular techniques.

- All preparation and demonstration techniques, as well as the process of waste disposal, demand due attention to health and safety legislation and require a full risk assessment. Risks must be controlled as far as is reasonably practicable.

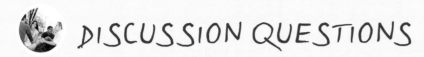

DISCUSSION QUESTIONS

1. Discuss the factors that govern the choice of fixative for cytology specimens. Why are cytological fixatives generally different from those used in histology?

2. Cytologists are trained to examine cell preparations under the microscope for alterations in cell morphology that might indicate disease. Cytologists are therefore *morphologists* by training. Suggest how the skills of the cytologist might need to change in the future, given that the range of cell demonstration techniques is becoming increasingly diverse.

3. Most cytology laboratories are built to containment level 2 standard. Discuss how this can be justified, given that the hazard group to which microorganisms in clinical specimens belong is usually not known.

Answers to the self-check questions, and tips for responding to the discussion questions, are provided in the book's Online Resource Centre.

 Visit www.oxfordtextbooks.co.uk/orc/shambayati/

3

The cervical screening process

John Crossley

There are many types of medical screening programmes currently employed throughout the world. These include screening for early stage disease, screening for predisposition to diseases, and screening for carriers of diseases. The purpose of this chapter is to introduce the principles and theory of medical screening programmes and to explain the terminology used when referring to their characteristics. We also explain the purpose and aims of cervical screening, and discuss in more detail the organization of the National Health Service (NHS) Cervical Screening Programme.

Learning objectives

By the end of this chapter you should be able to:

- Provide a definition of a medical screening programme.
- Confidently discuss and describe the factors which influence the characteristics of a screening programme.
- Outline the purpose and aims of the Cervical Screening Programmes.
- Define the target population and screening interval of the NHS Cervical Screening Programme.
- Briefly describe the role and activities of the different agencies involved in the NHS Cervical Screening Programme.

3.1 The basic theory of screening programmes

The definition of a screening programme can be given as 'the testing of an apparently normal population in order to detect the cases of a target condition'.

All successful screening programmes must adhere to the following basic principles:

- The target condition must be relatively common and disabling. If the condition is very rare or the clinical disease has very little consequence on the health of the affected individual, the benefits gained by screening will not justify the cost.

- The natural history of the target condition must have been studied and understood. If the timescale of the progression of the condition is unknown, there will be no evidence upon which to base **screening intervals** or treatment interventions.

- There must be a recognizable, treatable precursor or pre-symptomatic phase, as there is for cervical cancer, which will be discussed in more detail later in this chapter.

- The screening programme must be cost effective. The actual cost of a screening programme is the difference between the cost of the screening/treatment process and the cost of treating the clinical disease in the absence of screening. The number of disease-free years of life saved by the screening programme are referred to as **quality-adjusted life years**. It is an unfortunate fact of healthcare economics that quality-adjusted life years have a finite value beyond which the cost cannot be justified.

- The screening test must be reliable, valid, and repeatable to ensure standardization of the screening programme.

- The screening test must be acceptable and easy to perform, with a high **sensitivity** and **specificity**. This is discussed in more detail below.

- The treatments should be effective and available, with an agreed policy on who and how to treat.

Screening intervals
The period of time between each episode of testing.

Target population

The individuals at risk of having or developing the target condition are referred to as the **target population**. The proportion of the population that has the target condition represents the **prevalence** of the condition.

Prevalence
The proportion of the population that has the target condition.

CASE STUDY 3.1 *The National Health Service Breast Screening Programme*

The National Health Service Breast Screening Programme was set up by the Department of Health in 1988 and was the first national breast screening programme in the world. The programme uses mammography, an X-ray of each breast, to detect changes in the breast tissue which may indicate cancers that are too small for the patient or clinician to feel during manual examination. A small cancer is very often an early cancer and the early detection and treatment of breast cancer often leads to a better prognosis than if the cancer is left until clinical symptoms (e.g. a palpable lump) occur.

Screening is undertaken every three years (three-year screening interval) for women between the ages of 50 and 70. Women under 50 are not invited because the mammograms are not as effective at detecting small cancers in the denser breast tissue of pre-menopausal women.

Any changes detected in the breast tissue which are thought to be abnormal are sampled (biopsied) to provide a diagnosis. The diagnosed cancers are treated by surgical removal (excision) together with chemotherapy if necessary.

The individuals that have the target condition are referred to as the **positives**. The individuals that do not have the target condition are referred to as the **negatives**. Figure 3.1 shows an unscreened population of positive (red) and negative (blue) individuals.

The ideal screening process would accurately identify all the positive and negative individuals as in Figure 3.2.

However, no screening process is ideal. The factors which affect the success of screening are discussed below.

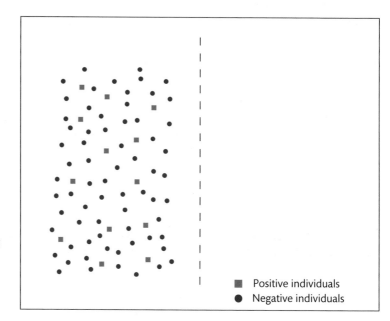

FIGURE 3.1

The broken central line represents a screening test. The left-hand side represents the target population before screening. The positives (red squares) remain undetected within the population of negatives (blue circles).

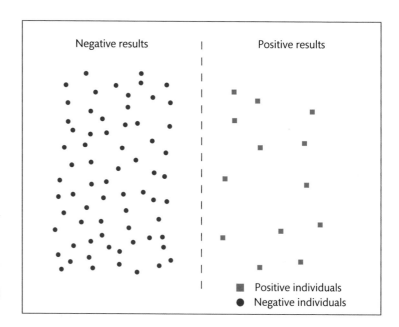

FIGURE 3.2

The broken central line represents the ideal screening test. The left-hand side represents those individuals testing negative and the right-hand side represents those individuals testing positive. As you can see, all the negatives (blue circles) tested negative and all the positives (red squares) tested positive.

Coverage

Coverage is the most important factor in influencing the effectiveness of a screening programme. The screening programme will fail if the coverage is low, irrespective of the effectiveness of the screening test used. Figure 3.3 shows a population screened with the ideal screening process but where coverage is limited to those individuals within the outlined area. When coverage is low, positive individuals remain undetected because they are not screened.

Coverage is dependent upon two factors:

- The proportion of the target population that has access to the screening process.
- Of those having access, the proportion that participate. This participation is referred to as **uptake**.

In order for a screening programme to be successful the coverage of the target population should be in the region of 70–80%.

When designing a screening programme it is important to qualify (who) and quantify (how many) the target population and, if necessary, set a screening interval appropriate to the natural history of the target condition. It is essential that there are sufficient skills and resources available to invite, screen, and perform the resulting treatments on the whole target population at the specified interval.

> **Coverage**
> The proportion of the target population that is screened.

Key Points

Successful screening programmes require sufficient resources to screen the entire target population and to treat all the positive individuals identified by the screening process.

Some programmes screen individuals who present in a clinical situation, for example antenatal screening. However, in screening programmes that require the invitation of individuals, there must be access to a database holding the names, addresses, and screening histories of the

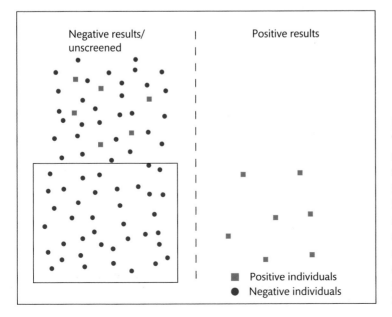

FIGURE 3.3
The broken central line represents the ideal screening test. The left-hand side represents those individuals testing negative and the right-hand side represents those individuals testing positive. However, only the proportion of the population within the square has been screened. As you can see, of those screened, all the negatives (blue circles) tested negative and all the positives (red squares) tested positive, but the unscreened positives remain undetected within the population of negatives.

target population. The invitation for screening should include information on the test, what it is for, where and how it will be performed, what the possible outcomes may be, and what the subsequent treatments/consequences will involve. An example of this information is the leaflet entitled *Cervical Screening—the Facts*, which is available in 19 different languages and accompanies all invitations for the National Health Service Cervical Screening Programme (NHSCSP).

Acceptance of the invitation will result in an appointment at which the individual must be counselled by the clinician responsible for testing/sample taking. The clinician should again inform the individual of what is involved in the test, how and when the results will be available, and give a brief explanation of the possible outcomes. They should also provide the individual with an opportunity to ask any questions they may have about the whole screening process. The individual must understand the sceening process before they can give **informed consent**.

One of the major factors in the individual's decision to consent to the screening process will be how the test is performed and whether the individual finds this acceptable. The quality of information provided about the test is crucial in this decision. It is a testament to the quality of information supplied by the cancer screening programmes in the UK and the healthcare professionals involved, that the cervical screening and breast screening programmes are successful and have a relatively high coverage.

Sensitivity

Sensitivity of a screening test
A measure of the ability of a screening test to identify the positives.

Sensitivity is a measure of the ability of a screening test to identify the positives, that is, the individuals within the target population that have the target condition. It is usually expressed as a percentage of positives correctly identified by the test out of the total positives actually tested.

The individuals within the target population that have the target condition and test positive when screened are referred to as **true positives**. Screening tests with a sensitivity of less than 100% will give rise to **false negatives**. False negatives are the individuals within the target population that have the target condition but have a negative result when screened. Figure 3.4 shows the result of population screening with low sensitivity. Note that some of the positive individuals have negative results and remain undetected.

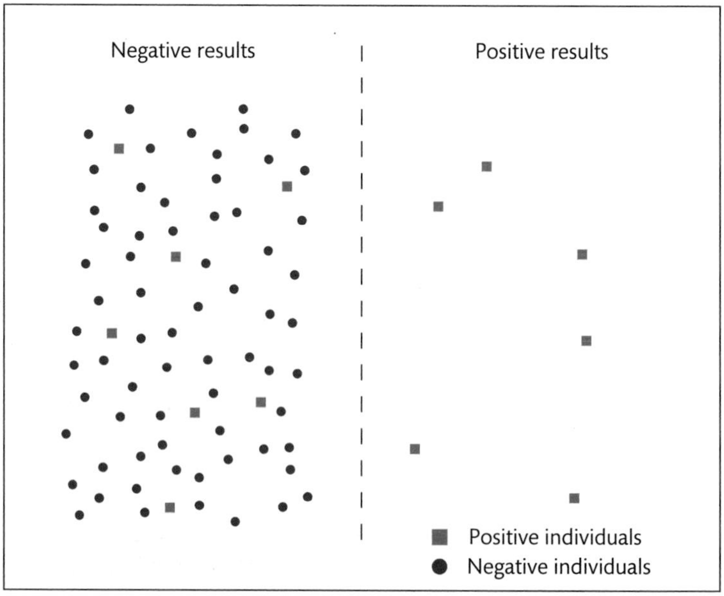

FIGURE 3.4

The broken central line represents a screening test with low sensitivity. The left-hand side represents those individuals testing negative and the right-hand side represents those individuals testing positive. As you can see, all of the negatives (blue circles) tested negative, but some of the positives (red squares) also tested negative.

Sensitivity can be calculated using the following formula:

$$\frac{\text{true positives}}{\text{true positives} + \text{false negatives}} \times 100\%$$

(Examples of the use of this and the following formulae are given below in Case Study 3.2.)

False negative results have a detrimental effect upon the effectiveness of, and the public confidence in, a screening programme. However, they are an inevitable consequence of virtually all screening processes as very few tests are 100% sensitive. False negative results can be minimized through quality assurance systems and procedures. Keeping the number of false negatives to a minimum helps to maintain the integrity of the programme but unfortunately does not prevent the potential devastation to the affected individuals and their families.

Specificity

Specificity is a measure of the ability of a screening test to identify the negatives, that is, the individuals within the target population that do not have the target condition. As with sensitivity, it is usually expressed as a percentage of negatives correctly identified by the test out of the total negatives actually tested.

The individuals within the target population that do not have the target condition and test negative when screened are referred to as **true negatives**. Screening tests with a specificity of less than 100% will give rise to **false positives**. False positives are the individuals within the target population that do not have the target condition but have a positive result when screened. Figure 3.5 shows the result of population screening with low specificity. Note that some of the negative individuals have positive results.

Specificity can be calculated using the formula:

$$\frac{\text{true negatives}}{\text{true negatives} + \text{false positives}} \times 100\%$$

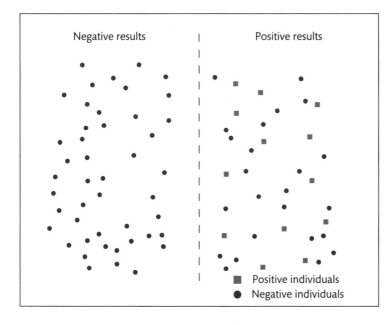

Negative results | Positive results

■ Positive individuals
● Negative individuals

FIGURE 3.5
The broken central line represents a screening test with low specificity. The left-hand side represents those individuals testing negative and the right-hand side represents those individuals testing positive. As you can see, all of the positives (red squares) tested positive, but some of the negatives (blue circles) also tested positive.

False positive results lead to unwarranted anxiety and unnecessary treatments, the latter of which will also add to the cost of the screening programme. Because sensitivity is calculated using true positives and false negatives, and specificity is calculated using false positives and true negatives, in statistical terms they are mutually exclusive. In practical terms, however, this is not always the case: attempts to minimize the number of false positives by increasing the threshold for a positive result (to therefore increase the specificity of the test) risks the possibility of increasing the number of false negatives and therefore decreasing the sensitivity of the test.

Positive predictive value

Positive predictive value (PPV) is a measure of the accuracy of a positive result. It expresses the proportion of the individuals testing positive for the target condition that actually have the target condition, usually as a percentage.

PPV can be calculated using the formula:

$$\frac{\text{true positives}}{\text{true positives} + \text{false positives}} \times 100\%$$

PPV is considered the 'clinician's gold standard', as it indicates the probability that a positive test result reflects the presence of the target condition. A high positive predictive value allows the clinician to treat individuals testing positive knowing there is a high probability the treatment is warranted and only a small risk that the treatment is unnecessary. Ideally, PPV would be 100%. However, this would require the number of false positives to be zero. Further, as with specificity, lowering the number of false positive test results can have a detrimental effect on the sensitivity of the test.

Negative predictive value

Negative predictive value (NPV) is a measure of the accuracy of a negative result. It expresses the proportion of individuals testing negative for the target condition that do not have the target condition, usually as a percentage.

NPV can be calculated using the formula:

$$\frac{\text{true negatives}}{\text{true negatives} + \text{false negatives}} \times 100\%$$

In programmes that employ repeat testing of individuals, the screening interval must allow testing to occur often enough to facilitate detection of the target condition before it can progress to a serious stage. If the negative predictive value is low, there is low confidence that a negative test result reflects the true absence of the condition. To offset the risk that a negative test result might not, in fact, be correct, the screening interval must be short so that testing occurs regularly to facilitate the detection of the target condition as early as possible. However, a high negative predictive value offers increased confidence that a negative test result reflects the true absence of the target condition. This increased confidence allows fewer episodes of testing and therefore extended screening intervals.

SELF-CHECK 3.1

How can altering the screening interval help to limit the consequences of a false negative test?

CASE STUDY 3.2 *Screening*

A screening programme operates to prevent disease X which has a 100% treatable, sub-clinical, precursor condition Y.

The target population is 125,000 and the programme has 80% coverage.
In one round of screening there are:

100,000 tests (80% coverage of the 125,000 target population)
79,200 true negatives (TN)
19,000 true positives (TP)
1,000 false negatives (FN)
800 false positives (FP)

$$\text{Sensitivity} = \frac{TP}{TP + FN} \times 100\% = \frac{19,000}{19,000 + 1,000} \times 100\% = 95\%$$

$$\text{Specificity} = \frac{TN}{TN + FP} \times 100\% = \frac{79,200}{79,200 + 800} \times 100\% = 99\%$$

$$\text{PPV} = \frac{TP}{TP + FP} \times 100\% = \frac{19,000}{19,000 + 800} \times 100\% = 96\%$$

$$\text{NPV} = \frac{TN}{TN + FN} \times 100\% = \frac{79,200}{79,200 + 1,000} \times 100\% = 99\%$$

Predictive value and prevalence

Bayes' theorem is a complex probability theory. However, in simplistic terms, Bayes' theorem dictates that, if sensitivity and specificity of a screening test remain constant, the positive predictive value of the test increases and the negative predictive value decreases as the prevalence of the target condition increases. This effect is demonstrated in Case Study 3.3.

Because of this relationship between predictive values and prevalence, differences in the predictive values of different screening processes may not relate to differences in their performances. For example, the breast screening programme cannot be compared to the cervical screening programme using predictive values alone. However, predictive values can be used to compare the performances of different screening tests designed to detect the *same* target condition within the *same* population, for example if we wished to compare mammography with manual palpation of breast tissue for the detection of breast cancer.

The success of any screening programme is dependent upon high coverage of the target population, acceptable and accurate detection of the target condition, and timely treatments with positive patient outcomes. However, poor performance in any one of the screening characteristics discussed above may result in failure. Consequently, all screening programmes must be continually developed in order to improve performance and to identify and accommodate any changes which may affect performance.

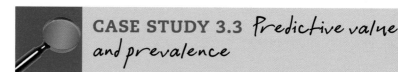

CASE STUDY 3.3 *Predictive value and prevalence*

Test A has a sensitivity of 90% and a specificity of 95% for the detection of condition B in a target population of 10,000.

If the prevalence of condition B is 5%:

Total positives = 500

Sensitivity of test A = 90% therefore:

True positives (TP) = 450
False negatives (FN) = 50

Total negatives = 9,500

Specificity of test A = 95% therefore:

True negatives (TN) = 9,025
False positives (FP) = 475

$$PPV = \frac{TP \times 100\%}{TP + FP} = \frac{450 \times 100\%}{450 + 475} = 48.6\%$$

$$NPV = \frac{TN}{TN + FN} \times 100\% = \frac{9,025}{9,025 + 50} \times 100\% = 98.9\%$$

If the prevalence of condition B is 25%

Total positives = 2,500
Sensitivity of test A = 90% therefore:
True positives (TP) = 2,250
False negatives (FN) = 250

Total negatives = 7,500

Specificity of test A = 95% therefore:

True negatives (TN) = 7,125
False positives (FP) = 375

$$PPV = \frac{TP}{TP + FP} \times 100\% = \frac{2,250}{2,250 + 375} \times 100\% = 85.7\%$$

$$NPV = \frac{TN}{TN + FN} \times 100\% = \frac{7,125}{7,125 + 250} \times 100\% = 96.6\%$$

3.2 Cervical screening

Cytology is the most commonly used primary screening tool for the prevention of **carcinoma** of the **uterine cervix** throughout the western world. Georgios Papanicolaou first recognized the ability of vaginal aspirate cytology to detect abnormal epithelial cells from cervical lesions in 1928. However, due to lack of acceptance by the establishment it wasn't until 1943 that he published, together with Herbert Traut, the paper 'Diagnosis of uterine cancer by the vaginal smear'. Since that time there have been many cervical screening programmes developed in numerous countries. All of these programmes are continually developing and improving the basic principles discovered by Papanicolaou.

Cross reference
Chapter 4, Section 4.4.

Rationale of cervical screening

The majority of cervical carcinomas are either squamous cell carcinoma derived from squamous epithelial cells as found on the ectocervix, or adenocarcinoma derived from the glandular epithelial cells of the endocervix. It is now widely accepted that the vast majority of these carcinomas have a recognizable precursor lesion.

These precursors are termed **intraepithelial** as the abnormal cells remain within the epithelium. Previously these lesions have been referred to as dysplasia and currently some countries, including the USA, use the term squamous intraepithelial lesion (SIL). In Britain, the squamous lesions are referred to as **cervical intraepithelial neoplasia (CIN)** and the glandular lesions are referred to as **cervical glandular intraepithelial neoplasia (CGIN)**.

Cross reference
We discuss CIN and CGIN in more detail in Chapter 5.

With no invasion through the basement membrane into the underlying tissues, these lesions are completely treatable using local excision or ablation methods.

SELF-CHECK 3.2

What other terms are used, both historically and internationally, for intraepithelial lesions of the cervix?

Papanicolaou's work demonstrated that cervical carcinomas and intraepithelial lesions shed cells from their surface. These cells can be collected, smeared on a glass slide, stained, and microscopically recognized to diagnose the lesion of origin. Papanicolaou's name and his method of cell sample preparation have given rise to two popular names for the cervical cytology test: the Pap test and the smear test.

The fact that many cervical carcinomas have a treatable precursor lesion, together with the ability to detect these lesions using cervical cytology, makes cervical carcinoma prevention an ideal candidate for screening.

SELF-CHECK 3.3

What are the popular names given to the cervical cytology test and how do you think they were derived?

The purpose of cervical screening

The purpose of all cervical screening programmes is to reduce the **incidence** and **mortality** of cervical cancer. To achieve this, the target population undergo cervical cytology testing to identify potential precursors of cervical carcinoma. These potential precursor lesions can then be managed and treated with respect to their perceived risk of progressing to invasive cervical carcinoma.

Throughout the world there will be variations in the infrastructure and organization of cervical screening programmes. The availability and prioritization of resources will differ but the purpose will be the same.

3.3 National Health Service Cervical Screening Programme

This chapter will now focus on the NHS Cervical Screening Programme in England which was established by the Department of Health in 1988. Scotland, Wales, and Northern Ireland have similar screening programmes but there are some variations between them all.

Target population

The target population of a screening programme may be best identified by excluding those not at risk. In the case of cervical cancer prevention the first exclusion is all those who do not have a cervix. Therefore all males and all women who have undergone a total hysterectomy for benign reasons are excluded.

Cross reference

Sections 4.7 and 13.3.

It is widely accepted that virtually all cervical carcinomas are caused by the **human papilloma virus** (**HPV**). There are more than one hundred different types of HPV. Many are transmitted through personal, non-sexual contact and cause either warts or sub-clinical infections. However, there are over 40 types that are sexually transmitted. These sexually transmitted HPV types are classified according to their risk of causing cervical carcinoma. Low-risk HPV types (e.g. types 6, 11, 42, 43 and 44) cause genital warts and CIN1, the lowest grade of CIN. High-risk HPV types (e.g. types 16, 18, 31, 33 and 45) can cause precursor lesions of cervical carcinoma. Most HPV infections, whether high risk or low risk, are transient and resolve within 6–18 months. However, persistent high-risk HPV infections increase the risk of the development of precursor lesions and subsequent cervical carcinoma.

SELF-CHECK 3.4

Can you list two low-risk HPV types and two high-risk HPV types?

From these facts it can be deduced that female virgins are at an extremely low risk of developing cervical carcinoma and can therefore be excluded from the target population. However, it is very difficult, if not impossible, to accurately identify these individuals. Therefore other age-based criteria must be applied.

The age of onset of sexual activity of individuals varies enormously. It varies between different countries and between different regions of the same country, and is also influenced by culture and religion. An overall average would probably be somewhere between 14 and 17 years. Therefore the earliest age the average woman is at risk of acquiring a high-risk HPV infection is between 14 and 17 years.

The development of cervical carcinoma is dependent upon:

- The acquisition of a high-risk HPV infection.
- Persistence of a high-risk HPV infection.
- Development of a high-grade precursor lesion.
- Invasion of the lesion through the basement membrane.

The natural history of the above progression is approximately 20 years on average (that is, it takes around 20 years from HPV infection to invasion of the lesion), with the longest period being between the development of the precursor lesion and invasion through the basement membrane. Therefore, despite a small number of cases that develop much more rapidly, cervical carcinomas in women below 30 years old are rare.

Because of the sexual attitudes and behaviour of young single people, and in contrast to the extremely small number of cervical carcinomas, there are a relatively high number of transient HPV infections in women under the age of 25 years. These transient HPV infections are the source of numerous inconsequential cytological 'abnormalities' detected during screening. These abnormalities resolve when the HPV infection resolves, but still burden the screening programme with early repeat samples and burden the patient with unnecessary anxiety.

The cost of screening, together with the unnecessary repeat samples and anxiety caused to many patients, is not justified by the prevention of the extremely small number of cancers occurring in young women. For this reason, in October 2003 the NHS Cervical Screening Programme changed the age of commencement of screening from 20 years to 25 years.

Cervical carcinoma is very rare in well-screened women. Patients with a long history of regular cervical cytology tests showing no evidence of a precursor lesion are at an extremely low risk of developing cervical carcinoma. For this reason women who are 65 years and above, and who have had three negative cervical cytology tests within the previous ten years are excluded from the NHS Cervical Screening Programme.

Key Points

The target population for the NHS Cervical Screening Programme can now be defined as 'All women between the ages of 25 and 64 years, with the exception of those that have undergone a hysterectomy for benign reasons'.

Screening interval

Some screening programmes are designed to detect conditions that are either present or absent, whereby the individual's status with respect to that condition doesn't change over time. The sensitivity and specificity of the test used will affect the reliability of the result but the actual status of the patient will remain constant throughout their lives. An example is a test for a genetic disorder: either the specified gene is defective or it isn't (or a specific allele is present or absent). Such conditions will only require one reliable test in the life of the patient.

Cervical screening is different as the target condition status of the patient can change. A negative cervical cytology result means one of two things. Either there is no precursor lesion present on the cervix (a true negative test result), or a precursor lesion is present but has not been detected by the test (a false negative test result).

If the result is a true negative, there is still the possibility that the patient will develop a precursor lesion in the future. For this reason subsequent testing is necessary. If the result is a false negative and an undetected precursor lesion is present, subsequent testing is necessary in an attempt to detect and treat the lesion before it progresses to cervical carcinoma.

The time interval between the necessary subsequent tests is called the screening interval. It would seem common sense to suggest that tests giving false negative results need repeating earlier, and therefore require a smaller screening interval, than true negative results. However, there is no way of determining which negative results are true and which ones are false. Therefore, the management of all negative results must allow for potential false negative results.

As stated previously, the natural history of the development of cervical carcinoma is approximately 20 years. The longest period of time within this progression is from the development of the high-grade precursor lesion to the time of invasion. This period is probably 10 to 15 years and is the window of opportunity for screening to detect the precursor lesion to allow appropriate treatment. The more screening tests that occur within this time period the greater the chance of detecting a precursor lesion.

The NHS Cervical Screening Programme has set the screening interval at three years for women aged 25 to 49 years. This screening interval allows four to six tests, that is, four to six attempts at detecting a precursor lesion within the 10–15-year window of opportunity. By contrast, the screening interval for women aged 50–64 years has been set at five years. This increased screening interval only allows three to four tests within the 10–15-year window of opportunity. But, as the women in the screening programme get older, their screening histories become longer and their risk of developing cervical carcinoma decreases. This decreased risk allows the screening interval to be increased. Women aged 65 years and above are only screened if their screening history is incomplete. They are screened until they have had three negative test results within the previous ten years.

Key Points

The screening interval for the NHS Cervical Screening Programme is three years for women aged 25 to 49 years and five years for women aged 50 to 64 years.

3.4 Agencies involved in the NHS Cervical Screening Programme

Call and recall

When the target population for screening has been identified and it has been decided how often to screen, a mechanism is needed by which individuals can be invited at the appropriate times. This requires a database containing all eligible individuals, their current address and their full screening history. There must also be an administrative centre responsible for communicating with patients regarding screening invitations and results, and with other health professionals in order to maintain the integrity of the database.

The database used by the NHS Cervical Screening Programme is the **National Health Application Infrastructure Services** (NHAIS), often referred to as the Exeter system. Although this system is a national database, it is operated and maintained by local call and recall offices using local data. These offices were previously referred to as family health service authorities, but the offices and responsibility have been taken over by primary care trusts (PCT).

The Exeter system is able to perform numerous data searches, referred to as analysis jobs, which can identify patients that fall within certain set criteria. One such job provides a **prior notification list** (PNL). The PNL is a list of all the women due for cervical cytology within the specified date parameters. The PNL should not be produced more than 12 weeks in advance of the due date. Each general practitioner (GP) practice receives a PNL for all their registered patients due for screening within the specified dates. The GP examines the list to verify the appropriateness of screening of all the individuals listed. There are three options given for each woman:

- Invite
- Postpone—the GP must specify a valid reason and give a new recall date
- Cease—the GP must specify a valid reason and sign the PNL

The GP may also use the PNL to communicate specific information about certain patients, for example learning disabilities or language barriers.

Once it has been established which women should be invited for screening from the returned PNLs, the invitation letters are generated using the Exeter system. The leaflet entitled *Cervical Screening—the Facts*, which includes information about the benefits and limitations of screening, must accompany all invitation letters. This leaflet is available in many languages and will help the patient make an informed choice about whether or not to accept the invitation for screening. Invitation letters may be sent by GP practices rather than the call and recall office. However, to prevent confusion and possible duplicate testing, women should not receive an invitation from more than one organization for the same test. Therefore, good communication between the call and recall office and GP practices is essential.

The call and recall office is responsible for recording all cervical cytology test results on the Exeter system. The laboratory report contains a standard action code which indicates the cytological findings and the appropriate recall interval for the patient. For example, a cytology report of negative, normal recall is coded 2A. Here, 2 is the code for the negative cytological findings and A is the code for a repeat test at the normal recall interval. Ideally, to prevent human data transfer errors and postal delays, there should be an electronic data link between the laboratory computer system and the Exeter system. If there is no electronic link this data is transferred from the laboratory to the call and recall office via the pink copy of the HMR 101 multi-copy request/report form. However, there are instances where some women registered within the PCT have their tests done in laboratories other than the linked laboratory and instances where patients move from one PCT to another. Therefore it is essential that data is transferred between the call and recall offices of different PCTs to ensure patients' screening histories are accurate and complete.

Although the sample taker is responsible for ensuring women are notified of their cervical cytology test result, arrangements for notification may differ between PCTs. Indeed the result letters are often generated by the call and recall office rather than the practice or clinic that has undertaken the screening sample. These letters inform the patient of the result of their test and indicate when their next test is due.

Key Points

The call and recall office is responsible for identifying and inviting women for screening at the appropriate interval. In the majority of cases the call and recall office also provides women with the result of their screening test.

The call and recall office is also responsible for producing the KC53 statistical return, which is used as an aid to monitor the effectiveness of the NHS Cervical Screening Programme.

Cross reference

See Chapter 6 for more information regarding the KC53 statistical return.

Sample takers and primary care

Women can choose to have their cervical cytology samples taken either at their GP practice by a GP or a practice nurse, or at a community clinic such as a contraception or a sexual health clinic. To ensure appropriate sampling of the cervix, all sample takers must be competent. 'Being competent' includes the completion of appropriate training to *attain* competence and regular attendance at update courses to *maintain* competence. In addition, the environment for sample taking must be suitable, with the necessary equipment, hygiene, and privacy facilities readily available.

Regular attendance of women for screening is fundamental to the coverage (and therefore success) of the cervical screening programme. If a woman has a bad experience during an episode of screening, whether due to the inconvenience of the time or place of the appointment, the conduct or competence of the sample taker, or the environment of the examination room, she may be disinclined to attend for subsequent testing.

Although the primary care team should encourage women to have regular cervical cytology tests, they must also ensure that each individual makes an informed choice about accepting the invitation to be screened. All women attending for cervical cytology sampling are made aware of the method of sampling, the limitations of the test, and when and how they will receive the result. Reassurance should be given that the majority of tests are 'normal', but the possible consequences of an 'abnormal' result must be explained.

Key Points

The sample taker is responsible for obtaining informed consent to the screening process from the woman prior to testing.

The sample taker is responsible for ensuring the sample is taken in the appropriate manner, for ensuring full and accurate completion of the request form, and for ensuring sufficient labelling of the sample to allow unique identification upon arrival at the laboratory. All the samples that have been sent to the laboratory must be recorded and there should be a system of ensuring that results have been received for each sample sent.

Finally, the sample taker must ensure the woman is informed of the result, provide an explanation of the result if necessary and ensure all recommendations contained in the report are appropriately actioned, for example early repeat tests or referral to the colposcopy clinic. Results that indicate the possible presence of invasive carcinoma must be communicated in person by the patient's GP or other responsible clinician.

The laboratory

The laboratory provides two services for the NHS Cervical Screening Programme: cytology and histopathology. To ensure appropriate quality of service, the laboratories must be accredited by Clinical Pathology Accreditation (UK) Ltd (CPA). Laboratories that are registered but not approved are closely monitored by the regional Quality Assurance Reference Centre (QARC) to ensure the quality of service provision.

Cytology

The processing of cytology samples, including cervical samples, is discussed in greater detail in Chapter 2. However, the general role of the laboratory will be outlined here. Cervical cytology samples are received from primary care hospital clinics and occasionally hospital in-patients. The samples are checked to ensure there are at least three matching patient identifiers (forename, surname, date of birth, NHS number) on the sample and accompanying request form. The samples are then allocated a unique request number and booked into the laboratory computer system.

SELF-CHECK 3.5

What is the minimum number of patient identifiers required for unique identification of a patient/sample?

All members of staff must be trained and hold the appropriate qualifications to demonstrate attainment of competence in the tasks they undertake. Further, all individuals undertaking screening and reporting must demonstrate continued competence by having regular update training, having their performance monitored internally, and by participating in the national **External Quality Assurance (EQA)** Scheme for Gynaecological Cytopathology. The laboratory must also participate in the national EQA Scheme for the Evaluation of Papanicolaou Staining.

All cytology reports must indicate the cytological findings (e.g. negative) and suggest an appropriate clinical management of the patient (e.g. repeat in six months). The clinical management is dependent upon the cytological findings and the patient's screening history. The report must be communicated to the sample taker, the patient's GP (if not the taker) and the call and recall office. This communication may be via an electronic link or via a paper copy of the report.

Key Points

The cytology laboratory is responsible for processing, screening, and reporting the cervical screening samples.

It is considered best practice that laboratories facilitate a **direct referral** system where patients requiring colposcopy or gynaecological investigations are contacted directly by the referral clinic to arrange an appointment. Direct referral helps to minimize the length of time between referral and appointment, allows the minimum of inconvenience of the appointment to the patient, and leads to higher clinic attendance rates. If direct referral is in place, the result must also be communicated to the appropriate referral clinic.

The success of the cervical screening programme is dependent upon the correct clinical management of the women screened. Although the cytology report contains the correct management for each woman, there must be procedures in place to ensure the management is undertaken. These procedures are referred to as **fail-safe**. The fail-safe procedures for early repeat tests are undertaken via the Exeter computer system in the call and recall office, but the laboratory has the responsibility of fail-safe for women referred for colposcopic examination or gynaecological investigations. These procedures will include sending questionnaires to clinicians regarding the management of patients and attempts to follow-up patients who fail or refuse to attend for assessment or treatment.

Histopathology

As well as providing cytology services to the NHS Cervical Screening Programme, the laboratory also provides **histopathology** services.

Patients who have been referred for abnormal cytology results or women referred for other clinical reasons, for example abnormal bleeding, may have a sample of their cervix removed for histological assessment. These samples are referred to as **biopsies**, and are taken in colposcopy clinics, other hospital outpatient clinics and operating theatres.

The biopsies are received in **formalin** fixative, which is a 10% solution of formaldehyde in water. The samples must be appropriately sliced and sampled, a process known as **dissection**, before being processed to paraffin wax (see below). However, some small biopsies are processed whole and therefore do not require dissecting. The sampled tissues are placed in a tissue cassette (see Figure 3.6) which is labelled with the unique laboratory number to ensure the chain of custody (knowing which patient the biopsy came from) for the sample is maintained during processing.

Cross reference
Fundamentals in Biomedical Science, Histopathology.

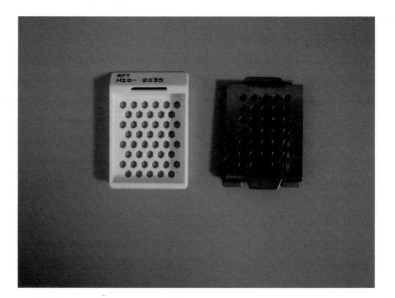

FIGURE 3.6
On the left is a tissue processing cassette. The sampled tissue is placed within the cassette and the metal lid (on the right) is snapped into position to enclose the tissue. The cassette and lid have holes to allow the passage of processing solvents and solutions. As you can see, the specimen number is printed on the cassette to enable identification.

Processing to paraffin wax involves total dehydration of the sampled tissue using alcohol, replacement of the alcohol with a clearing solution such as xylene, and finally impregnation of the tissue with paraffin wax. This process is undertaken in a programmable, automated, processing machine. The processing of the tissues is completely enclosed within the processing machines in order to contain the solvent fumes. Optimal processing is time consuming and is usually done overnight.

Following processing, the tissues are embedded in paraffin wax using appropriately sized moulds to give a wax block of tissue. The tissues must be orientated correctly in the mould to ensure appropriate cross sectioning at **microtomy**, described below. The labelled tissue cassette is incorporated into the block to maintain the chain of custody of the tissue.

The purpose of impregnating and embedding the tissues in paraffin wax is to facilitate microtomy, the cutting of very thin slices or **sections**, from the block of tissue. Microtomy is undertaken on a machine called a microtome, shown in Figure 3.7. The sections that are produced are usually between 2–5 micrometers (μm) thick. They are floated on warm water (in a water bath) and transferred to an appropriately labelled glass slide. The slide and section are allowed to dry and can then be stained to enable the tissues to be visualized during microscopy. The usual routine stain in the histology laboratory is the haematoxylin and eosin (H&E) stain.

All histology samples must be examined and interpreted by an appropriately qualified histopathologist. The report should make reference to the adequacy of the biopsy and whether the biopsy explains any cytology findings during screening. For example, 'This is an adequate cervical biopsy showing no abnormality and offers no explanation to the origin of the severely dyskaryotic cells present in the cervical cytology screening test.' There will be cases where the cytology result and histology result do not correlate; these must be discussed between histology, cytology, and colposcopy colleagues to decide the appropriate clinical management of the patient.

Key Points

The histology laboratory is responsible for processing, interpreting, and reporting cervical biopsies.

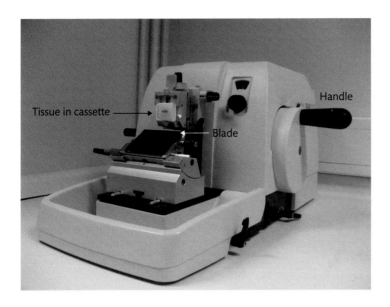

Tissue in cassette

Handle

Blade

FIGURE 3.7

This is a microtome. It is used to cut very thin slices (sections) of tissue biopsies. One turn of the handle advances the blade toward the tissue by a predefined, adjustable distance. The section thickness is equal to the distance of advancement.

The laboratory is responsible for producing the KC61 statistical return which is used as an aid to monitor the cytology reporting profile of the laboratory and the outcomes of referrals to colposcopy/gynaecology.

Cross reference

See Chapter 6 for more information regarding the KC61 statistical return.

Colposcopy services

Colposcopy is undertaken using a specialized microscope called a **colposcope**. Colposcopy is often facilitated by the application of a weak acetic acid solution or an iodine solution to the surface of the tissue to be examined.

The colposcopy services are usually provided within a hospital's gynaecology and/or genito-urinary medicine departments. However, there are also community-based colposcopy clinics. Colposcopy clinics receive referrals from primary care and may also operate a direct referral system for patients with abnormal cervical cytology results.

Colposcopy

The illuminated, magnified examination of the cervix, vagina, or vulva.

SELF-CHECK 3.6

Can you name two solutions that are applied to the cervix during colposcopy in order to identify abnormal tissue?

All colposcopists must be appropriately trained and must maintain their skills by attending update training and by examining the recommended number of new and follow-up patients.

During the colposcopic examination, the colposcopist must diagnose the degree of any disease identified during the procedure and take biopsies to confirm the diagnosis histologically. The biopsies may be small (less than 1 cm) punch biopsies taken for confirmation of the colposcopic diagnosis only, or they may be larger biopsies which provide both a diagnosis and treatment by excision of the diseased area. Cases where the patient receives excisional treatment at the first examination are referred to as 'see and treat' cases.

The treatment for diagnosed lesions is dependent upon the degree of abnormality and the preference of the treating clinician. Low-grade lesions (CIN1) can be treated using ablative treatment methods, which destroy the diseased tissue on the cervix. Examples of this are cold coagulation, where a heated probe is used to destroy the tissue, and laser treatment, which vaporizes the abnormal tissue. High-grade lesions should be treated using excisional treatment methods to allow excision margins to be assessed. Excision margins are the edges of the biopsies and are important in assessing whether all the diseased tissue has been removed and therefore whether the treatment has been successful. If diseased tissue is present at the excision margins, there is a possibility that there is part of the lesion still on the cervix. Examples of excisional treatment methods are large loop excision, where a hot wire loop is used to excise the lesion, and cone biopsy, where a knife is used to excise a cone of tissue containing the lesion.

SELF-CHECK 3.7

Why are excision margins important?

The colposcopy services will also perform the initial follow-up of patients they have treated for CIN, performing colposcopic examinations and cytology sampling. At the appropriate time, which is dependent upon the initial diagnosis and follow-up findings, the patients are discharged back to the primary care providers and the call and recall office.

Key Points

The colposcopy clinic is responsible for examining, diagnosing, and treating the lesions detected by cervical screening.

Cross reference
See Chapter 6 for more information regarding the KC65 statistical return.

Colposcopy clinics must complete a KC65 statistical return each annual quarter. The KC65 allows direct monitoring of the colposcopy services with respect to waiting times for appointments, method of treatment, and outcomes.

The National Office

The National Office of the NHS Cancer Screening Programmes is responsible for monitoring and improving the overall performance of the NHS Cervical Screening Programme. The National Office was set up in 1994 and aims to develop systems, guidelines, and policies which will assure a high quality of cervical screening throughout the country.

 SUMMARY

Screening

Success depends upon:

- High coverage
- A test that is reliable with a high sensitivity and a high specificity
- Proven methods of treatment/counselling
- Cost effectiveness

NHS Cervical Screening Programme

This aims to reduce the incidence and mortality of cervical cancer by detecting and treating potentially precancerous lesions of the cervix. It is delivered by multiple services including:

- Call and recall
- Sample takers and primary care
- Cytology and histology laboratories
- Colposcopy

All of these are of equal importance for the success of the programme.

 # FURTHER READING

- **Raffle AE and Muir Gray JA (2007)** *Screening–Evidence and Practice.* **Oxford University Press, Oxford.**

- **Wilson JMG and Jungner G (1968)** *Principles and Practice of Screening for Disease.* **Public Health Paper No. 34. WHO, Geneva.**

- **Information on and publications by the NHS Cervical Screening Programme are available from: http://www.cancerscreening.nhs.uk/cervical (accessed 2010).**

- **Information on and publications by Cervical Screening Wales are available from: http://www.screeningservices.org.uk/csw (accessed 2010).**

- **Information on and publications by the cervical screening programme in Scotland are available from: http://www.healthscotland.com/topics/health-topics/screening/cervical.aspx (accessed 2010).**

- **Information on and publications by the cervical screening programme in Northern Ireland are available from: http://www.cancerscreening.n-i.nhs.uk/cervical/toc.html (accessed 2010).**

 # DISCUSSION QUESTIONS

1. Discuss the basic factors which must be considered when configuring and implementing a screening programme.

2. Describe the aims of cervical screening and the rationale of how this is achieved.

3. Describe the roles of the various agencies involved in the NHS Cervical Screening Programme.

Answers to the self-check questions, and tips for responding to the discussion questions, are provided in the book's Online Resource Centre.

 Visit www.oxfordtextbooks.co.uk/orc/shambayati/

4

Normal cervical cytology

Andrew Evered

Cytological screening for **precancerous** lesions of the cervix has been undertaken in the UK since the early 1960s. Cervical cytology accounts for the bulk of the workload of most cytology laboratories in this country.

> ## Key Points
>
> Cervical precancer is an abnormal growth of tissue in the cervix that may, in time, develop into cancer. Cancer is a complex and often fatal disease characterized by the uncontrolled growth of cells that invade surrounding tissues and spread to distant body sites.

This chapter aims to build on the principles outlined in Chapter 3. By now you should have a good understanding of the rationale for cervical screening and know how samples are collected and processed in the laboratory. We will now turn from these general principles to the specifics of cervical cytology. After an introductory description of the anatomy, physiology, and histology of the female genital tract we will examine the cytological features of normal cervical epithelial cells and other normal constituents of cervical samples. We will also look at the cytological features of inflammation, cervical infections, metaplasia, and iatrogenic changes, each of which can so often confuse the inexperienced cytologist.

Learning objectives

By the end of this chapter you should be able to:

■ Describe the normal anatomy, physiology, and histology of the female genital system.

■ Describe the cytological features of normal epithelial cells.

■ Explain the hormonal variations observed in cervical samples.

■ Describe the variation in cellular appearances caused by inflammation, metaplasia, and iatrogenesis.

■ List the non-epithelial cell types encountered in cervical samples.

■ List the pathogenic and non-pathogenic microorganisms that can be seen in cervical samples.

■ Define the criteria for judging the adequacy of cervical samples.

4.1 Normal anatomy and histology of the female genital system

The female genital system comprises the vulva, vagina, cervix, uterus, fallopian tubes, and ovaries. These structures are depicted in Figure 4.1. The histological appearance of the cervical **epithelium** is incorporated in the diagram.

The **vulva** protects the openings of the **vagina** and **urethra**, and consists of two pairs of skin folds, the **labia majora** and **labia minora**. These skin folds are lined by **keratinizing stratified squamous epithelium**.

Key Points

The vulva is the only part of the female genital system that normally develops keratin. This protein provides a tough layer of protection for the delicate internal structures.

Key Points

Stratified squamous epithelium contains multiple cell layers and is highly specialized for providing protection. It may be keratinized, as in the vulva, or non-keratinized such as in the vagina and cervix.

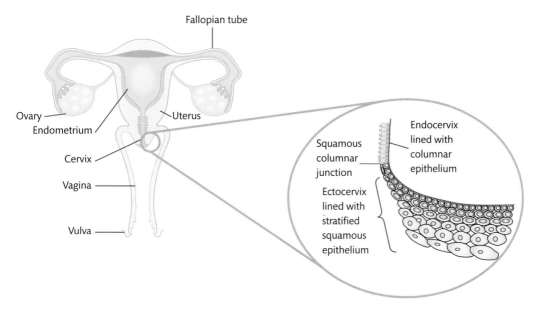

FIGURE 4.1
Anatomy of the female genital system.

The vagina is a muscular canal leading from the vulva to the **cervix**. It forms the lower part of the birth canal and is approximately 10 cm long in adulthood. The cells that line the vagina show cyclical variations according to circulating levels of the sex hormones oestrogen and progesterone. Under normal conditions the lining of the vagina consists of **non-keratinizing stratified squamous epithelium**.

The cervix is the neck of the uterus and forms the first part of the so-called birth canal. It is about 4 cm in length and consists of an inner canal (the **endocervix**) lined by a single layer of **columnar epithelial cells**, and an outer portion (the **ectocervix**) lined by non-keratinizing stratified squamous epithelium.

Columnar cells
So named because they are tall and narrow.

The point at which the endocervical canal opens out into the vagina is called the **external os** (to distinguish it from the **internal os**, which describes the opening from the endocervical canal into the uterus). The stratified squamous epithelium of the ectocervix is divided into four layers of cells: the **basal**, **parabasal**, **intermediate**, and **superficial** layers, as shown in Figure 4.2.

Key Points

Basal cells are the least mature of the squamous epithelial cells in the cervix, followed by parabasal cells and intermediate cells. Superficial cells are the most mature surface cells of stratified squamous epithelium.

Undifferentiated cells
Cells that completely lack specialization and have an ill-defined morphology.

The columnar cells of the endocervix, shown in Figure 4.3, are derived from **reserve cells**, which lie beneath and between the columnar cells. Reserve cells are **undifferentiated cells**, characterized by their ability to divide and differentiate into a number of different cell types, depending on genetic cues and influences from the local tissue environment. In the cervix, reserve cells usually differentiate into endocervical cells, but in certain circumstances (such as **squamous metaplasia**—see Section 4.3 below) they can develop into squamous epithelial cells.

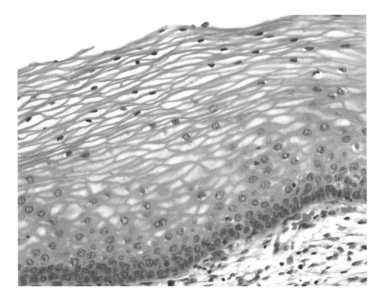

FIGURE 4.2
Stratified squamous epithelium of the ectocervix showing the basal, parabasal, intermediate, and superficial layers.
Reproduced with kind permission from Public Health Wales.

Appropriate sample processing is essential to ensure the resulting cellular preparation is representative of the submitted sample and is suitable for screening. Unrepresentative or unsuitable cellular preparations may result in false negative results and are therefore detrimental to patient care.

SELF-CHECK 6.5

How is the training for sample preparation delivered?

Primary screening

Primary screening is the term used to describe the initial, complete microscopic analysis of a cellular preparation. All individuals undertaking primary screening must:

- Hold the City & Guilds Diploma in Cervical Cytology (or equivalent, e.g. NHS Cervical Screening Programme certificate of competence). This evidences the attainment of the required skills and knowledge to practice.
- Have successfully completed an LBC conversion course if their initial training was undertaken in conventional cytology.
- Screen a minimum of 3,000 cellular preparations per annum to ensure the necessary exposure to cellular material to facilitate the maintenance of competence to practice.
- Achieve a primary screening sensitivity of 90% for all grades of abnormality and 95% for high-grade abnormalities (see later) to demonstrate competence in the detection of cellular abnormalities.
- Participate in the NHS Cervical Screening Programme EQA Scheme for Gynaecological Cytopathology (see later) to demonstrate competence to practice.
- Undertake three days of recognized update training every three years to ensure their knowledge and skills are appropriate for current practices.

Primary screening has three possible outcomes:

- No abnormal cells present (negative).
- Inadequate sample for screening purposes.
- Potentially abnormal cells present.

Negative and inadequate samples are reported by the primary screener, but potentially abnormal samples are referred for checking by a member of staff with enhanced skills. The primary screener's opinion must be recorded and available to the checker to facilitate the identification of gaps in the knowledge and skills of the primary screener, and therefore identify and address training requirements.

Rapid review/preview

All samples reported as negative or inadequate at primary screening must be partially re-screened by a qualified individual, other than the primary screener. This partial re-screen is referred to as rapid review. In some departments this additional partial screen is undertaken prior to the primary screen and is referred to as rapid *preview*. Partial screening involves screening the sample at a normal pace but not covering all of the cellular material as intensively as in primary screening. Therefore the 'rapid' description refers to the time it takes to complete the task rather than the speed at which the cellular preparation is moved during screening. If the rapid (p)review outcome is in agreement with the primary screening report, the report is issued. Samples with discrepancies between the primary screening report and the rapid (p)review outcome must be referred for checking.

Rapid (p)review is a quality control (QC) method, and is designed to detect errors which may occur during primary screening. If the error consists of abnormal cells which have not been detected at primary screening, the sample is referred to as a primary screening false negative. However, because the error has been detected, the sample will not be reported as negative and is therefore not classed as a false negative in the wider context of the screening programme. (False negatives within the programme are those screening errors which remain undetected, and sampling errors.)

Cross reference

See Chapter 3 for the formula used to calculate the sensitivity.

Abnormal samples accurately identified at primary screening are referred to as primary screening true positives. Having qualified the false negatives and true positives, the sensitivity of primary screening can be calculated.

Because rapid preview is undertaken before primary screening, all abnormal samples are rapid previewed. Just as abnormal samples missed at primary screening but detected at rapid (p)review enable the sensitivity of primary screening to be calculated, the abnormal samples missed at rapid preview but detected at primary screening enable the sensitivity of rapid preview to be calculated. A similar process is not possible with rapid review because the only abnormal samples that are reviewed are the primary screening false negatives. Therefore rapid preview allows a measure of effectiveness of the QC method that rapid review cannot provide. However, there are no national quality standards regarding the sensitivity of rapid preview, and there are no recognized teaching methods to improve the sensitivity of rapid preview, making any requirement for remedial action difficult to prove or achieve.

SELF-CHECK 6.6

What are the three possible outcomes of primary screening?

Checking

Samples identified as containing potentially abnormal cells, and samples with discrepancies between the primary screening report and the rapid (p)review outcome, must be referred for checking. Checking is a secondary complete screen of the cellular preparation and is undertaken by individuals who have demonstrated enhanced knowledge and skills beyond those required for primary screening. The enhanced knowledge can be demonstrated by additional qualifications such as a Master's degree (MSc), or the Institute of Biomedical Science (IBMS) higher specialist diploma. The enhanced skills will be demonstrated through records of professional development and audit. All individuals undertaking checking must:

- Hold the City & Guilds Diploma in Cervical Cytology (or equivalent, e.g. NHS Cervical Screening Programme certificate of competence). This evidences the attainment of the required skills and knowledge to practice.

- Have successfully completed an LBC conversion course if their initial training was undertaken in conventional cytology.

- Check a minimum of 750 cellular preparations per annum to ensure the necessary exposure to cellular material to facilitate the maintenance of competence to practice.

- Participate in the NHS Cervical Screening Programme EQA Scheme for Gynaecological Cytopathology (see later) to demonstrate competence to practice.

- Undertake three days of recognized update training every three years to ensure their knowledge and skills are appropriate for current practices.

Reporting abnormal samples

Samples deemed to be negative or inadequate by the checker are reported as such. Samples which are very difficult to interpret or confirmed as containing potentially abnormal cells are referred to a consultant cytopathologist or an appropriately qualified biomedical scientist for reporting. All individuals undertaking reporting of abnormal samples must:

- Have successfully completed an LBC conversion course if their initial training was undertaken in conventional cytology.
- Report or review a minimum of 750 cellular preparations per annum to ensure the necessary exposure to cellular material to facilitate the maintenance of competence to practice.
- Participate in the NHS Cervical Screening Programme EQA Scheme for Gynaecological Cytopathology (see later) to demonstrate competence to practice.

In addition to the above, all biomedical scientists undertaking reporting of abnormal samples must:

- Hold the City & Guilds Diploma in Cervical Cytology (or equivalent, e.g. NHS Cervical Screening Programme certificate of competence).
- Hold the advanced specialist diploma in cervical cytology. This evidences the attainment of the required skills and knowledge to practice.
- Report a minimum of 500 abnormal samples within the annual minimum of 750.
- Undertake three days of recognized update training every three years to ensure their knowledge and skills are appropriate for current practices.

Table 6.1 shows the possible outcomes for cervical cytology following screening, checking, and reporting. Also included are how the patients are clinically managed for each report category and the reasons for this.

Fail-safe

The cytology laboratory must have a fail-safe mechanism which ensures that all patients referred for colposcopic/gynaecological assessment are offered an appointment and that sufficient measures are undertaken to encourage the patient to attend. The laboratory liaises with the sample taker and/or GP, who in turn contacts the patient, initially in writing and then face to face. All correspondence must be recorded for audit purposes. Ensuring patients attend for the appropriate investigations and treatment helps to maintain the effectiveness of the screening programme. More details can be obtained from the NHS Cervical Screening Programme publication number 21.

SELF-CHECK 6.7

What is the purpose of fail-safe?

Monitoring laboratory effectiveness

The cytology laboratory is responsible for monitoring its own effectiveness within the screening programme. This is facilitated by the Department of Health requirement for the laboratory to produce the annual KC61 statistical return which is made up of three parts:

A   (1) Breakdown of tests by result and source.

     (2) Turnaround times for tests.

     (3) Number of tests sent to/received from another laboratory.

TABLE 6.1 This table shows the possible outcomes of cervical cytology and the patient management associated with each outcome.

Outcome	Clinical management and reason
Inadequate	**Early repeat** (3 months). The sample is deemed unrepresentative of the disease status of the cervix, therefore a repeat is required.
Negative	Either **normal recall**. Or **early repeat** in accordance with screening history (6 or 12 months).
Borderline changes	Either **early repeat** (6 months). Or **refer for colposcopic assessment** if: • 3rd occurrence. • High-grade disease is suspected. • The changes are in endocervical cells. The significance of the cell changes present is uncertain. The changes may regress (repeat samples to monitor) or may be the manifestation of a high-grade disease (refer). If repeat samples are requested, the patient can only be returned to normal recall after 3 consecutive negative results.
Mild dyskaryosis	**Early repeat** (6 months) is acceptable practice for the 1st occurrence as the cell changes may regress. **Referral for colposcopic assessment** is seen as best practice for the 1st occurrence but is mandatory for the 2nd occurrence. Up to 20% of patients referred for mild dyskaryosis have high grade disease. If repeat samples are requested, the patient can only be returned to normal recall after 3 consecutive negative results.
Moderate dyskaryosis	**Referral for colposcopic assessment**. Approximately 75% of patients with this result will have high-grade disease.
Severe dyskaryosis	**Referral for colposcopic assessment**. Approximately 75% of patients with this result will have high-grade disease.
Severe dyskaryosis/ ? invasive carcinoma	**Urgent referral for colposcopic assessment**. The patient is suspected of having cervical carcinoma.
? Glandular neoplasia	**Urgent referral for colposcopic/gynaecological assessment**. The patient is suspected of having carcinoma (cervical, uterine or extra-uterine).

B Breakdown of GP and community tests by result and age.

C Breakdown of referrals to colposcopy/gynaecology by test result and outcome. This also includes calculating the positive predictive value (PPV—see Chapter 3) of moderate dyskaryosis or worse resulting in cervical intraepithelial neoplasia grade 2 (CIN 2) or worse, and calculating the percentage of patients lost to follow-up.

The laboratory quality standards that are derived from this statistical return are:

• Inadequate results as a percentage of all samples.

• Borderline changes plus mild dyskaryosis as a percentage of all adequate samples from women aged 25–64 screened at GP and community clinics.

• Moderate dyskaryosis or worse as a percentage of all adequate samples from women aged 25–64 screened at GP and community clinics.

• PPV (as above).

The national quality standards are published annually as standard ranges within which the laboratory performance is expected to lie. The ranges reflect the 10th to 90th percentile ranges of the nationally submitted KC61 returns from the previous year. Laboratory performance which is outwith these ranges must be investigated as it may indicate over/under interpretation of the samples, resulting in inappropriate clinical management of patients and decreased effectiveness of the screening programme.

External quality assurance schemes

As mentioned above there are two mandatory national EQA schemes in the NHS Cervical Screening Programme. Although national schemes, they are both locally run and overseen by the regional Quality Assurance Reference Centre.

The EQA Scheme for the Evaluation of Papanicolaou Staining in Cervical Cytology was introduced in 2004, with an aim to optimize and standardize the staining quality of cervical screening samples throughout the programme. The scheme is run quarterly and all laboratories must submit four slides, two of which are randomly selected for evaluation by a panel of trained assessors. The staining characteristics of the nuclei and cytoplasm of the cells are numerically scored in accordance with their correspondence to set criteria. The total score provides one of four possible outcomes:

- Good
- Acceptable
- Marginal
- Substandard

Outcomes of 'substandard' will result in action points either within the assessment panel, at a local level, or at a national level, dependent upon the number and frequency of such outcomes. (More details can be obtained from the NHS Cervical Screening Programme publication number 19.)

The EQA Scheme for Gynaecological Cytopathology was introduced in 2003 and replaced the proficiency testing scheme which had run since the early 1990s. The aim of the scheme is to identify and remedy potential individual poor performance within the NHS Cervical Screening Programme. Initially the scheme was run annually but since 2008 it has been run twice per year. All laboratories are asked to submit cellular preparations with specific outcomes and consensus opinions at screening, checking, and reporting levels. These submitted preparations are then distributed to an assessment panel to ensure the appropriateness of their inclusion into the scheme, and to exclude examples that do not fulfil the specified criteria.

Slide sets of ten preparations which satisfy the panel are compiled and delivered to each laboratory. All individuals undertaking screening, checking, or reporting of cervical screening samples must examine the slides in the set and submit responses for each preparation. The submitted responses for the slide sets are used to obtain a peer consensus opinion for each slide. These consensus opinions are deemed the correct responses, and are used to mark the submitted responses of each individual. Scores below the bottom 2.5th centile in the range of marks within each peer group are deemed poor performance, as are clinically serious errors (missed high-grade abnormalities). Persistent poor performance is defined as two episodes of poor performance in any three rounds, and results in action points for the re-education of the individual concerned. More details can be obtained from the NHS Cervical Screening Programme publication number 15.

SELF-CHECK 6.8
What does the acronym EQA stand for?

Systemic cytology

Unlike cervical cytology, which employs one method of sampling and only two methods of sample preparation (see Chapter 2), there are a variety of methods employed in systemic cytology to prepare the numerous different sample types for analysis. These will not be detailed here but include direct spread, LBC methods (as in cervical cytology), and cyto-centrifugation (whereby the material is transferred to a glass slide by centrifugation). Whichever method is employed (and this will vary between departments and between different specimen types within the same department) the basic quality principles remain the same.

The sample must be labelled with at least three patient identifiers and be accompanied by a completed matching request form stating the date and time of sampling and the investigative procedures required to be undertaken. Correct labelling and identification of samples is vitally important (see cervical cytology above). Recording the date and time of sampling, and the date and time of reception in the laboratory, facilitates auditing of the time taken for the sample to reach the laboratory. This is important for two reasons. First, unlike cervical cytology, systemic cytology samples are often received unfixed and therefore susceptible to deterioration of the cellular material. This deterioration can lead to over or under interpretation of any cell changes present, resulting in incorrect reports and inappropriate clinical management of patients. Second, because systemic cytology samples are often diagnostic samples, the timeliness of the report is important, as delayed treatments can be detrimental to patient outcomes.

Specific analyses may be required on some systemic cytology samples. Examples of these include differential cell counts (the proportion of different types of cells present in the sample) and the presence of certain crystals. These analyses must be specifically requested by the clinician to ensure the final report contains all the relevant clinical information required to provide the optimal clinical management of the patient.

Sample taking for some systemic cytology samples is an invasive procedure, often uncomfortable for the patient and requiring considerable clinical resources. For this reason, it is important that the initial sampling contains adequate material for diagnostic purposes. In an attempt to ensure adequate sampling, many departments attach an appropriately trained/qualified member of staff to the sampling clinic. Rapid processing and analysis of directly spread samples enables trained biomedical scientists to confirm sample adequacy, or enables consultant cytopathologists to report the sample in the clinic.

As with cervical cytology, all staff processing systemic cytology cellular samples must be competent in the systems being employed to produce the cellular preparations. The training required to ensure this competence must either be delivered by the manufacturers of the system (as in LBC systems) or by validated, in-house competency training with demonstrable outcomes. The quality of the preparation and staining of the samples must be checked and recorded by an appropriate member of staff, usually an experienced biomedical scientist.

All systemic cytology samples must be reported by appropriately qualified staff, which includes medically qualified staff and appropriately trained biomedical scientists, depending upon the type of sample. However, in many departments the samples are pre-screened, usually by a biomedical scientist, prior to reporting. This is important for two reasons. First, it gives trainees and potential trainees exposure to these samples, thus facilitating training. (The screener's opinion should be recorded to allow feedback.) Second, it means the sample has been screened twice before reporting (once by the screener and once by the reporter), helping to reduce the risk of erroneous reports due to detection errors.

There should be regular auditing of the cytology reports against patient diagnoses and outcomes to highlight areas of potential poor sample preparation and inaccurate interpretation of the cellular material. Through these audits, changes in practice and additional education and training can be initiated to improve the quality and therefore clinical value of the service.

6.6 Quality issues in colposcopy services

A description of colposcopy services is provided in Chapter 3. The quality issues in colposcopy will be discussed here. All individuals undertaking colposcopy must hold the Royal College of Obstetricians and Gynaecologists (RCOG)/British Society for Colposcopy and Cervical Pathology (BSCCP) Certificate in Colposcopy. To be awarded this certificate they must pass an examination after completing a structured theoretical and practical training programme that involves undertaking:

- Fifty colposcopy cases under direct supervision, of which twenty must be new cases and ten must be high-grade disease.
- One hundred colposcopy cases under indirect supervision, of which 30 must be new cases and 15 must be high-grade disease.
- Completion of a training log book.
- Microscopy training sessions in histopathology and cytopathology.

All trained colposcopists must attend at least one BSCCP recognized colposcopy meeting every three years, and see at least 50 new abnormal cytology referrals each year to ensure the necessary exposure to cervical disease to facilitate the maintenance of competence to practice. Over- or under-interpretations of colposcopic appearance can lead to over- or under-treatment of patients.

Key Points

Accurate colposcopy assessments help to ensure diagnoses and treatments are appropriately undertaken in a timely manner, minimizing the discomfort and anxiety caused to patients.

Women are referred to colposcopy because of cytology results or for other clinical reasons. The reason for referral must be available at all colposcopic examinations, and the actual grade of cytological abnormality must be known in at least 90% of cytological referrals. Knowledge of the grade of the cytology abnormality is helpful, because the grade can indicate to the colposcopist the site and type of colposcopic appearances to expect. At least 90% of women referred for abnormal cytology results should be seen in colposcopy clinic within eight weeks of the referral. However, of those women referred with possible invasive disease (reports of severe dyskaryosis/? invasive carcinoma and ? glandular neoplasia), 90% must be seen within two weeks of the referral.

Key Points

Timely diagnoses and treatments help to ease the anxiety of patients and help to prevent further progression of the disease following cytological identification.

As with sample taking, counselling of patients and the gaining of consent for colposcopy and treatments is vitally important. Women must fully understand what to expect before, during, and after the process, and what the possible outcomes and consequences are likely to be.

If women are diagnosed and treated at the first visit to colposcopy, there must be histological evidence of cervical intraepithelial neoplasia (CIN) or cervical glandular intraepithelial neoplasia (CGIN) in more than 90% of cases. This ensures that treatments are not being undertaken unnecessarily, causing needless discomfort and distress to patients and wasting valuable resources within the screening programme.

Following the visit to colposcopy, the patient and the referring practitioner must be informed of the results of the investigations within four weeks (best practice) or eight weeks (minimum standard). The timely communication of outcomes reduces patient anxiety, improves patient experience and encourages compliance with subsequent follow-up and screening.

The colposcopy department is responsible for monitoring its own effectiveness within the screening programme. This is facilitated by the Department of Health requirement to produce the quarterly KC65 statistical return which is made up of five parts:

A—Breakdown of women referred to colposcopy by referral indication and result of referral.

B—Breakdown of women referred to colposcopy by result of referral smear and time from referral to first appointment.

C—Breakdown of first attendances by type of procedure and result of referral.

D—Breakdown of biopsies, by time from biopsy to informing patient of result.

E—Breakdown of biopsies by type and outcome.

6.7 Quality issues in histology services within the Cervical Screening Programme

A description of histology services is provided in Chapter 3. The quality issues in histology will be discussed here. The purpose of histology is to provide a report on the tissue samples (biopsies) taken from the cervices of women identified by cytology and/or colposcopy, as having potentially significant cervical disease. This is important because the histological outcome of colposcopically directed biopsies is often seen as the 'gold standard' against which the accuracy of the cytology report is measured. It is also important because the histological findings (presence and degree of CIN or CGIN) dictate the clinical management of patients:

- No CIN—complete the appropriate follow-up for the referral cytology result and then return to normal recall.
- CIN 1–2-year negative cytology follow-up before returning to normal recall.
- CIN 2, 3 or CGIN—10-year negative cytology follow-up before returning to normal recall.

The biopsies are received in the laboratory in a pot of fixative, accompanied by a request form. The pot and form must have a minimum of three matching patient identifiers. The processes involved in producing the report require transfer of part or the whole biopsy from the labelled pot of fixative to a plastic cassette. Following impregnation of the biopsy with paraffin wax, a thin section is cut and transferred to a glass slide. It is vitally important that accurate labelling is used during these transfers to safeguard the identity of the samples, therefore ensuring patients receive the correct report and subsequent clinical management.

The tissue sections produced are stained with haematoxylin and eosin (H&E) to provide contrast for microscopy (see Chapter 2). The stained tissue sections are checked for suitable quality by a senior biomedical scientist prior to reporting by a histopathologist.

All histopathology specimens must be interpreted and reported by appropriately qualified medical staff. Wherever possible, any relevant cytology should be reviewed when the biopsy is reported. If the initial histopathological findings do not correspond with those predicted by the cytology and/or colposcopy appearances, further tissue sections should be cut deeper into the biopsy to ensure the material examined is representative of the whole biopsy. The histology report should state the adequacy of the biopsy and whether the findings account for the cytological/colposcopic appearances.

As with cytology and colposcopy, the histology report is an individual's subjective opinion of the microscopic appearances of the sampled tissue. And, as with all opinions, these are subject to over- and under-interpretation.

Key Points

Correlation between cytology, colposcopy, and histology is an important audit to optimize the clinical management and outcomes of patients (see multidisciplinary meetings below).

All histology laboratories must participate in the appropriate EQA schemes to ensure the technical preparation and microscopic interpretations of the biopsies are of the appropriate standard. The production of poor sections and poor staining can result in inaccurate interpretations, leading to inappropriate clinical management of patients, which in turn is detrimental to the effectiveness of the screening programme.

SELF-CHECK 6.9

What is the 'gold standard' against which the accuracy of the cytology report is measured?

6.8 Multidisciplinary meetings

It is important that the outcomes of patients referred for colposcopic or gynaecological assessment are monitored. Discrepancies between cytological, colposcopic, and histological interpretations must be discussed at multidisciplinary meetings to decide the appropriate clinical management of the patient. All three disciplines must be represented at these meetings by staff with an appropriate level of knowledge and responsibility, for example consultant medical and consultant non-medical staff. The cases should be reviewed prior to the meeting and the findings of the review presented. The cases discussed, the clinical management decisions made, and those in attendance at these meetings must be documented.

Cross reference
Chapter 7, Section 7.5.

6.9 Quality assurance reference centre

The regional directors of public health are responsible for the coordination and quality of the Cervical Screening Programme within the NHS. Each government office region has a Quality Assurance Reference Centre (QARC), which is an NHS body lead by the regional quality assurance director who is accountable to the regional director of public health. The purpose of the QARC is to:

- Ensure the national standards set for the screening programme are met, by monitoring the statistical returns from each agency within the region.
- Coordinate the national EQA schemes at a regional level.
- Coordinate the national cervical cancer audit (see below) at a regional level.
- Undertake quality assurance visits to agencies within the programme, to ensure facilities and resources are suitable for service delivery.
- Maintain close liaison with key professionals, to facilitate clear lines of communication within the screening programme.
- Facilitate the introduction of new developments, e.g. LBC.
- Identify and investigate areas of poor performance.
- Identify, and encourage the adoption and transfer of good practice.

By facilitating the maintenance of minimum standards and helping to improve the performance of all aspects of cervical screening, the QARCs ensure that all women have access to high quality cervical screening services.

6.10 Cervical cancer audit

The purpose of audit is to monitor the effectiveness of the screening programme and to identify areas of good practice and areas where improvements can be made. It has been a requirement for many years that all women identified as having cervical cancer must have their screening history audited and the result of this audit made available to the woman, if she so wishes. However, until 2006 there was no national protocol for conducting these audits. Consequently, there was no uniformity nationally, and no method of collating and analysing the information derived at local levels.

NHS Cervical Screening Programme publication 28 *Audit of Invasive Cervical Cancers* sets out a protocol by which all women who develop cervical cancer have all aspects of their screening history audited. This includes:

- A review of the call and recall invitations sent to the patient.
- A review of the woman's uptake of the invitations.
- A full review of all available cytology samples and suggested clinical management.
- A review of all colposcopy/gynaecology invitations, attendances, colposcopic findings, treatments, and follow-up.
- A full review of all available histology samples.

For each woman developing cervical cancer, two women that have not developed cervical cancer, but who match for age and area of residence are selected as controls. The selection of these controls is done automatically by the NHAIS system. Comparing the screening histories of the women who have developed cervical cancer with the controls may help to identify areas within the programme that have failed or areas that require improvement.

On a local level, the audits are coordinated by the Hospital Based Programme Coordinator (HBPC). Regionally the audits are overseen and data collated by the QARC. The national office of the NHS Cervical Screening Programme works in conjunction with the regional QARCs, cancer registries, and Cancer Research UK to collate and analyse the data to produce information about the performance of the national screening programme. This information should allow the identification of best practice and highlight areas where improvements can/must be made.

 # SUMMARY

- The appropriate training and education of the staff responsible for the delivery of the component services underpin the quality of a cervical screening programme.

- Individual competencies must be achieved and maintained, and agencies must record all procedures and activities undertaken to facilitate this.

- The quality of performance of the individuals and agencies involved in cervical screening can be monitored by their adherence to the quality standards set by the programme.

- Adherence to the quality standards does not guarantee acceptable performance 100% of the time, but demonstrates a commitment to achieve and maintain the appropriate quality of the service provided.

- Monitoring of the quality standards, multidisciplinary meetings, and the findings of clinical audit, allow the cervical screening services provided to individuals and the population as a whole, at local and national levels, to be continually improved.

- A perfect screening programme may be unobtainable, but high quality screening services delivered to the target population at the appropriate screening interval are extremely effective at reducing the incidence and mortality of cervical cancer.

 # FURTHER READING

Information on and publications by the NHS Cervical Screening Programme are available from http://www.cancerscreening.nhs.uk/cervical. Particular attention to quality is given in:

- **Publication No. 1**
  Achievable Standards, Benchmarks for Reporting, and Criteria for Evaluating Cervical Cytopathology
  May 2000

- **Publication No. 12**
  Qualifications and Training for Non-medical Laboratory Staff Working in the UK Cervical Screening Programme
  January 2000

- Publication No. 15
  External Quality Assessment Scheme for Gynaecological Cytopathology
  March 2009

- Publication No. 18
  Cervical Screening Call and Recall: Guide to Administrative Good Practice
  February 2004

- Publication No. 19
  External Quality Assessment Scheme for the Evaluation of Papanicolaou Staining in Cervical Cytology, Protocol, and Standard Operating Procedures
  April 2004

- Publication No. 20
  Colposcopy and Programme Management: Guidelines for the NHS Cervical Screening Programme
  May 2010

- Publication No. 23
  Taking Samples for Cervical Screening—a Resource Pack for Trainers
  April 2006

- Publication No. 27
  Improving the Quality of Written Information Sent to Women about Cervical Screening: Guidelines on the Content of Letters and Leaflets
  December 2006

- Publication No. 28
  Audit of Invasive Cervical Cancers
  December 2006

- Juran JM and Blanton Godfrey A (1999) *Juran's Quality Handbook*, fifth edition. McGraw-Hill, New York.

- *Recommended Code of Practice for Laboratories Participating in the UK Cervical Screening Programmes 2010.* British Society for Clinical Cytology. http://www.clinicalcytology.co.uk/resources/cop/BSCC_COP_2010.pdf (accessed 2010).

DISCUSSION QUESTIONS

1. Discuss the term 'quality' and what is meant when it is related to laboratory medicine and cervical screening.
2. Discuss the use of quality standards in cervical screening.
3. Describe the process of cervical cancer audit and how this can benefit a cervical screening programme.

Answers to the self-check questions, and tips for responding to the discussion questions, are provided in the book's Online Resource Centre.

 Visit www.oxfordtextbooks.co.uk/orc/shambayati/

Diagnostic cytopathology

Andrew Evered

In Chapters 4 and 5 you learned that cervical cytology is an excellent example of the application of cytological principles. However, the cytological identification of cancer and precancerous conditions is not limited to the female genital tract. As early as 1860, sputum cytology was used for the identification of cells from lung cancer. In fact, cells can be removed from any suspected **malignant** site, spread on a glass slide or collected into a fluid fixative, processed by the laboratory, and examined by the cytologist to reach a quick diagnosis.

Diagnostic cytopathology is the diagnosis of disease at the cellular level. It involves the preservation, staining, and microscopic examination of cells that have been extracted from the body. A variety of human diseases can be diagnosed in this way, including **infections**, **crystal deposition diseases**, and **benign tumours**, but the primary aim of our discipline is the diagnosis of **cancer** and **pre-cancerous conditions**.

Key Points

Diagnostic cytology deals primarily with the diagnosis of cancer.

This chapter will prepare you for what is to come in the remainder of the book. In subsequent chapters you will be taken on a remarkable cytological journey that will take in virtually all body sites and a multiplicity of tumour types.

Learning objectives

By the end of this chapter you should be able to:

- Discuss the general principles of diagnostic non-cervical cytology.
- Discuss the role of cytology in the diagnosis of neoplasia.
- Describe the nomenclature and classification systems for neoplastic conditions.

- Outline the mechanisms of tumour genesis and metastasis.
- Outline the mechanisms involved in a range of benign cellular processes.
- Explain why both benign and malignant conditions can sometimes confuse cytological decision making.
- Outline the role of the cancer multidisciplinary team.

7.1 The general principles of diagnostic, non-cervical cytology

Neoplasm
An abnormal growth of new tissue.

Cytology can be performed on virtually any type of tissue specimen to look for **neoplastic** cells. The existence of cervical screening programmes worldwide is testament to the great utility of cytology for the detection and diagnosis of neoplasia. It is therefore no wonder that many clinicians and cytologists recognize that the same principles can be applied to a wide variety of bodily specimens.

The previous chapters provide us with a useful platform from which to explore the field of diagnostic cytology. Briefly, we can summarize the underlying principles of diagnostic cytology as follows:

1. Cancer cells lose their cohesive properties early in the disease process. This loss of cell cohesion encourages the shedding of cells from tissue surfaces, thereby increasing the likelihood that cancer cells will be detected.

2. Specimen collection is normally simple, painless, and rapid.

3. In many clinical cases, a cytological diagnosis can remove the need for exploratory surgery.

4. As discussed in Chapter 2 with cervical cytology, our discipline can be effectively applied to the screening of high-risk groups of individuals for specific diseases.

5. At present there is no other comparable test, in terms of clinical effectiveness and cost effectiveness, for the detection of unsuspected and asymptomatic tumours.

6. In addition to the early detection of cancer, cytology can also be used for diagnosing infections and a variety of other non-cancerous conditions.

SELF-CHECK 7.1

What makes cytology such an appealing diagnostic test for the clinician who suspects cancer in a patient?

Screening versus diagnostic cytology

The fundamental difference between screening and diagnostic cytology lies with the individuals from whom specimens are taken. Cytological screening aims to detect neoplasia in an essentially healthy population group as early in the disease process as possible, before any symptoms become apparent. In such populations disease prevalence will be relatively low and achieving high test **sensitivity** is of paramount importance. The aim is to detect as many cases of disease as possible whilst minimizing the occurrence of positive tests in individuals who do not have the disease in question (**false positives**) and of negative test results in individuals who do have the disease (**false negatives**). In reality, cytological screening is acceptable for only a few types of cancers.

Key Points

The **sensitivity** of a test for the detection of a disease is a way of comparing the number of cases correctly identified with the number which are missed. It is calculated by dividing the number of correctly identified cases by the total number of cases in the tested population. The fewer the number of missed cases, the higher will be the test sensitivity.

Cross reference

Chapter 3, Section 3.1.

SELF-CHECK 7.2

Can you think of any cancers for which attempts have been made to employ cytology as a screening test, other than cervical neoplasia?

In diagnostic cytology the focus is on symptomatic patients in whom there is a clinical suspicion of cancer. Disease prevalence in this group will be relatively high and the maintenance of a high level of test **specificity** is crucial. The false positive test result in diagnostic cytology is an anathema. A false negative cytology result, although unfortunate, is unlikely to affect patient management in an adverse way.

Key Points

The **specificity** of a test is the probability that a negative test result represents the genuine absence of the disease in question. It is calculated by dividing the number of correctly identified negative test results by the total number of individuals without disease in the tested population. The fewer number of false positive cases the higher will be the test specificity.

SELF-CHECK 7.3

Try to think of the detrimental effects of false positive and false negative cytology test results in both the screening and diagnostic scenarios. How do the two scenarios differ?

BOX 7.1 The benefits and hazards of screening: a non-cytological example

Breast cancer provides a useful example to explain the benefits and potential harm that can occur as a result of screening. Screening trials in which patients are randomly allocated to receive mammographic screening or to serve as unscreened controls have been undertaken in a number of countries. These studies show a clear reduction in deaths from breast cancer in women aged over 50 who were screened. This has been attributed to the earlier detection of small non-invasive breast tumours which would otherwise have had a high chance of progressing to invasive breast cancer if left untreated. There is also evidence that screen detected cancers are of a less aggressive type than those which are clinically detected. An interesting observation, however, is that screened women under the age of 50 have a higher mortality rate for breast cancer than unscreened women. As yet, there is no clear explanation for this phenomenon, but what *is* clear is that screening can cause harm as well as good if the wrong individuals are screened or if an unsuitable test is employed.

Having established the differences between screening and diagnostic cytology we can now discuss the principles which bind these two modalities together. Regardless of whether a cytological specimen originates from a patient with symptoms of cancer or from an individual who is being screened for early cancer, the *morphological rules of thumb* are the same. Chapter 1 provides an overview of the cytological features of normal and neoplastic cells. Cancer cells are derived from normal cells and will inevitably share with them many morphological features. There are two implications of the previous statement. First, *cancer cells can mimic normal cells*, sometimes almost to perfection. This can give rise to false negative reporting. Second, *normal cells can mimic cancer cells*, leading to the problem of false positive cytological reports. The minimization of false positives and false negatives is a huge challenge in cytology, and only comes with years of experience and good *pattern recognition skills* on behalf of the cytologist.

The differences between cancer cells and normal cells are based on *cell size, cell shape, the interrelationship of cells, and characteristics of the nucleus*.

Key Points

Cytologists are *morphologists*. They analyse and interpret the shape, size, and interrelationship of cells.

Detailed morphological descriptions will be given in subsequent chapters under the headings of specific tumour types and body sites.

SELF-CHECK 7.4

To what extent do morphological features reflect the biological behaviour of malignant cells? (Hint: think at the molecular level.)

Tumour marker

A substance that is associated with cancer and can be used to indicate the presence of the disease.

Despite the emphasis that is placed on cell morphology it is sometimes impossible to reach a cytological diagnosis based on the size and shape of cells and their nuclei. This is where **tumour markers** are useful. Certain tumours secrete products, display antigens, or express genetic anomalies that can be detected using clever laboratory tests such as special stains, immunocytochemistry, and molecular techniques. These methods are described in detail in Chapters 2 and 13 so no more will be added here. Table 7.1 shows a few of the most popular tumour markers that can be detected, not only to aid diagnosis but also to help in the selection of appropriate therapy and to predict disease **prognosis**.

7.2 Nomenclature and classification of neoplastic conditions

We will keep this section short and uncomplicated, for the simple reason that nomenclature and classification systems for neoplastic conditions can be confusing to the uninitiated. Most **tumours** are named on the basis of their *tissue of origin* and on their likely *behaviour* (benign or malignant).

Tumour

An alternative term for a neoplasm, i.e. it is an abnormal mass of tissue.

From reading Chapter 5 you will already be familiar with most tumours of **epithelial** tissue. However, the human body is made up of several different types of tissue, any of which can succumb to neoplastic disease. Table 7.2 shows a very simple version of the classification system for human epithelial and **connective tissue** tumours.

TABLE 7.1 Examples of tumour markers.

Type of marker	Example of marker	Example of tumour identified
Secretory product	Acid phosphatase Alpha foetoprotein Hormone products	Prostate cancer Hepatocellular carcinoma Endocrine tumours
Cell-bound antigen	Calretinin Carcinoembryonic antigen (CEA) Chromogranin	Mesothelioma Carcinoma Small cell carcinoma of the lung
Gene or chromosome anomaly	Aneuploidy of chromosomes 3, 7, 17 Mutated B-type RAF (BRAF) gene Mutated epidermal growth factor receptor (EGFR) gene	Urothelial carcinoma Thyroid cancer Lung adenocarcinoma

TABLE 7.2 Classification of epithelial and connective tissue tumours.

Tissue	Normal cells involved	Benign neoplasm	Malignant neoplasm
Epithelium	Squamous Glandular Transitional	Papilloma Adenoma Papilloma	Squamous cell carcinoma Adenocarcinoma Transitional cell carcinoma
Connective tissue: • Fibrous • Adipose • Bone • Smooth muscle • Skeletal muscle	 Fibrocyte Fat cell (adipocyte) Osteocyte Smooth muscle Striated muscle	 Fibroma Lipoma Osteoma Leiomyoma Rhabdomyoma	 Fibrosarcoma Liposarcoma Osteosarcoma Leiomyosarcoma Rhabdomyosarcoma

Key Points

Epithelial tissue lines the surfaces of organs, whereas connective tissue supports and binds different tissues together.

Unfortunately, many tumours do not fit neatly into this classification system, including tumours of **embryonal** tissue, **nervous** tissue, and the **haematopoietic** system. Examples of such tumours are **lymphoma**, **leukaemia**, and **melanoma**. Yet other tumours are named after the person who first recognized them, such as **Hodgkin's disease**, **Ewing's sarcoma**, and **Burkitt's lymphoma**. Suffice to say that in total there are about 200 different types of human cancer. In many instances a biopsy is necessary for definitive histological diagnosis. Nevertheless, many tumours have been well characterized cytologically. Part of the skill of a cytologist is in recognizing when the limitations of cytology as a diagnostic tool have been reached.

We will now look more closely at the biology of neoplastic disease.

7.3 The mechanisms of tumour genesis and metastasis

Having set the scene with the range of tumour types that can be observed in cytological samples, we are now in a position to explore the concepts of tumour development and the morphological characteristics of cancer cells. The molecular biology of cancer is complex. Although cytologists can get by without a deep understanding of tumour biology, an appreciation of cancer at the molecular level assists in the understanding of cell morphology and will certainly prepare you for what is to come in Chapter 13.

The basic concept of neoplasia was outlined in Chapter 1. Here we extend the discussion to include the mechanisms of malignant tumour growth and the spread (or **invasion**) of cancer cells from their primary site to distant organs and tissues, a process known as **metastasis**.

SELF-CHECK 7.5

What is the most reliable feature that differentiates malignant from benign tumours?

Cancer is a genetic disease. No discussion of tumour development is complete without an explanation of the genetic basis of cancer. The idea that cancer is a **clonal** disorder derived from a single **transformed** cell has already been discussed in Chapter 1. The role played by **oncogenes** and **tumour suppressor genes** as the responsible agents for the initial transforming event was also covered in that chapter.

With molecular events at the centre of human cancer development, we can list five fundamental properties of malignant tumours. These are:

1. Cell **immortalization**

2. Evasion of **apoptosis**

3. Resistance to growth inhibition

4. Independence from **mitotic** control

5. **Angiogenesis**

Oncogenes
These are normal genes that have the potential to become cancer-initiating genes.

Tumour suppressor genes
These are normal genes whose job is to inhibit cell division and therefore help protect against tumour formation.

Key Points

The capacity for unlimited cell growth (immortalization) and the ability to evade the normal pathways that lead to cell death (apoptosis) are the key properties of a malignant tumour.

The whole process of tumour initiation, invasion, metastasis, and secondary tumour deposition is difficult to convey in a simple diagram, but an attempt is made in Figure 7.1.

To explain the growth of a malignant tumour it might be helpful to borrow from evolutionary theory. Just as populations of animals and plants will change according to environmental conditions, so a tissue containing a single transformed cancer cell will change to accommodate the new tumour cells. There are, of course, many differences between the evolution of multicellular organisms and tumour growth, but perhaps the most notable is the timescale involved. Multicellular organisms can take millions of years to evolve. A tumour may take

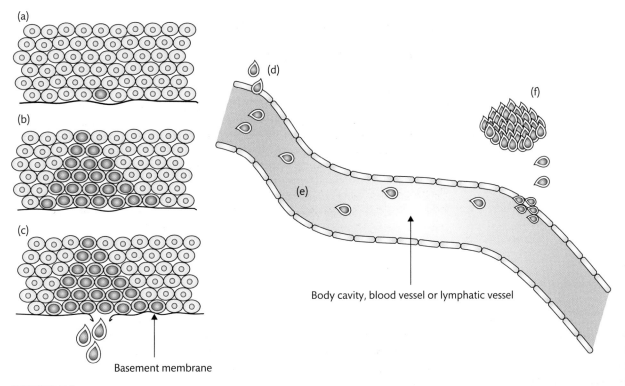

FIGURE 7.1

Diagrammatic illustration of the sequential steps involved in tumour genesis and metastasis. (a) Transformation of a normal cell into a tumour cell (indicated in red). (b) Clonal expansion. (c) Loss of contact inhibition and breach of basement membrane. (d) Invasion of tumour cells into body cavities, bloodstream, or lymphatic vessels. (e) Metastasis. (f) Secondary tumour deposits.

months, weeks, or even days to completely replace a normal population of cells. With their unlimited capacity for growth and their greed for nutrients and space, tumour cells have a distinct survival advantage over their normal counterparts. The result is a **clonal expansion** of the cancer cells at the expense of normal tissue, which often degenerates and dies.

The growth of a tumour often pursues an erratic, unpredictable course. Although most tumours progressively enlarge, some may suddenly shrink as a result of **ischaemic necrosis** caused by the tumour outstripping its own blood supply. Yet others may remain dormant for a considerable length of time.

So, we have learned that cells of malignant tumours multiply continuously, uncontrollably, and often very quickly. One of the unique properties of a malignant tumour is its ability to metastasize. In the process of metastasis there is an initial invasion of the malignant cells into surrounding tissues. As the cancer grows, it expands and begins to compete with normal tissues for space and nutrients. The invasiveness of the malignant cells may be related to mechanical pressure of the growing tumour, motility of the malignant cells, and enzymes produced by the malignant cells. Also, malignant cells lack what is called **contact inhibition**. When non-malignant cells of the body divide, their widespread migration is inhibited by contact on all sides with other cells. Malignant cells do not conform to the rules of contact inhibition; they have the ability to invade healthy body tissues with very few restrictions.

Clonal expansion

An increase in the number of genetically identical cells arising through cell division.

Ischaemic necrosis

The localized death of tissue caused by a lack of blood and oxygen supply.

BOX 7.2 *Moving towards molecular cytology*

With the success of the Human Genome Project and other research into the molecular basis of cancer, many researchers are turning their attention to the application of this knowledge to the development of techniques aimed at the diagnosis, prognosis, and even treatment of human cancer. Although the techniques of molecular biology have had, thus far, a relatively small impact on diagnostic cytology, this role can only increase with the passage of time. Among the developments are the identification of genes and their products using hybridization techniques, gene and tissue microarrays, and the detection of viruses and other agents that may play an important role in the genesis and natural history of human cancer.

Monoclonal tumour
A population of genetically identical neoplastic cells.

Phenotype of a tumour
This refers to the observable physical and biochemical characteristics of the tumour.

Tumour heterogeneity
This refers to the phenotypic variation observed within a tumour.

Karyotype of a tumour
This is the number, form, size, and arrangement of chromosomes within the constituent nuclei.

Mutation
An alteration in a DNA sequence of a gene.

Despite the postulated **monoclonal** origin of tumours, an interesting observation at the **phenotypic** level is the extreme **heterogeneity** exhibited by some malignant tumours. Attributes such *as invasive potential, growth rate, metastatic ability, karyotype, hormonal responsiveness, and susceptibility to anticancer drugs* can be quite variable in different portions of the tumour. A suggested mechanism to explain tumour heterogeneity relates to the inherent *genetic instability* of its constituent cells. The relatively high rate of random, spontaneous **mutations** during clonal expansion gives rise to subpopulations of cells with different phenotypic attributes.

Following invasion, some of the malignant cells may detach from the **primary tumour** and invade a body cavity or enter the bloodstream or lymphatic vessels. This can lead to widespread metastasis. Those malignant cells that survive in the blood or lymph invade adjacent body tissues and establish **secondary tumours**. In the final stage of metastasis, the secondary tumour becomes vascularized, a process known as angiogenesis. Angiogenesis is the development of new networks of blood vessels that provide nutrients for the further growth of the tumour. Some malignant tumours secrete proteins that serve as chemical triggers for blood vessel growth. Such **angiogenic growth factors** may also be derived from inflammatory cells (e.g. macrophages).

7.4 Benign processes affecting cell morphology

You may reasonably ask why cytologists are so interested in benign processes such as **inflammation**, tissue repair, cell death, and **metaplasia**. After all, these processes rarely harm the patient or affect clinical management. The explanation for our fascination lies with the *interpretive difficulties* that can arise from these processes. Indeed, tissue injury followed by cell **degeneration** and **regeneration**, sometimes occurring concurrently with cell death, metaplasia, or neoplasia, are a common source of frustration and interest for the cytologist.

Injury causes inflammation

Cells and tissues can be injured by numerous agents. Injurious agents may be broadly classified as *physical* (e.g. heat, cold, radiation), *chemical* (e.g. acids, alkalis, drugs), and *microbiological* (e.g. bacteria, viruses, fungi, protozoa). Inflammation is the response of tissues to injury and is outwardly manifested by the *redness, swelling, pain, and heat* which are often apparent at the site of tissue damage. At the cellular level the tissues mount an inflammatory response which usually includes an increase in the number of *inflammatory cells* (e.g. neutrophils, lymphocytes, macrophages) at the site.

Inflammation: the changing balance between cell degeneration and repair

Perhaps of greater interest and significance to the cytologist than the non-epithelial reaction to tissue injury are the changes that occur in the epithelial cells. Figure 7.2 illustrates some of the morphological effects on cells following injury. At first the cells exhibit changes associated with degeneration, which may include *blurring* of the nuclear chromatin, nuclear **pyknosis** and *wrinkling* of the nuclear membrane. Cytoplasmic changes may include *vacuolation* and *swelling*. If the cell dies these cellular alterations become extreme and widespread. At the same time, surviving cells begin to show the features of regeneration and repair. An increase in protein synthesis is reflected in *generalized nuclear enlargement, multinucleation, chromatin coarsening, nuclear hyperchromasia*, and an *increase in the number and size of nucleoli*. An *undulated nuclear membrane* may also be apparent during the repair process. Cytologists often refer to these morphological alterations collectively as **reactive changes**.

As inflammation is a continuous process, the prevalence of each of the cytological features described above will depend upon the snapshot in time at which a cell sample is taken and whether the injurious events persist or fade away.

SELF-CHECK 7.6

Why are the morphological changes in epithelial cells of greater significance than the non-epithelial response to injury?

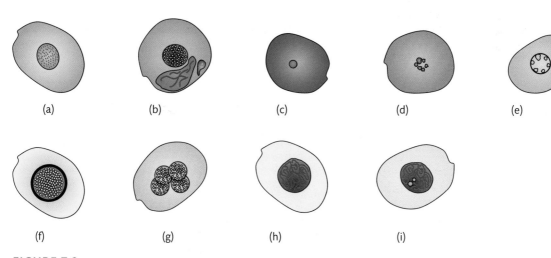

(a)   (b)   (c)   (d)   (e)

(f)   (g)   (h)   (i)

FIGURE 7.2

Diagrammatic representations of cells undergoing degeneration (a–e) and repair (f–i). (a) Normal cell. (b) Blurring of chromatin, breakdown of nuclear membrane, cytoplasmic vacuolation. (c) Nuclear shrinkage and condensation of chromatin (karyopyknosis). (d) Nuclear fragmentation and agglutination of chromatin (karyorrhexis). (e) Almost complete dissolution of nucleus (karyolysis). (f) Nuclear enlargement, prominent nuclear border. (g) Multinucleation. (h) Slight coarsening of chromatin, slight nuclear hyperchromasia, undulations in the nuclear membrane. (i) Prominent nucleoli.

Two types of cell death: apoptosis and necrosis

Death is an inevitable event in the life cycle of a cell. It may occur as a result of unplanned and irreversible damage to the cell, a phenomenon known as **necrosis**, or through a more naturalistic process termed **apoptosis** (programmed cell death). Although the features of necrosis are fairly non-specific, most cytologists will be able to recognize it by assessing the 'background' information in cell samples. Some of the features of necrosis are shown in Figure 7.3 and include cytoplasmic vacuolation, dense nuclear pyknosis and disintegration of the cell membrane.

Apoptosis, on the other hand, is characterized first by cytoplasmic shrinkage or *blistering*, followed by condensation and fragmentation of the nuclear chromatin into small granules resembling mercury droplets. Various stages of apoptosis can be seen in Figure 7.4.

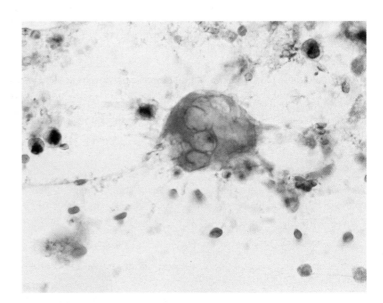

FIGURE 7.3
Cellular features of necrosis.
Note the cytoplasmic vacuolation and breakdown of the cell membrane. The nuclear chromatin has condensed to such a degree that it has become practically invisible.

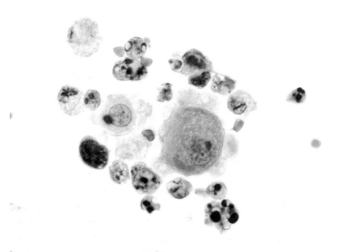

FIGURE 7.4
Apoptosis.

The nuclear fragmentation seen in Figure 7.4 is termed **karyorrhexis**. In contrast to necrosis, apoptotic cells remain relatively intact but gradually shrink to become what are known as **apoptotic bodies**. Apoptotic bodies are often identified as part of the background in normal and abnormal cytological samples.

Iatrogenesis: another source of interpretive difficulty

As if the myriad changes associated with inflammation weren't enough to confuse the microscopic interpretation of cell samples, **iatrogenesis** can certainly confound the most experienced of cytologists. Iatrogenesis may be simply defined as *damage or disease caused unintentionally by medical treatment or any other form of medical intervention*. In cytological terms, 'iatrogenic changes' refer to the cellular alterations and potential misdiagnosis resulting from medical interventions. Iatrogenic changes may take the form of cell damage or destruction, disorders of cell growth and maturation, or an increased risk of opportunistic infection.

Most descriptions of the cellular alterations associated with iatrogenesis are anecdotal. Perhaps the most carefully characterized are the changes seen in cervical specimens. A whole section of Chapter 4 is devoted to the iatrogenic changes commonly found in cervical samples. Broadly speaking we can classify the cytological patterns according to the intervention modalities that cause them. These are grouped as follows and discussed further in the text below:

Cross reference
Chapter 4, Section 4.9.

1. Irradiation
2. Chemotherapy
3. 'Physical' intervention (e.g. surgery, urinary tract catheterization, endoscopic examination, the intrauterine contraceptive device)
4. Exogenous hormones

The cellular effects of **radiation treatment** are in essence those of the degeneration and repair that follows any other kind of injury to cells. The changes can be quite variable and depend mainly on the dose and period of exposure. Not all post-radiation samples will show radiation changes; in others the changes may persist for many years. Figure 7.5 shows the *marked nuclear and cellular enlargement* and *cytoplasmic vacuolation* that are common in radiation-induced changes.

BOX 7.3 Iatrogenesis: a broader description

Iatrogenesis, or physician-induced damage, is by no means restricted to cytological misdiagnosis. There are many sources of clinical iatrogenesis, including medical error, drug interactions, adverse side effects of drugs, drug interactions, hospital acquired infections, physician-assisted suicide, and even medical torture! Iatrogenic illness, or even death, is in many cases an unfortunate and unintentional side effect of modern medical practice. In contrast, deliberate negligence and avoidable errors are in direct conflict with the important medical mandate 'first do no harm'. Breaches of this principle can result in medical negligence claims, a problem that is sadly on the increase in developed nations.

Nuclear membrane *wrinkling*, *nuclear* **hyperchromasia**, **multinucleation**, and infiltration of the cell by polymorphs may also be seen. All of these effects can mimic malignant disease almost to perfection. To complicate matters further, post-radiation malignancy can be almost impossible to distinguish from benign radiation change. The cells are more likely to be malignant if they show marked **pleomorphism**, *nuclear crowding*, or **mitotic figures**.

Iatrogenic changes associated with **chemotherapy** are less well described than those induced by radiation treatment. The changes induced by cytotoxic drugs are often indistinguishable from neoplasia and are a rare but important pitfall in cytology. An example of the effects of chemotherapy is shown in Figure 7.6.

Broadly speaking, the effects are similar to radiation changes but may also include *enlargement of nucleoli*, *perinuclear haloing* and *irregular condensation of chromatin*. In addition to these changes, cytotoxic drugs increase the risk of opportunistic infections by bacteria, viruses, and fungi.

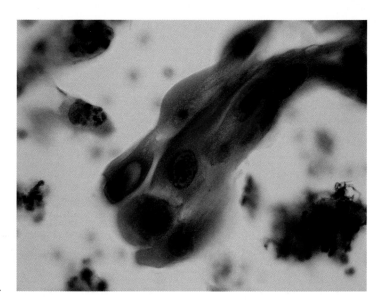

FIGURE 7.5
Radiation changes.

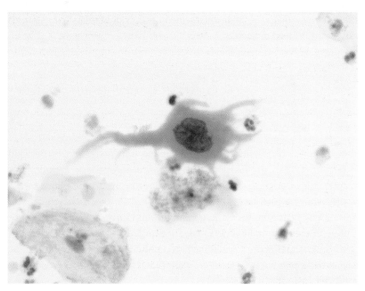

FIGURE 7.6
Chemotherapy-induced changes.

CASE STUDY 7.1 Iatrogenic changes associated with cyclophosphamide therapy

Cyclophosphamide is an agent used for the treatment of a number of malignant diseases. Following metabolism of the drug in the liver, the urinary products can cause marked inflammation and necrosis in the lining of the bladder. The clinical effects include a haematuria (blood in the urine) and bladder pain, which can be relieved by hydration of the patient, but may result in the need for surgery.

A 52-year-old woman is receiving cyclophosphamide chemotherapy for ovarian cancer and develops dysuria (difficulty in urination). Further tests reveal gross haematuria. A urine sample is sent to cytology. Discuss the cytological appearances that may be expected. What diagnostic pitfalls must the cytologist be aware of?

'Physical' intervention is the term we use to describe any form of medical intervention resulting in physical trauma to cells and tissue. These are many and varied, and may include surgery (e.g. excision of cervical intraepithelial neoplasia), ablation (e.g. laser destruction of cervical intraepithelial neoplasia), and instrumentation (e.g. urinary tract catheterization, endoscopic examination, the intrauterine contraceptive device, etc.). As with radiation treatment and chemotherapy, the iatrogenic changes associated with physical intervention follow the pathway of acute damage and degeneration of cells and a longer phase of repair and regeneration. Epithelial repair and regeneration is usually complete within one to two months following physical intervention. It is therefore prudent to wait at least two months before taking follow-up samples for cytology. A curious iatrogenic effect following physical intervention is that of **inappropriate re-epithelialization** of damaged tissues. For example, in 30–70% of cervices subjected to surgery the cervix becomes lined by epithelium resembling that of the fallopian tube and/or endometrium. This condition, known as **tubo-endometrioid metaplasia**, can give an erroneous impression of neoplasia. The cytological findings in tubo-endometrioid metaplasia are described in Chapter 4.

Cross reference
Chapter 4, Section 4.9.

The final example of iatrogenesis we will discuss is the effect of **exogenous hormones**. The epithelial cells lining the urinary tract and female genital tract are sensitive to administered steroid hormones, particularly oestrogen and progesterone. The cytological changes in cervical samples associated with steroid hormones are discussed in Chapter 3. Occasionally, the presence of glycogen-laden squamous cells in the urinary sediment of prostate cancer patients is the result of oestrogen treatment for prostatic carcinoma.

Cross reference
Chapter 3.

7.5 The role of cancer multidisciplinary teams

Section 7.3 exposed cancer as a complex disease process at both the molecular and cellular level. Cancer is also a complex disease clinically. A patient with, or suspected of having, cancer should not be viewed as the simple sum total of his or her symptoms but rather as a unique human being who requires the attention of a variety of healthcare professionals. The modern delivery of healthcare for patients with cancer relies on different specialists working together to provide integrated care. This, in essence, is the role of the cancer **multidisciplinary team** (MDT).

In recent years there has been significant growth in MDT working as a result of increasing specialization, advances in medical technologies, and recommendations by respected agencies. Multidisciplinary teams and their meetings now occupy a central role in developed health systems.

The core membership of a cancer MDT will generally include the following specialists:

1. **Clinician**
2. **Surgeon**
3. **Radiologist**
4. **Oncologist**
5. Histopathologist
6. Cytopathologist
7. Specialist nurses and other specialists involved in the patient's care

Ancillary roles of the multidisciplinary teams

While the primary purpose of an MDT is to help team members resolve difficult cases, teams often fulfil a variety of additional functions:

1. They promote communication and coordination between diverse specialties.
2. They provide a forum for clinical audit.
3. They can identify service gaps and breakdowns in coordination or communication between individual specialists.
4. They enhance the professional skills and knowledge of individual team members.
5. They can generate innovation to ensure progress in practice and service effectiveness.

What makes a successful multidisciplinary team?

Putting people together in a group representing many disciplines does not in itself guarantee efficient and effective patient care. The essential elements of a successful MDT are listed below:

1. Personal commitment

Individual team members must be personally committed to attend meetings and participate fully.

2. A common goal

The central aim of the MDT meeting is to help ensure the best possible care for cancer patients.

3. Clarity of roles and communication

The person with overall responsibility for the care of the patient will normally act as chair of the MDT meeting and ensure that discussions are focused and concise.

4. Institutional support

Healthcare agencies across the UK recognize that the work of the MDT is positively related to the quality of clinical care, and support MDT working accordingly.

What inhibits multidisciplinary teamwork?

Teamwork in all areas of professional life can be hampered in the following situations:

1. Logistics

A commonly stated problem is that of 'actually getting people together'. Careful planning of timetables and other practicalities must be considered.

2. Resources

Insufficient time, physical space, or a lack of specific equipment is detrimental to MDT working.

3. Attitudes of team members

Effective MDT working requires a mutual understanding between, and tolerance of, different professional groups.

Which patients should be discussed?

All patients with a diagnosis of cancer should be discussed at MDT meetings. In the UK this is a **clinical governance** requirement.

The MDT should be consulted at the following key stages of progress through cancer diagnosis and treatment:

1. At early diagnosis (to agree **staging** investigations and discuss any further investigations).

2. At definitive diagnosis (to agree a treatment plan).

3. Following initial treatment (to determine if further treatment is required).

4. If/when condition changes (to discuss follow-up and further management).

> **Clinical governance**
> The system through which health organizations are accountable for continuously improving the quality of their services and safeguarding high standards of care.

What actually happens at a multidisciplinary team meeting?

For all meetings, a list of patients to be discussed is circulated in advance to team members. Patient samples and images are located along with reports, for review by pathologists and radiologists, prior to the meeting. At the meeting the responsible clinician will present the patient's details and lead the discussion. The radiologist, histopathologist, and/or cytopathologist will each present images and slides where relevant to the discussion. Other team members contribute to the discussion and collectively agree the patient's management plan, ensuring that the patient's preferences are represented and taken into account. Once a decision is made, the responsible clinician will summarize the planned course of action, and this is documented in the patient's notes.

CASE STUDY 7.2

A 62-year-old male smoker presents to his GP with shortness of breath and productive cough. The GP refers the man for a chest X-ray which shows an opacity in the lower lobe of his left lung. Discuss the key events in the subsequent clinical management of this patient.

SUMMARY

- Diagnostic cytopathology is the diagnosis of disease at the cellular level. It can be used in two modalities; as a screening test for the early detection of precancerous lesions in asymptomatic individuals, and as a diagnostic test for the identification of cancer cells in symptomatic patients. Whichever modality is used, cytology is seen as a simple, inexpensive, and effective means of detecting neoplastic conditions.

- The biology of tumour growth is complex. Cytologists are better able to interpret cell samples and have a fuller appreciation of diagnostic pitfalls when they understand the basic mechanisms and natural history of cancer development.

- The microscopic interpretation of cells taken from the human body is often not straightforward. Inflammatory disease, cell degeneration, regenerative process, and a bewildering array of benign conditions all represent pitfalls which can entrap the unwary or inexperienced cytologist.

- Cancer patients require the attention of a number of different specialists. In ensuring the best possible care for cancer patients, teams consisting of surgeons, oncologists, pathologists, radiologists, and specialist nurses come together to decide on the best possible care for individual patients. Whenever cytology plays a part in the diagnosis and/or follow-up of cancer patients, the participation of the reporting cytologist in multidisciplinary meetings is essential.

DISCUSSION QUESTIONS

1. The microscopic interpretation of cells can be limited by several factors. Discuss the nature of this problem and suggest ways in which diagnostic reliability can be improved, now and in the future.

2. The most common question asked of the cytopathology department is 'are malignant cells present?' What other questions might the clinician be interested in, and to what extent can cytopathology answer these?

3. Discuss the contribution made by cytologists in screening for and diagnosis of neoplastic conditions.

Answers to the self-check questions, and tips for responding to the discussion questions, are provided in the book's Online Resource Centre.

 Visit www.oxfordtextbooks.co.uk/orc/shambayati/

Cytology of urine

Behdad Shambayati

The cytological examination of a urine specimen is a simple and inexpensive method of assessing patients who present with **haematuria** (blood in urine). However, haematuria is a common presentation for a variety of non-malignant conditions, hence cytological yield for cancer is low.

Cytological evaluation of urine is useful for follow-up of patients who have been treated for **urothelial carcinoma**, as these patients are at risk for recurrence.

Urinary cytology can be used as a screening test for individuals who are at higher risk of developing urothelial cancer, such as workers exposed to carcinogens, including aniline dyes, and workers in the rubber industry.

Haematuria
The presence of red blood cells in urine is called haematuria. There are many causes for haematuria, including trauma, infections, stones, benign enlargement of prostate gland, and malignancy.

Urothelial carcinoma
This is a malignancy arising from epithelium lining the bladder, ureters, or renal pelves.

SELF-CHECK 8.1

List the main indications for urinary cytology.

Learning objectives

By the end of this chapter you should be able to:

- Describe the normal anatomy of the urinary tract.
- Describe the histology of urothelium.
- List the common specimen types and preparatory methods.
- Describe reporting terminology in urine cytology.
- Describe the cytological constituents of voided urine.
- Explain aetiology, symptoms, and treatment of bladder cancer.
- Describe common cytological findings in neoplasia.
- Describe the effects of therapy on the urothelium.

Before we embark on the cytology, we must make ourselves familiar with the basic anatomy and histology of the urothelium.

8.1 **Anatomy and histology**

Renal calyces

Systems of ducts which carry the urine.

Renal pelvis

The central area in the kidney that urine passes before it is funnelled into the ureter.

Prostate gland

This is a gland in men that produces an alkaline fluid which contributes to the fluid component of semen.

The urinary tract consists of two kidneys, two ureters, a urinary bladder, and a urethra. The kidneys produce and excrete urine by filtering blood. Urine passes from the **renal calyces** via the **renal pelvis** into the **ureters** and is stored in the bladder. It is voided via the **urethra**. In men, the urethra also passes through the **prostate gland** (Figure 8.1). The renal pelvis, the ureters, the bladder, and parts of the urethra are lined by a specialized type of epithelium known as **urothelium** (Figure 8.2). Small portions of the urethra are lined by squamous epithelium.

Urothelium was originally called **transitional epithelium** as it was thought to be a transition between squamous and glandular epithelium. This term is still in use. Urothelium is impermeable to urine and does not allow the noxious substances in urine to cross back into the blood. The other feature which makes this epithelium unique is that it has specialized structures within, which allows the urothelium to expand and contract depending on the volume of urine in the bladder.

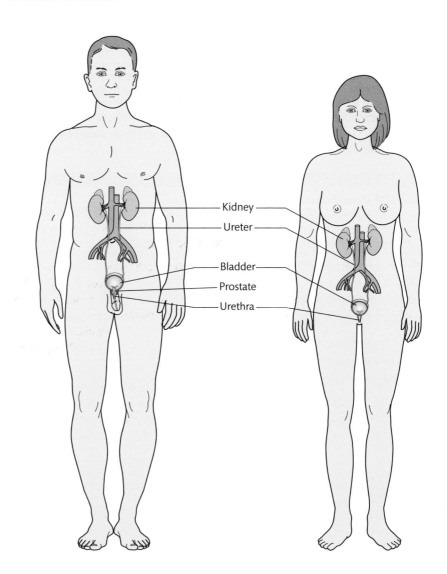

Kidney
Ureter
Bladder
Prostate
Urethra

FIGURE 8.1

Line drawing of urinary tract. Urine is produced in the kidneys. It is collected in the bladder and finally voided via the urethra.

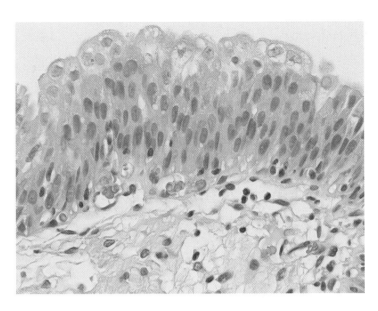

FIGURE 8.2
Normal urothelium-histology section.

> **Key Points**
>
> Urothelium is a highly specialized epithelium. It has two important roles: it is imperme-
> able and does not allow the noxious substances in urine to cross back into the blood,
> whilst it is able to expand to accommodate large volumes of urine in the bladder.

Urothelium is a multilayered epithelium: in some respects, urothelium is similar to the cervical epithelium that you learned about in Chapter 4. We arbitrarily divided squamous epithelium into layers: a basal layer, a parabasal layer, an intermediate layer, and a superficial layer. However, compared to cervical cytology, these cell layers are not so easily identified in urothelium and most cytologists refer to them as deeper layer cells and superficial cells.

Cross reference
Chapter 4, Section 4.4.

Figure 8.3 is a diagrammatic representation of these arbitrary cell layers. We will look at the cytology of urine later on but, for clarity, let us look at some common cell types now. The deeper layer cells are similar to parabasal squamous cells and have dense staining cytoplasm with a sharp border (Figure 8.4). The superficial layer cells are fairly large cells, and may have a single nucleus, two nuclei (binucleated), or be multinucleated and cover intermediate cells underneath. For this reason, you may hear them referred to as **umbrella cells**. Figure 8.5 shows a binucleated and multinucleated superficial/umbrella cell.

Unlike squamous and other epithelia, the urothelium is a very low turnover/exfoliation epithelium that takes almost a year to shed. In common with other epithelia, the urotheli-um can also undergo the process of **metaplasia**. Common forms of metaplasia occurring in the urinary tract include squamous and glandular metaplasia.

Cross reference
Chapter 4, Section 4.3.

8.2 Specimen types

Before looking at the cells in detail, we must make ourselves aware of different specimen types and preparation methods, as they may affect the appearance of the cells. Most labora-tories receive three main types of urine samples: **voided**, **ileal conduit**, and **catheter** urine.

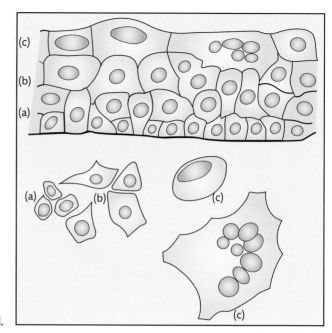

FIGURE 8.3
Diagrammatic drawing of urothelial cell layers.
(a) Basal and parabasal. (b) Intermediate. (c) Superficial.

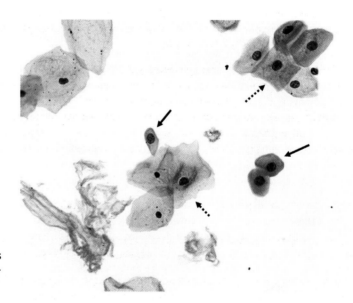

FIGURE 8.4
Normal deeper layer urothelial and squamous cells. Solid arrows are pointing at deeper layer cells, dashed arrows show squamous cells.

Bladder washings and **brushings** are more specialized techniques that are not as commonly encountered, but which we will cover briefly here, too.

When designing a cytopathology request form (either paper or electronic), it is important to create mandatory field boxes to ensure that the correct specimen type (i.e. voided, catheterized, brush, etc.) is stated by the requesting doctor.

Voided urine

Voided urine is the most common sample type, for which the patient is asked to produce a sample into a pot. Voided urine samples undergo degeneration very quickly so it is advisable to collect the sample in a pot containing a preservative. Commercially available preservatives

(a)

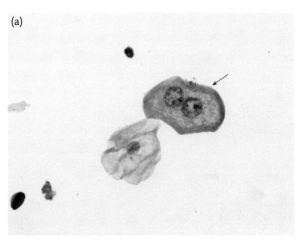

(b)

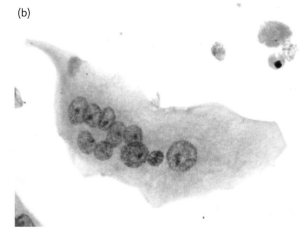

FIGURE 8.5
(a) Superficial urothelial (arrow) and superficial squamous cell. Superficial urothelial cells (umbrella cells) are often bi- or multinucleated and have abundant cytoplasm.
(b) Multinucleated superficial urothelial cell.

have been tested and found to work very well in reducing cellular degeneration and stopping bacterial growth. Urine pots containing 30 ml of preservative can be supplied to urology clinics, to which 30 ml of urine should be added.

There has been a lot of debate amongst cytologists on collecting voided urine samples. It is thought that urothelial cells are often passed at the beginning or end of voiding, and therefore mid-stream urine is unsuitable for cytology. Time of collection is also important: though early morning samples are quite cellular, the cells show marked deterioration as they have been degenerating for several hours overnight in the bladder. This results in inferior morphology. The second sample of the day (or a sample produced later in the day) is therefore preferred.

Ileal conduit urine

This specimen type is useful for follow-up of patients who have been treated for bladder cancer, as they have an increased risk of developing cancers in the ureters or the kidneys.

An ileal conduit is a 'urinary diversion' that is constructed from a section of ileum when a patient has had their bladder removed during surgery for bladder cancer.

Cross reference
Treatment for bladder cancer
page 196, Chapter 8.

Urine samples from these patients have a quite varied appearance: they may show large numbers of degenerating intestinal epithelial cells, macrophages, inflammatory cells, and mucus. These samples require careful screening, as occasional malignant cells could easily be overlooked amongst the busy background. Figure 8.6 shows some of these features.

Catheter urine

A urinary catheter is a short piece of thin plastic tubing that is inserted into the bladder to allow urine to drain freely into a bag.

Urine from catheter specimens is often collected from a bag which has been at room temperature for several hours, and so cells may show degenerative changes.

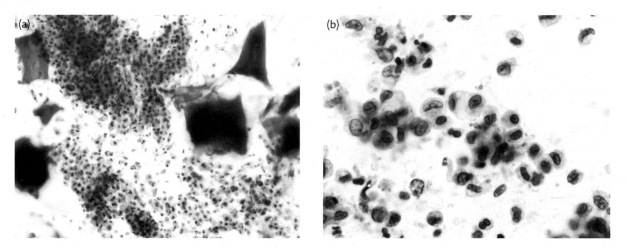

FIGURE 8.6
Ileal conduit urine samples. (a) Mucus and degenerate intestinal cells. (b) Vacuolated intestinal cells.

Bladder washings

Bladder washings are not a common source of urine sample as the procedure is invasive: it involves injecting 50–100 ml of saline through a catheter and aspirating the fluid before collection. This procedure is sometimes done before a biopsy and produces a high cell yield of freshly exfoliated cells.

Upper urinary tract brush cytology

This specialized technique may be employed when malignancy of the upper urinary tract such as in the renal pelvis or ureters is suspected. Using endoscopic and X-ray control, a fine brush is passed into the ureter. Once the brush has reached the suspicious area, the brush is moved back and forth to dislodge cells. These cells adhere to the bristles of the brush and can be spread onto a slide. As this is a difficult sample to obtain and the opportunity for repeat samples is limited, it is advisable that a biomedical scientist is available at the time of sample collection to prepare optimal quality smears.

In catheterization, the cellularity of bladder washing and brushing samples is generally quite high as the instrument mechanically scrapes the epithelial surface. These specimen types may include cells from the top superficial layers through to the basal layer. The cytologist must be familiar with this spectrum to avoid making an incorrect diagnosis. It is important that the sample type is indicated on the request form. Figure 8.7 shows some of these features.

Key Points

The method used to obtain a sample of urine has a significant influence on cytological appearances, and the cytologist must familiarize themselves with the varied picture that is seen in different sample types.

(a) 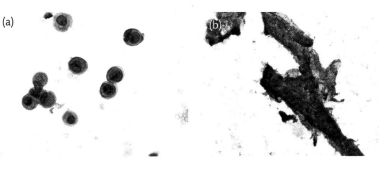 (b)

(c)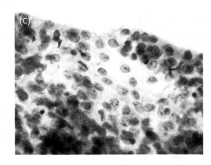

FIGURE 8.7

Normal urothelial cells. (a) This photomicrograph shows numerous deeper layer cells with high nuclear to cytoplasmic ratio (ureteric brushings). (b) A sheet of normal urothelial cells that have been forcefully removed. This is a common finding in catheterized samples (catheter urine). (c) High power, same case as (b). These cells are very regular in size and have normal chromatin pattern.

SELF-CHECK 8.2

List some of the advantages and disadvantages of different urine specimen types.

8.3 **Sample processing**

General sample processing was discussed in more detail in Chapter 2. However, we will briefly mention common preparation techniques in this chapter.

The main challenges in preparing urine samples are to:

- Concentrate the low cellular sample.
- Preserve morphology.
- Ensure adequate adhesion of cells to the glass slide.
- Produce material that covers only a small area of the glass slide to allow screening in a reasonable time.

The sample can be concentrated in a variety of ways, including centrifugation, cytocentrifugation, large volume cytocentrifugation, and liquid based cytology (LBC). Common fixatives in use are alcohol based, which provide good preservation of morphology and allow staining with Papanicolaou stain. The cell adhesion can be improved by using glass slides that have been treated with an adhesive agent; this can be done in house or obtained commercially.

Cross reference
Chapter 2.

Centrifugation and direct preparation

This is the basis of most cytological techniques, where the sample is spun and a preparation is made from the deposit. The preparation is fixed in alcohol and stained with Papanicolaou stain. This preparation method is very easy and fast to produce, but the material may be spread over a large area, which makes screening difficult. Cellularity may be low due to cell loss during the fixation step, unless slides have been treated with an adhesive agent.

Cytocentrifugation and large volume cytocentrifugation

This popular method takes slightly longer than the direct method but produces very good preparations. The sample is first centrifuged and the deposit is appropriately diluted and placed in the cytocentrifuge chambers. The fluid specimen is spun directly onto a glass slide. This produces a monolayer in a defined area on the glass slide.

Liquid based cytology

LBC has been mentioned in earlier chapters with regard to cervical screening. The same methodology can be employed in the preparation of urine samples. Two technologies are in common use: Tripath and Hologic LBC systems. These systems produce monolayers which are concentrated in a defined slide area. Both methodologies are semi-automated and more time consuming than the direct preparation method. However, the cellular preservation can be good if the sample is collected into the proprietary preservative at the time of collection. Samples processed using the Tripath system can be collected in **BD CytoRich™ Non-Gyn Blue Preservative** and for the Hologic system in **PreservCyt® or in CytoLyt®**.

SELF-CHECK 8.3

List some advantages and disadvantages of methods employed in sample processing.

8.4 Terminology used in reporting urine samples

It is important to use simple and concise terminology when reporting urine samples. There are no fully agreed definitions, but the categories Unsatisfactory, Negative, Atypical, Suspicious, and Malignant are commonly used:

Unsatisfactory: currently there are no agreed criteria for assessing adequacy in urine samples. However, a sample that is completely obscured by inflammatory cells or consists only of blood should be reported as unsatisfactory. Normal urine is sparsely cellular, but should contain a few urothelial cells to ensure that the sample is representative.

Negative: no malignant cells seen.

Atypical: used sparingly in urine cytology when changes present are not quite normal, but fall well short of malignancy. Such cells are said to show **atypia**.

Suspicious: a suspicious report is issued when cells include some features of malignancy, but the cells may be small in number, or where degenerative changes are present.

Malignant: a sample should only be reported as malignant when there are indisputable cell changes consistent with malignancy.

Atypia
This refers to deviation from normal, or not 'typical' of what is normally accepted. Its use in cytology refers to changes that are short of malignancy but raise some concern.

8.5 Normal voided urine

Normal voided urine generally contains very little cellular material. However, as stated earlier, different methods of specimen collection will affect cellular content. When examining a urine sample, you may encounter other cells or non-epithelial elements in addition to urothelial cells.

Urothelial cells

In voided urine, cells usually occur as isolated single cells: clusters are uncommon. It was said before in Section 8.1 that urothelium is a multilayered epithelium. The lower level urothelial cells have dense cytoplasm with well-defined borders, and stain green/blue with Papanicolaou stain. Their nuclei are round, and the chromatin has a finely granular pattern. Small nucleoli may also be present.

The superficial urothelial cells or umbrella cells are larger than deep layer cells and have abundant cytoplasm; their nuclei may also contain nucleoli. Their low nuclear to cytoplasmic ratio helps to confirm their benign nature. Figure 8.8 shows a deeper layer cell and binucleated superficial cell.

Urothelial cells undergo degeneration and the nuclei can become hyperchromatic. Some may also contain a red inclusion body. The nature of these inclusion bodies is not clear but they have no diagnostic significance. Figure 8.9 shows a degenerated urothelial cell with the red inclusion body.

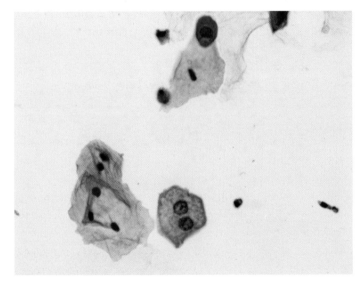

FIGURE 8.8
Normal urine.

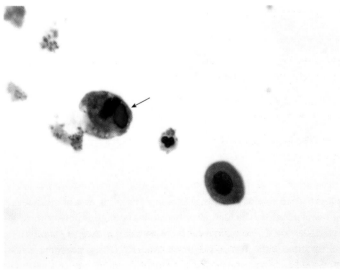

FIGURE 8.9
Normal urine. Degenerate urothelial cell with red inclusion body.

Other cells and non-cellular elements

You learnt the morphology of squamous cells in Chapter 4. In urine samples from women squamous cells can outnumber the urothelial cells. The squamous cells may arise from the bladder, or commonly from urethra, vulva, or vagina. Columnar cells may be seen as a result of metaplasia or contamination from the female genital tract.

In men, spermatozoa (Figure 8.10a), **seminal vesicle** cells (Figure 8.10b), **corpora amylacea** (Figure 8.10c), and **prostatic** cells may also be present.

The occasional spermatozoon may be present as a contaminant. However, spermatozoa in large numbers could be due to a condition known as **retrograde ejaculation**, where the semen, which should normally exit via the urethra, is directed to the bladder. Seminal vesicle cells may have large pleomorphic nuclei with hyperchromatic nuclei. Care should be taken

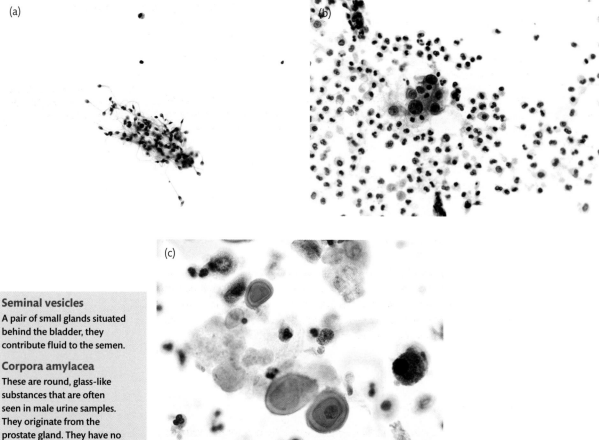

Seminal vesicles

A pair of small glands situated behind the bladder, they contribute fluid to the semen.

Corpora amylacea

These are round, glass-like substances that are often seen in male urine samples. They originate from the prostate gland. They have no diagnostic significance. Corpora amylacea are also rarely seen in respiratory samples.

Retrograde ejaculation

This is the entry of semen into the bladder instead of going out through normal ejaculation route via the urethra.

FIGURE 8.10

Other cells present in male urine samples. (a) Spermatozoa-voided urine. (b) Seminal vesicle cells. A group of cells presumed to be of seminal vesicle origin. The cytoplasm of these cells contains the brown pigment lipofuscin. The background includes numerous polymorphs. (c) Corpora amylacea-voided urine. Corpora amylacea are laminated structures, staining blue-green here; they occur more with advancing age and are derived from degenerate cells.

FIGURE 8.11
Endometrial cells–voided urine. Groups of epithelia cells presumed to be of endometrial origin in voided urine of a 37-year-old female. The follow-up samples were negative for malignancy.

interpreting these features as they could be mistaken for malignancy, but one clue to their benign nature is the presence of the brown pigment **lipofuscin** in their cytoplasm.

Other cells commonly seen in urine include neutrophils and red blood cells. Other epithelial cells from the female genital tract may also be identified in urine samples; Figure 8.11 shows a cluster of endometrial cells which were seen in an otherwise unremarkable urine sample.

Lipofuscin
This is a brown substance found in various cells, associated with breakdown of fats and proteins.

SELF-CHECK 8.4

List cells that may be seen in a normal urine sample.

8.6 Benign findings

Urine cytology is a useful test in identification of **high-grade urothelial cancer**. This will be discussed later in Section 8.7 of this chapter. In this section we will look at some benign changes that are important to recognize because they may give additional information to the clinician to help in the management of the patient (as in the case of **urinary casts**) or may lead to false positive diagnosis (as with **polyomavirus** infection).

Infections

The bladder could be infected by bacteria, fungi, viruses, or parasites.

Bacterial infections are relatively common, and may cause a prolific neutrophil response. Bacteria may also be seen under the microscope, but require microbiological culture for correct identification.

The most common fungus seen in urine samples is **candida**. In female patients it could be due to contamination from the vagina.

Parasitic infections include the protozoan ***Trichomonas vaginalis*** and the trematode **Schistosoma**. Schistosoma are rarely seen in urine samples in developed countries: Case Study 8.1 is included here as it describes the interesting life cycle of this parasite.

Cross reference
Chapter 4, Section 4.7 for description of morphology of candida and *Trichomonas vaginalis*.

CASE STUDY 8.1

An 18-year-old male of Egyptian origin attended his GP having recently arrived in the UK. He complained of pain when passing urine and having the urge to urinate more often. The GP performed a dipstick test which showed presence of blood and infection. He was prescribed antibiotics, and urine samples were submitted for microbiology and cytology.

The cytology specimen consisted mostly of polymorphs. However, on close examination, large oval structures were seen amongst the polymorphs, and these did not take up the stain. On closer examination these were seen to have terminal spines, and were recognized as eggs of the **trematode** *Schistosoma haematobium*. Figure 8.12 shows a typical egg surrounded by polymorphs.

caused by three main species: *Schistosoma haematobium*, *Schistosoma mansoni*, and *Schistosoma japonicum*. The other two species, *Schistosoma mekongi* and *Schistosoma intercalatum* cause infections less frequently. *Schistosoma haematobium*, causes infection of the urinary tract and is most commonly found in Africa and the Middle East in areas of poor sanitation.

This parasite has an interesting life cycle that involves both humans and an intermediate host, a fresh water snail. Humans are infected by the free swimming larval stage called **cercaria**, which burrows into skin when humans come into contact with contaminated water. The larva then travels to the liver via the bloodstream and matures into the adult form. After a period the adults migrate

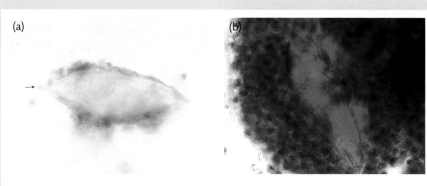

FIGURE 8.12

Schistosoma haematobium. (a) Ova of the *Schistosoma haematobium*. Voided urine. This photomicrograph shows the ova of the *Schistosoma haematobium*. A terminal spine is just visible (arrow). (b) *Schistosoma haematobium*. An ova is seen amongst the inflammatory cells. The terminal spine is not visible.

This finding was also confirmed by the microbiology department upon analysis of a wet preparation specimen of urine sample. He was initially treated with **praziquantel**, but the patient did not return for follow-up appointments, so his response to therapy is not known.

Schistosomiasis (also known as **Bilharzia**, after the German physician Theodor Bilharz who first described the cause of urinary schistosomiasis in 1851) is a parasitic disease caused by species of the fluke worm Schistosoma. It is thought that up to 200 million people are infected worldwide. There are five different species. The majority of infections are

to the bladder to copulate and lay eggs. The eggs are passed out with the urine and hatch into the larval stage **miracidia**, which enters a fresh water snail. In the snail the larva develops further into the free swimming larval cercaria form, and so the life cycle continues.

Schistosoma haematobium if left untreated can cause chronic inflammation in the bladder. It is associated with increased risks for squamous carcinoma of the bladder.

The treatment, which is effective against the adult stage of all types of human schistosomiasis, is praziquantel.

Viral infections of the urinary tract include **human papilloma virus (HPV)**, **herpes simplex virus (HSV)**, **cytomegalovirus (CMV)**, and **polyomavirus**.

Cytological changes due to infection by HPV and HSV were described in Chapter 4.

Cytomegalovirus is a member of the herpes virus family that is acquired by most people during their lives, but rarely causes disease unless the person is **immunocompromised**. Like other members of the herpes virus family it can remain dormant and reactivation can occur, particularly in people with a compromised immune system. It is rarely seen in urine samples of patients receiving chemotherapy or in transplant patients who are receiving **immunosuppressive therapy**. In cytology the effect of the virus on the cells can be seen as a large **intranuclear** inclusion.

The human polyomavirus will be described in detail as it is a frequent cause of false positive diagnosis because the infected cells resemble degenerate malignant urothelial cells.

There are five types of polyomavirus that have been found in humans. Of these the BK and JC viruses infect the urinary tract. They were named BK and JC after the initials of the patients in whom they were identified in the 1970s. The BK virus was seen in a renal transplant recipient and JC virus was in a patient with Hodgkin's disease. The infection with these viruses is widespread but there are no significant consequences of infection, except in immunosuppressed or immunocompromised patients. The photomicrographs in Figure 8.13 show the effect of polyomavirus in urothelial cells.

Infected cells may become quite large in size, and the nucleus is usually eccentric (placed on one side of the cell). The nuclear outline becomes thickened, and the chromatin appears cloudy, glassy, or opaque such that the typical chromatin pattern disappears.

The changes are sometimes so dramatic that the cells can be confused with malignant cells. For this reason they have been given the name **decoy cells** by cytologists. Close examination of these cells will reveal that there are, however, some differences that should allow for the correct interpretation of these cells. For example, the nuclear contour is almost always smooth and the nucleus is round, unlike malignant cells that can have an irregular nuclear outline. The infected cells also present as single cells, in contrast to malignant cells that appear both as single and loose clusters. Lack of visible chromatin is also unlike malignant cells, which show coarsened chromatin.

Cross reference
Chapter 4, Section 4.7.

Immunocompromised
This describes the state of a person's immune system that is less able or has lost the ability to fight infections. Examples of immunocompromised people include those with AIDS, and those receiving chemotherapy or drugs to suppress the immune system after organ transplantation.

Immunosuppresive therapy
Drugs given to patients who have had an organ or bone marrow transplant to suppress their immune system to prevent rejection.

Intranuclear
Within the nucleus of a cell.

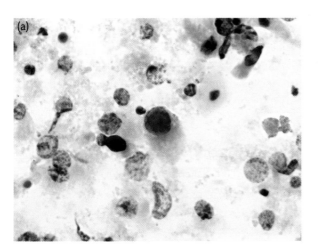
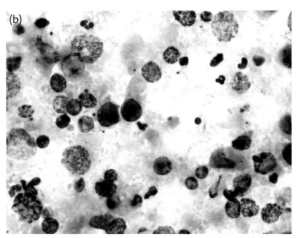

FIGURE 8.13
(a) Polyoma virus infected cell. The cell in the centre of the field has an eccentrically placed nuclei, the chromatin of the cell appears glassy. (b) Polyoma virus infected cells.

CASE STUDY 8.2

End-stage renal disease

The complete failure of the kidneys to remove waste and produce urine. Dialysis or kidney transplantation is the only treatment for end-stage renal disease.

Creatinine

This is a waste product of creatine phosphate metabolism in muscle tissue. Measurement of serum creatinine level is used to monitor renal function.

A 38-year-old female, who had received a kidney transplant for **end-stage renal disease** six years previously, was referred to a transplant clinic for evaluation following an increase in her serum **creatinine** level from 120 λmol/l to 170 λmol/l over the preceding six months. (Normal range 40–120 λmol/l for women.) She was on combination immunosuppressive therapy.

The patient submitted urine for cytology. The urine cytology was reported as 'specimen includes numerous enlarged cells which have eccentric nuclei. The chromatin appears glassy and featureless. The features are consistent with polyomavirus infection.' Figure 8.14 shows a typical field of view from this patient's urine.

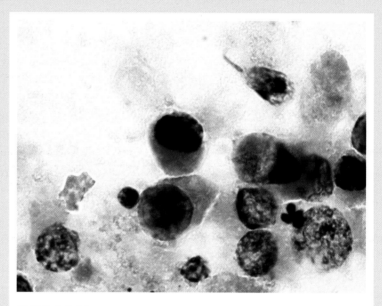

FIGURE 8.14
Polyoma virus infected cell. The enlarged, eccentric nuclei of the cells in the middle of field are virtually replaced by glassy, homogenous inclusions. The background includes numerous degenerate urothelial cells. Also note the round contour of nuclear outline.

Interstitial nephritis

This is inflammation of the kidney that involves the tissues surrounding the renal tubules.

Cytopathic

Characteristic pathological changes in a cell.

Nephropathy

Disease of the kidney.

The patient later had a renal biopsy which showed advanced **interstitial nephritis** and nuclear inclusions in tubular cells, with prominent nuclear **cytopathic** changes, compatible with polyomavirus **nephropathy**. The patient's immunosuppressive therapy was modified and, in the months that followed, serum creatinine levels stabilized at 140 λmol/l.

Measurement of serum creatinine level is a simple test to monitor renal function. This case demonstrates that urine cytology is a useful non-invasive screening tool when polyomavirus infection is suspected. The definitive diagnosis of polyomavirus nephropathy in this case was made on a renal biopsy.

(a)
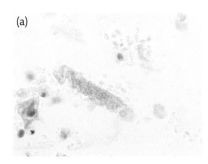

(b)
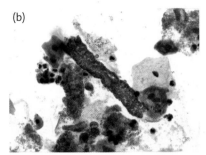

(c)
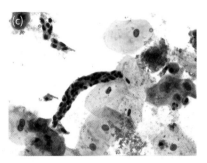

FIGURE 8.15

Urinary casts. (a) Granular cast-voided urine. Granular casts are fairly common, possibly formed from breakdown of cells, or aggregates of proteins. (b) Red blood cell cast–voided urine. A typical red blood cell cast from a patient who had glomerular nephritis. (c) Epithelial cast–voided urine. This is a well-preserved example of renal epithelial cast.

Casts

Urinary casts are cylindrical structures that form in the tubules and collecting ducts of the kidney. The casts dislodge and pass into the urine. Casts are usually reported by microbiologists when they examine unstained wet preparations of urine samples, but may also be detected in Papanicolaou stained cytological preparations.

Casts can be divided into two groups: acellular (containing no cells) and cellular. Acellular casts include **hyaline** and **granular casts** (Figure 8.15a) which are often seen in normal individuals and are of little clinical significance. Cellular casts include **white blood cell casts**, **red blood cell casts** (Figure 8.15b), and **epithelial casts** (Figure 8.15c).

Red blood cell casts are quite significant and are characteristically found in **glomerular disease**. White blood cell casts may suggest inflammation or infection. Epithelial casts are seen in various conditions that result in epithelial cell death, including **tubular necrosis**.

Calculi

Urinary calculi, more commonly known as urinary stones, are solid structures formed from minerals present in urine. They are most often due to deposition of calcium oxalate or uric acid crystals. The formation of calculi is called **lithiasis**. Although calculi are not diagnosed by the cytologist, they can affect the presentation of the cells and we should be aware of their effect on urinary epithelium to avoid a false positive diagnosis.

Urinary calculi usually form in the renal pelvis and can vary in size. If they are very small, they may pass without causing any symptoms, but once they reach a size of 2 mm, they can cause obstruction in the ureters. The resulting obstruction with dilatation causes severe pain. The other presentation is haematuria due to the mechanical damage caused to the urothelium.

Diagnosis of calculi is usually based on clinical presentation and imaging. Some calculi are large enough to be seen by plain X-ray, followed by **intravenous urogram (IVU)** (see Figure 8.16) and CT scanning. Cytological specimens from these patients may be quite cellular: the background could include red blood cells, inflammatory cells, and clusters of urothelial cells, which may be papillary in appearance. These clusters are difficult to differentiate from a low-grade urothelial carcinoma based on light microscopy (see Figure 8.17).

Hyaline casts
Formed by deposition of glycoprotein in the tubules, they can be seen in concentrated urine of normal individuals.

Granular casts
These are either formed from depositions of proteins or are broken down cellular casts.

Glomerular disease
The term given to diseases that affect the filtration units in the kidneys, the glomeruli.

Tubular necrosis
This is death of tubular cells that form the tubules in the kidney. It is one of the causes of renal failure.

Cross reference
Section 8.8.

Intravenous urogram (IVU)
Also known as intravenous pyelogram, this is a radiological procedure for visualizing the kidneys, ureters, and the bladder. Contrast media which is injected into the patient's vein is excreted by the kidneys, and is passed via the ureters into the bladder. X-rays taken before and at specific intervals after the injections show up the contrast media and help visualize the urinary tract.

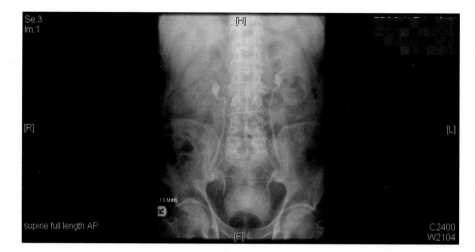

FIGURE 8.16
Normal intravenous urogram (IVU).

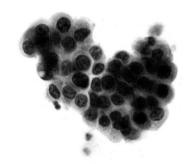

FIGURE 8.17
Normal urothelial cells. A papillary cluster of urothelial cells with rounded edge is often seen in patients with proven ureteric calculi.

SELF-CHECK 8.5

What is the significance of calculi to the cytologist?

Crystals

Most crystals seen in urine samples do not have major clinical significance; their formation depends on pH, the temperature that the urine was stored in prior to processing, and the chemical nature of the urine. Crystals are usually reported by microbiology departments on wet preparation samples. However, many survive cytological processing and can be seen in Papanicolaou stained preparations. These include triple phosphate acid crystals (Figure 8.18a) and uric acid crystals (Figure 8.18b).

Other non-neoplastic findings

Enterovesical fistula
Normally there is no connection between the urinary system and alimentary canal. An enterovesical fistula is an abnormal connection between bladder and alimentary system.

There are rare occasions in which the identification of foreign matter or contaminants in urine may aid in correct diagnosis and management of patients. One such instance is recognition of faecal matter. Faecal matter may be present in urine samples due to **enterovesical fistula**. This uncommon condition is due to an abnormal connection between the bladder and any segment of the bowel.

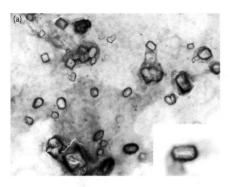

FIGURE 8.18

Examples of crystals seen in urine samples. (a) Triple phosphate crystals. Inset shows a high-power view of typical coffin lid appearance of triple phosphate crystals—voided urine. (b) Uric acid crystals—voided urine. Uric acid crystals showing large variation in appearance.

Pneumaturia
The passage of gas from the urethra during urination.

Faecaluria
The mixing together of urine with faeces which is normally due to presence of a fistula.

Cystoscopy
This is an endoscopic procedure that is used to examine the bladder. A cystoscope (a type of endoscope) is passed via the patient's urethra. It is also possible to take biopsies during cystoscopy. See Figure 8.20.

CASE STUDY 8.3

A 57-year-old male patient with a history of recurrent urinary tract infection was referred to the urology clinic by his GP.

At the urology clinic a full history was taken. He had previously been diagnosed with diverticular disease. He complained of pain near the pubic bone, and also mentioned that he could see gas bubbles in his urine and occasionally noted gas being released from his urethra whilst urinating (**pneumaturia**).

The urologist examined his urine sample, which was dark in colour, and noted the presence of faecal material (**faecaluria**). Suspecting the possibility of an enterovesical fistula, the urologist submitted urine cytology for faecal matter microscopy and carried out cystoscopy as a first-line investigation.

Urine cytology showed extensive vegetable matter (see Figure 8.19a) and partially digested meat fibres (Figure 8.19b). This was considered consistent with faecal material.

Cystoscopy (Figure 8.20) showed areas of inflammatory mucosal changes, and biopsies were obtained to exclude malignancy. Mucus was also observed in the bladder. Histological examination of the biopsies showed chronic inflammation.

The patient later had a barium enema and underwent CT scanning.

The tests confirmed presence of a **colovesical fistula** (connection between colon and bladder). The patient underwent surgical treatment: the affected segment of colon was resected and the defect in the bladder was closed up.

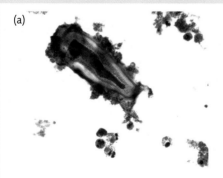

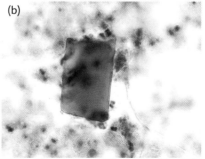

FIGURE 8.19
(a) Vegetable matter. The double wall cellular structure is consistent with a plant cell.
(b) Meat particle. A partially digested meat particle showing muscle striation.

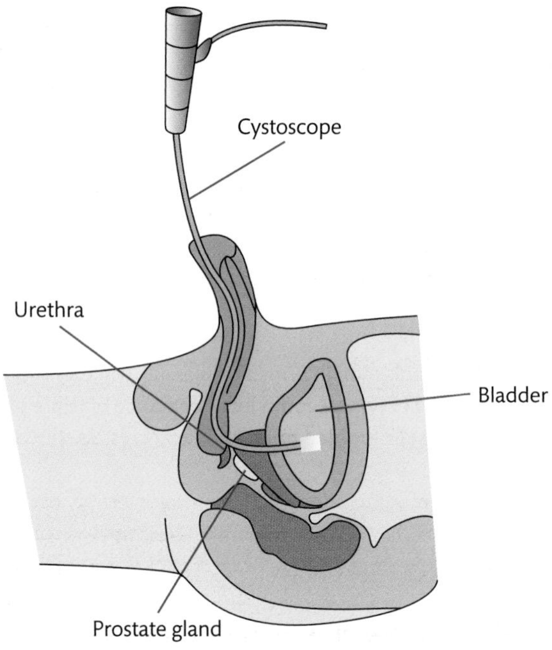

FIGURE 8.20
Cystoscopy involves using a fibre optic endoscopic device to examine the bladder and the urethra under local or general anaesthesia. The image is normally viewed on a monitor. During cystoscopy the urologist may also carry out therapeutic procedures.

Diverticular disease
This is a condition where the inner, lining layer of the colon bulges out through the outer muscular layer. These outpunchings are called diverticula. Infection of a diverticulum is called diverticulitis, which may cause tears in the colon, resulting in various complications including fistula formation and inflammation in the abdominal cavity.

The most common cause of enterovesical fistula is **diverticular** disease of the colon. Other causes include Crohn's disease, colonic cancer, or previous radiation therapy. In a Papanicolaou stained preparation, faecal matter appears as partially digested vegetable matter and muscle fibres. The background usually includes bacteria and debris.

8.7 Urothelial neoplasms

Both benign and malignant tumours may arise in the urinary tract. However, the cytological recognition of benign lesions is not possible using light microscopy because benign tumours do not have consistently recognizable diagnostic features. For this reason, we will discuss only the recognition of malignancy during the rest of this chapter.

Before we embark on the cytology of urothelial carcinoma, it is important to be familiar with the basic relevant clinical information, which will be helpful when assessing a sample.

Incidence and mortality statistics

Cross reference
Case Study 8.1.

The commonest malignancy in the urinary tract is carcinoma. In developed countries around 90% are usually of urothelial cell origin. The remainder are of squamous cell and adenocarcinoma origin. Urothelial carcinoma is often referred to as **transitional cell carcinoma** or **TCC**. In countries where schistosomiasis is endemic the usual cancer is squamous carcinoma. Urothelial carcinoma can arise anywhere in the urinary tract, but commonly it occurs in the bladder. Worldwide, bladder cancer is more common in the industrialized nations, the exception being Egypt, which has the highest incidence in the world (37 per 100,000) due to endemic schistosomiasis.

TABLE 8.1   Statistics on UK incidence and mortality for bladder cancer.

	Incidence (2006)		Mortality (2007)	
	Male	Female	Male	Female
Total	7,307	2,957	3,283	1,635
Age standardized rate (per 100,000 population)	19.4	5.9	8.1	2.8

For the remainder of this chapter we will limit our discussion to diagnosis and management of bladder cancer.

Table 8.1 summarizes some available statistics on UK incidence and mortality for bladder cancer.

Risk factors

Urothelial carcinoma is mainly caused by carcinogens ingested and excreted in urine. **Aromatic amines**, which are present in cigarette smoke, are known to be carcinogenic to the urothelium. Aromatic amines were also used in many industries in the manufacture of dyes and pigments for textiles, paints, plastics, paper, and hair dyes. They were also used in the production of drugs, pesticides, and in the rubber industry. Once the carcinogenic effect of aromatic amines was proven, they were withdrawn from these industries. In the UK this occurred in the 1950s.

Other agents reported to be involved in carcinogenesis of urothelial carcinoma include the **cytotoxic** drug **cyclophosphamide** and the analgesic **phenacetin**. The ova of *Schistosoma haematobium* has already been mentioned in its association with the development of squamous carcinoma of the bladder.

SELF-CHECK 8.6

What are the risk factors for urothelial neoplasms?

Presentation

The most common presenting symptom is haematuria. This is often painless and intermittent. Other symptoms include an increase in the frequency of urination and pain. Occasionally, bladder cancer presents with recurrent infection.

Investigation of patients with suspected bladder cancer

Patients should be managed by a urologist who is part of a **multidisciplinary cancer network team**.

The patients usually have a cystoscopy (Figure 8.20), and the urologist may also send a urine sample for cytology and microbiology.

Radiology plays an important role in the diagnosis of cancer in the urinary tract. Intravenous urogram (IVU) (see Figure 8.16) is the standard imaging technique in patients with haematuria having suspected bladder cancer.

Age standardized rate
The statistical method of comparing cancer rates and mortality between different geographical populations. This calculation removes the age bias by adjusting the rate to take into account how many people of different ages are in the population being looked at.

Cross reference
Case Study 8.1.

Aromatic amines
Organic molecules which contain one or more benzene rings.

Cytotoxic
Any agent that is toxic to cells and causes cell death.

Cyclophosphamide
A chemotheraputic agent used in combination with other drugs in treatment of many cancers including lymphoma and leukaemia. It is also used in treatment of some autoimmune disorders.

Phenacetin
A pain killer, which has similar uses as paracetamol and was initially banned from general use in 1968 as it was linked to bladder and kidney cancer. The ban was later revoked but its use is highly restricted.

Multidisciplinary cancer network
A group of medical professionals who are specialists in cancer diagnosis and treatment. The team follow nationally agreed guidelines to ensure they deliver the best care for cancer patients.

Cross reference
Chapter 7.

Low-grade and high-grade tumours

This refers to how the cells appear under the microscope. Histopathologists use various cytological and architectural criteria to determine the grade of a tumour. In a nutshell, in a low-grade tumour the cells look like normal cells and are 'well differentiated'. Low-grade tumours generally grow slowly. Cells in a high-grade tumour appear very abnormal and 'poorly differentiated'.

Diagnosis and grading

Definitive diagnosis prior to treatment is made on biopsy. Urothelial carcinomas are clinically separated into **non-invasive papillary** (Figure 8.21a), flat growing **carcinoma *in situ* (CIS)** (Figure 8.21b), and **muscle invasive carcinoma** (Figure 8.21c).

Non-invasive papillary carcinomas are tumours that do not invade underlying tissue and have a papillary (nipple-like) growth pattern that projects into the bladder. Papillary carcinomas tend to recur after treatment and up to 8% of these tumours may progress to invasive carcinoma. These tumours are histologically graded as either **low grade** or **high grade**. Low-grade lesions are less likely to progress to invasive carcinoma and spread. High-grade lesions have a higher tendency to become invasive. Another category of papillary lesions was defined by the World Health Organization (WHO) in 1994 to categorize a group of papillary lesions that

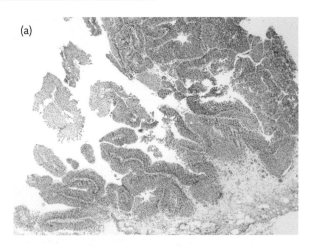
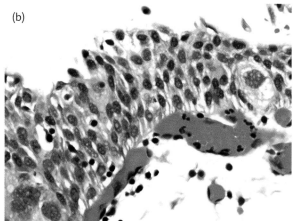
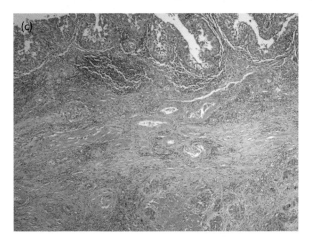

FIGURE 8.21

Examples of urothelial carcinomas. (a) Papillary carcinoma, histology section. This low power photomicrograph show finger-like projections of multilayered urothelial cells, which show minimal cytological atypia. (b) Carcinoma *in situ*—histology section. The full thickness of the epithelium is composed of neoplastic cells. These cells have not invaded the basement membrane. They show polymorphism and hyperchromasia. (c) Invasive carcinoma—histology section. Tumour cells are invading the muscle.

are slow growing and unlikely to spread. These lesions were called papillary urothelial lesions of low malignant potential or **PUNLUMP**. As the name suggests, these lesions are unlikely to progress to invasive carcinoma. Papillary lesions are generally easy to see at cystoscopy due to their growth pattern.

Carcinoma *in situ* has a flat growth pattern that spreads along the surface of the bladder. These tumours are histologically graded as high grade and the majority will eventually become invasive. Carcinoma *in situ* can develop alone or in association with papillary lesions. Due to their flat growth pattern their cystoscopic appearance is different to papillary lesions and cystoscopic assessment may not always be diagnostic.

Muscle invasive tumours are those that at presentation have invaded the **muscularis propria**. These tumours are high grade and behave aggressively. After invasion through the muscle layers, the tumour cells may spread to lymph nodes and eventually to bone, liver, and lung.

Muscularis propria
The deep muscle layer in the bladder that contracts during urination.

Treatment of bladder cancer

Superficial bladder cancers (that is, tumours that are confined to mucosa and do not invade the muscle) are often treated by **transurethral resection of bladder tumour (TURBT)**.

TURBT is an endoscopic procedure. Under general anaesthesia the cystoscope is passed through the urethra and after visualization of the tumour a special electrically heated wire loop is used to remove the tumour and seal the blood vessels. Papillary lesions may recur, and in these patients and those with high-grade disease chemotherapy into the bladder is given. This is called **intravesical therapy**. **Mitomycin C**, a type of intravesical chemotherapy drug, is often used. Another form of intravesical therapy is BCG.

Mitomycin C
This is a chemotherapeutic agent that is used in treatment of many cancers, including breast, rectal, and bladder cancer. It is used intravesically in treatment of bladder cancer.

The **Bacillus Calmette-Guerin vaccine** (BCG) which is used to vaccinate against tuberculosis, is commonly used for treatment of urothelial carcinoma. BCG is particularly useful in treatment of carcinoma *in situ* and high-grade urothelial carcinoma. BCG vaccine is prepared from a strain of the attenuated (weakened) live bovine tuberculosis bacillus. BCG has been used as a form of **immunotherapy** in treating bladder carcinoma since the 1980s. This effective treatment appears to work by initiating a local immune reaction against the tumour. The vaccine is injected into the bladder once a week for 6–12 weeks. Patients undergoing this treatment are routinely monitored for tumour recurrence and treatment response. Patients with carcinoma *in situ* who do not respond to BCG therapy are treated with surgery.

Immunotherapy
Any form of therapy that treats the disease by activating, amplifying, or suppressing the immune response.

Invasive tumours are usually treated with **radical cystectomy** (removal of bladder). Radical cystectomy in both sexes is an extensive operation but it remains the standard procedure for determining accurate staging, minimizing the risk of tumour recurrence, and optimizing curative potential. This operation differs in men and women: in women it is referred to as **anterior pelvic clearance**, and involves removing the bladder, urethra, lower end of the ureters, the front wall of the vagina, uterus, Fallopian tubes, and ovaries. Internal lymph nodes that lie inside the pelvis are also removed. In younger women the ovaries may be preserved. In men it is called **cystoprostatectomy**, and involves removal of the bladder, prostate, lower end of ureters, and sometimes the urethra.

When the bladder is removed, the surgeon also needs to carry out a further operation to allow for urine collection. There are several possible techniques available; the most common method is a procedure called **urostomy** or ileal conduit. In this procedure the ureters are disconnected from the bladder and joined to one end of a segment of bowel that is isolated from the rest of the intestine. This section of bowel is then brought to the skin surface on the abdomen. The end of the bowel that opens onto the abdominal wall is known as a **stoma**. The urine then empties through this stoma into a small bag.

The other surgical procedure which is suitable for some patients involves bladder reconstruction. This involves using a section of the small bowel to make a new reservoir. The ureters are implanted into this new reservoir, which is then sewn onto the urethra. The benefit of this method is that it allows the patient to pass urine in much the same way as they would normally and thus improves the quality of life after cystectomy.

Patients who are unfit for surgery may be treated with radiation therapy or a combination of chemotherapy and radiotherapy.

8.8 Cytology of urothelial carcinoma

Anaplastic cells

These display significant pleomorphism, hyperchromasia with high nuclear to cytoplasmic ratio. Anaplastic cells do not show any differentiation.

Urothelial carcinomas show a continuous pattern of growth from the relatively benign looking low-grade lesions to **anaplastic** looking high-grade carcinomas. Cytology has low sensitivity in diagnosis of low-grade urothelial carcinoma as low-grade lesions do not exhibit significant nuclear abnormality. Figure 8.22a is a histological section from a low-grade urothelial carcinoma. The cells look deceptively like normal cells (compare this to Figure 8.2), and the histopathological diagnosis is made on architectural features. Figure 8.22b is a histological section from a high-grade urothelial carcinoma and cells that exfoliate from this lesion will show sufficient nuclear abnormality to be recognized as abnormal.

Low-grade urothelial lesions

These lesions encompass papillary urothelial neoplasm of low malignant potential and low-grade urothelial carcinoma. For ease and practicality of cytological reporting we will discuss both these lesions under one heading. For more in-depth discussion refer to the further reading list at the end of this chapter.

It was mentioned earlier that these lesions are very difficult to diagnose with routine cytology as the cells closely resemble normal urothelial cells (Figure 8.23a). The cellular content of voided urine may be higher, and it may contain an increased number of cells. More often, however, cell clusters are present in larger numbers. Some of the clusters may be loosely cohesive (Figure 8.23b) and papillary looking fragments may also be seen (Figures 8.23c and d). On close high-power examination, these cells show mild variation in nuclear shape and size,

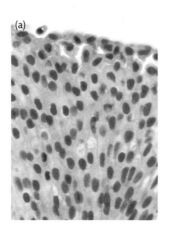
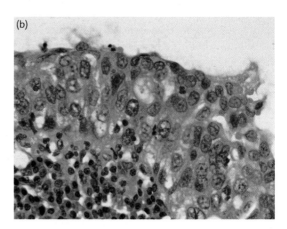

FIGURE 8.22

(a) Low-grade urothelial carcinoma–histology section. (b) High-grade urothelial carcinoma–histology section.

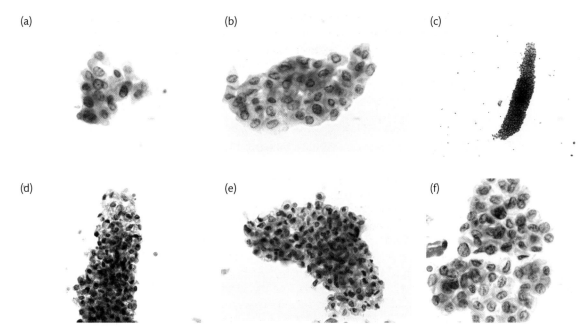

FIGURE 8.23

Atypical cells (confirmed low-grade urothelial carcinoma on biopsy). Voided urine.
(a) These cells show minimal cytological abnormality. There is mild variation in nuclear size
and variation in chromatin pattern. (b) Another example of atypical cells exhibiting loosely
cohesive clusters. (c) This low-power view shows a papillary-like group of urothelial cells.
(d) High-power view of (c) shows nuclear crowding and variation in size and shape. (e) In
this tissue fragment cells are disorganized and have overlapping nuclei. (f) Cells in this field
show a lot of variation. The nuclei are varied in shape and size, have irregular outline with
varied chromatin pattern.

increased nuclear to cytoplasmic ratio, and irregular nuclear outline, and the chromatin pattern may be varied when compared to neighbouring cells (Figures 8.23e and f). These architectural and cytological changes are, however, very similar to those seen in calculi (see Figure 8.17) and cannot be used to diagnose malignancy reliably. These changes are often reported as atypical or suspicious of malignancy.

Key Points

Summary of cytological features raising the suspicion of a low-grade lesion

- Increase in cellular content in voided urine samples.
- Increase in number of single cells and cell clusters.
- Papillary clusters.
- Mild variation in nuclear size and shape.
- Slight increase in nuclear to cytoplasmic ratio.
- Irregular nuclear outline.
- Chromatin pattern varied amongst neighbouring cells.

High-grade urothelial carcinoma

These include cells exfoliated from high-grade papillary urothelial carcinoma and carcinoma *in situ*. Generally, a high-grade urothelial carcinoma is readily identifiable in urine samples as there are reliable criteria for its identification.

Voided urine samples may be quite cellular (Figure 8.24a) and isolated single cells and clusters are quite common (Figures 8.24b and c); these may show marked **pleomorphism** (Figure 8.24d). Cells show a high nuclear to cytoplasmic ratio (Figure 8.24e) and marked nuclear hyperchromasia (Figures 8.24f and g). The chromatin may be coarsely granular (Figures 8.24h and i) and the nuclear outline may be quite irregular (Figure 8.24j). Some cases show enlarged nucleoli (Figure 8.24k). **Squamous differentiation** is frequently seen in high-grade urothelial carcinoma (Figure 8.25a). The background may include **necrotic debris**, inflammatory cells, and blood (Figure 8.25b). It is important to note that these background features could be removed by the processing methods employed in the laboratory, particularly the LBC technology.

Squamous differentiation

The process where fully differentiated epithelium takes physical characteristics of squamous cells. This sometimes occurs in malignant tumours, including urothelial, breast, and endometrial carcinoma.

Necrotic debris

The remains of dead cells.

Key Points

Summary of cytological features of high-grade urothelial carcinoma

- High nuclear to cytoplasmic ratio.
- Single and clusters of abnormal cells.
- Occasionally very marked hyperchromasia.
- Irregular nuclear outlines.
- Coarsely granular chromatin.
- Large nucleoli.
- Dirty background.

It is important to avoid a false positive cytology report as the patient may be subjected to unnecessary investigations. Most false positive diagnoses are due to misinterpretation of changes due to polyoma virus, calculi, and treatment changes, and the over-interpretation of benign cell groups as malignant. Case Study 8.4 highlights some of the issues that can arise from a false positive diagnosis malignancy.

Key Points

Most false positive diagnoses are due to misinterpretation of changes caused by polyomaviruses, calculi, and variations in treatment, or to the over-interpretation of benign cell groups as malignant.

Other malignant lesions

Earlier in the chapter it was mentioned that pure primary squamous carcinoma is rare and is usually associated with *Schistosoma haematobium* infection. Squamous carcinoma could also been seen in the bladder as result of direct spread of the tumour from the cervix or vagina. Primary adenocarcinoma of the bladder is also a rare finding.

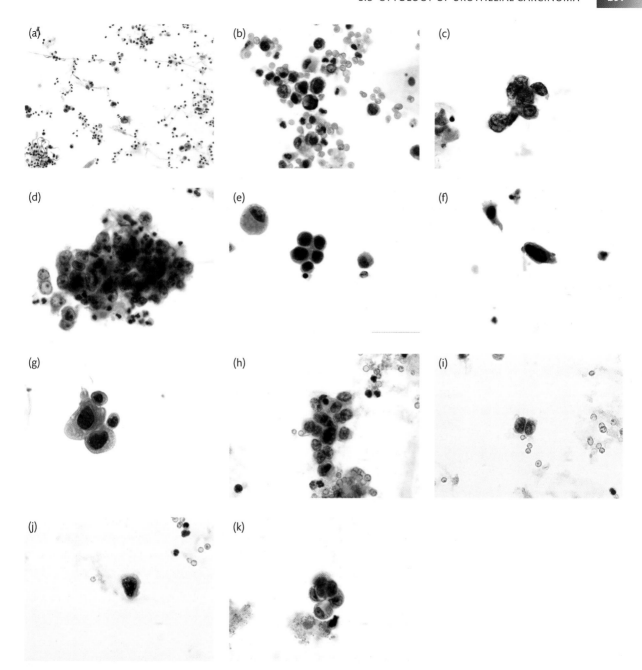

FIGURE 8.24

High-grade urothelial carcinoma—voided urine. (a) Low-power view showing increase in cellularity. (b) Isolated malignant cells. (c) Cluster of malignant cells with hyperchromatic nuclei. (d) Cluster of malignant cells exhibiting many diagnostic features including pleomorphism. (e) Malignant cells in this field have a very high nuclear to cytoplasmic ratio. The nucleus has occupied most of the area of the cell. (f) Note the marked nuclear hyperchromasia in this high-grade urothelial carcinoma. (g) Another example of marked nuclear hyperchromasia. (h) These malignant cells have coarse chromatin pattern. (i) These two cells also have coarse and irregular chromatin pattern. (j) In addition to other features of malignancy mentioned earlier, this cell also has an angular nuclear outline. (k) Prominent nucleoli are also seen in some cases.

(a)

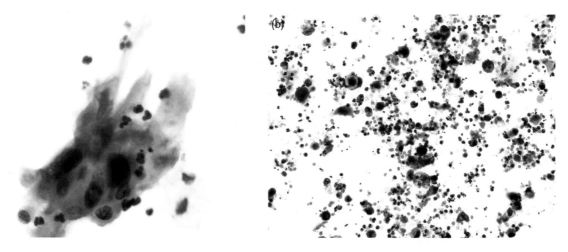

FIGURE 8.25
High-grade urothelial carcinoma with squamous differentiation-voided urine. (a) These malignant cells are keratinized. Elsewhere in the same slide there were malignant cells with more usual features of urothelial carcinoma. (b) This low-power field shows the 'dirty' background that could be seen in a malignant urine sample. This includes blood, inflammatory cells, and cell debris. These useful clues may be removed during processing.

CASE STUDY 8.4

A 67-year-old male patient with a history of haematuria was referred to the urology clinic by his GP.

At the urology clinic he was asked to submit a urine sample, and underwent cystoscopy. The cystoscopic findings were not significant but random biopsies were taken.

The biopsies did not show any significant abnormalities. However, the urine sample included single enlarged cells that had high nuclear to cytoplasmic ratio and abnormal chromatin pattern (Figures 8.26a and b). The sample was reported as 'suspicious of high-grade urothelial carcinoma'.

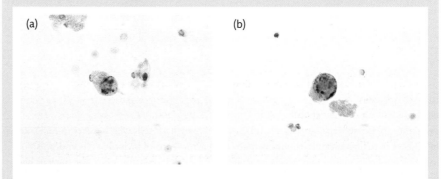

FIGURE 8.26
Polyoma virus-voided urine. (a) This cell with high nuclear to cytoplasmic ratio has a chromatin pattern in keeping with polyoma virus infection. (b) This cell has a perfectly round and smooth outline. The chromatin also has a glassy appearance.

The patient had an ultrasound and an IVU. Further urine samples were also submitted for cytological examination. The IVU and ultrasound did not show any abnormality, but the urine samples still included abnormal cells similar to the first sample. The sample was reported as 'atypical degenerate urothelial cells present suspicious of urothelial carcinoma'. The patient was discussed at the MDT and a further investigation of the upper urinary system was recommended. A **retrograde ureteropyelography was** carried out to allow for investigation of the ureters. This also did not show any defects.

The patient was re-discussed at the MDT. Slides were critically reviewed in the light of negative cystoscopy, IVU, retrograde ureteropyelography, ultrasound, and biopsies. It was decided that the cells that were originally reported as atypical probably represented polyomavirus infection.

The patient was discharged and followed up periodically; he remained disease free.

This case of polyomavirus infection was in a patient who did not have any history of immunosupression. However, transient and asymptomatic reactivation of polyomaviruses can be seen in some individuals and observed in urine samples. Therefore cytologists should be aware of the cytological changes that accompany this infection. Looking critically at Figures 8.26a and b we can see that although these cells have high nuclear to cytoplasmic ratio, the nuclear outline is perfectly round. The chromatin also appears opaque and glassy. These features are more in keeping with polyomavirus infection than neoplasia.

Retrograde ureteropyelography

Sometimes called retrograde pyelogram this is a combination of cystoscopy and X-ray examination, which allows visualization of the bladder, ureters, and renal pelvis. During a cystoscopy, contrast dye, is introduced into the ureters via a catheter and X-ray images taken before and after this instillation allow visualization of any abnormal areas.

Cells from prostatic adenocarcinoma may occasionally be seen in a urine sample, either as the result of direct spread to the bladder or due to the invasion of the urethra. While the primary diagnosis is usually known in most cases, cytology can occasionally suggest the diagnosis if characteristic features are present.

Case Study 8.5 is about a patient who had been previously treated for prostatic carcinoma and presented with haematuria.

Distant metastatic spread can occasionally be seen from primary breast carcinoma, lung, or malignant melanoma. Case Study 8.6 is an unusual presentation of a primary tumour first manifesting itself in the bladder.

CASE STUDY 8.5

A 56-year-old male who had been treated with radiotherapy for locally advanced prostate cancer presented with haematuria. The urologist asked the patient to submit a urine test for cytology and microbiology.

The urine sample was highly cellular and included clusters of cells with hyperchromatic nuclei. Some of the cells appeared to be forming gland-shaped structures (Figure 8.27a). A cell block of the urine sample was prepared for immunocytochemistry. Cells stained strongly positive for prostatic specific antigen (PSA) confirming their prostatic origin (Figure 8.27b).

The patient had cystoscopic biopsies which confirmed infiltration of the bladder wall with prostatic adenocarcinoma.

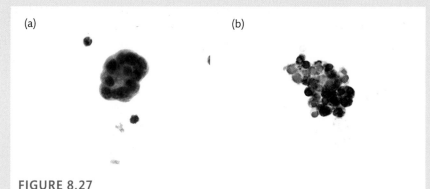

FIGURE 8.27
Prostatic carcinoma. (a) Prostatic carcinoma—voided urine. This group of malignant cells have formed an acinar structure and have round nuclei with smooth nuclear outlines. (b) The malignant cells show positive staining with PSA.

Cross reference
Chapter 4, Section 4.9.

8.9 Iatrogenic changes

Iatrogenic changes are caused by diagnostic or therapeutic procedures. Iatrogenesis was discussed in Chapter 4. Recognition of iatrogenic changes in the urinary tract is important to the cytologist as some of these changes can mimic malignancy. It is important to remind the clinicians of the importance of providing a full clinical history on request forms to enable the cytologist to assess the cytological changes correctly.

CASE STUDY 8.6

A 56-year-old smoker attended his GP complaining of weight loss, tiredness, and haematuria. The GP asked the patient to submit a urine sample for cytology and referred the patient urgently to the urology clinic.

The urine was bloodstained and included small, round, isolated single cells with hyperchromatic nuclei and scanty cytoplasm (Figure 8.28a). The other striking feature was the presence of cell clusters which showed 'nuclear moulding', where the tumour cells mould or shape around each other to form a distinctive structure (Figure 8.28b) (see also Chapter 10, Section 10.8). These features are suggestive of small cell anaplastic carcinoma.

The cytology was reported as: 'Atypical cells present which have high nuclear to cytoplasmic ratio. There is evidence of nuclear moulding in the cell clusters. Features are suggestive of small cell anaplastic carcinoma. Suggest further investigation.'

The GP requested a chest X-ray. This showed a right hilar mass which was suggestive of lung neoplasm. The patient was therefore referred to the chest physician urgently. The patient had good lung function and had a bronchoscopy which confirmed the presence of a tumour. Bronchial brushings confirmed the presence of small cell anaplastic carcinoma. The patient was discussed at MDT and treated with combination therapy.

Small cell anaplastic carcinoma of neuroendocrine origin may very rarely originate from the bladder, but a lung primary metastasis is most likely and should be excluded first.

Cross reference
Chapter 10, Section 10.7.

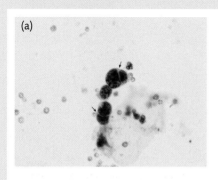
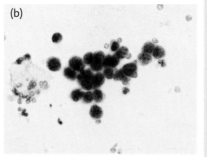

FIGURE 8.28
Small cell anaplastic carcinoma-voided urine. (a) These hyperchromatic malignant cells have a coarse chromatin pattern and very high nuclear to cytoplasmic ratio. Also note nuclear moulding (arrows). (b) This cluster of malignant cells also has a coarse chromatin pattern, high nuclear to cytoplasmic ratio, and the nuclei are moulding.

The effect of catheterization and changes due to an ileal conduit were mentioned in Section 8.2. In this section other changes induced by therapy, such as radiotherapy, chemotherapy, and immunotherapy were discussed.

Effects of radiotherapy

Bladder tumours may be the intended target for radiotherapy or the bladder may be irradiated when other pelvic tumours are being treated. Normal urothelium has an extremely slow turnover rate: the early radiation induced changes are initially mild and may not become recognized for 6–12 months.

Voided urine samples may show increase in exfoliation of urothelial cells (Figure 8.29a). Cell size may increase, but normal nuclear to cytoplasmic ratio is preserved (Figure 8.29b). Other changes include multinucleation and hyperchromasia (Figure 8.29c). The nuclear chromatin may be featureless (bland) and nuclear outline remains smooth. Cytoplasmic changes include vacuolation (Figure 8.29d).

Effects of chemotherapy

There are numerous chemotherapeutic agents in use for the treatment of cancer. The bladder epithelium may be affected by agents given to treat malignancies of other organ systems as well as for treating bladder cancer. Some of these chemotherapeutic treatments have been in use for many years, and the changes on the urothelium are better recognized. One such drug, cyclophosphamide, and its effect on the urothelium is well documented. The active metabolite of this drug concentrates in the urine and causes urothelial cell death. As such, these changes closely mimic urothelial carcinoma. The nuclei may exhibit hyperchromasia, the nuclear cytoplasmic ratio may be increased, and their nuclear outlines may become irregular. In addition,

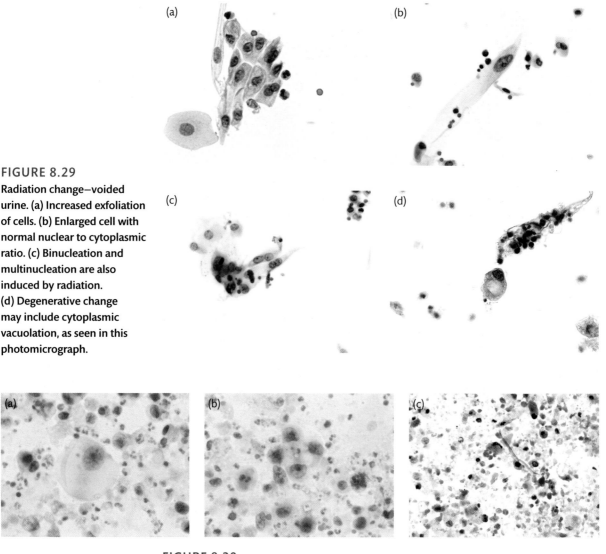

FIGURE 8.29
Radiation change–voided urine. (a) Increased exfoliation of cells. (b) Enlarged cell with normal nuclear to cytoplasmic ratio. (c) Binucleation and multinucleation are also induced by radiation. (d) Degenerative change may include cytoplasmic vacuolation, as seen in this photomicrograph.

FIGURE 8.30
Chemotherapy effect (cyclophosphamide)–voided urine. (a) This cell is enlarged, the nucleus is eccentric with low nuclear to cytoplasmic ratio. (b) The changes seen in this field are very similar to urothelial carcinoma and without appropriate clinical details would result in false positive diagnosis of carcinoma. (c) The cells are degenerate, but show keratinization and hyperchromasia. The background shows necrotic debris. Without clinical history, these changes could be erroneously reported as urothelial carcinoma.

one or more nucleoli may be visible. The background may show evidence of necrosis, such as presence of cell debris (Figures 8.30a and b). Other changes include nuclear pyknosis, keratinization, and increased exfoliation (Figure 8.30c).

You learned earlier about features of urothelial carcinoma. Changes due to iatrogenesis are very similar to urothelial carcinoma and without adequate clinical history drawing a distinction between the effect of chemotherapy and malignancy is not possible.

(a)   (b)   (c)

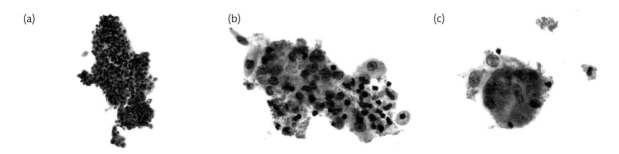

FIGURE 8.31
BCG therapy-voided urine. (a) BCG may cause exfoliation of a large number of urothelial cells. These may shed as single cells, clusters or in sheets as in the photomicrograph. (b) This field includes urothelial cells, polymorphs, and elongated histiocytes. (c) Multinucleated histiocytes indicating a granulomatous response in the bladder.

Effects of Bacillus Calmette-Guerin

BCG immunotherapy is widely used in management of carcinoma *in situ* and the cytologist should be aware of the effects of this therapy on urothelium.

The cytological picture could be varied: during the initial phase of therapy a neutrophil response may be seen, with increased exfoliation of epithelial cells (Figure 8.31a). Later in the course of therapy, numerous **epithelioid histiocytes** will be seen; these may be single, multinucleated or form aggregates (Figures 8.31b and c). The epithelial cells may show degenerative changes, and some may show an increase in hyperchromasia. These changes are usually well short of neoplasia; however, if there are any concerns that the cells are neoplastic, comparison with previous cytology and histology is very helpful.

Epithelioid histiocytes
Histiocytes (macrophages) that look like epithelial cells. Epithelioid histiocytes are seen in granulomatous inflammation.

Cross reference
Epithelioid cells in Chapter 10, Section 10.6.

SELF-CHECK 8.7

List some of the iatrogenic changes that a cytologist reporting urine samples should be aware of to avoid a false positive diagnosis.

8.10 **Ancillary techniques**

It was explained earlier in the chapter that light microscopy has its limitations and does not detect all urothelial carcinomas. Immunocytochemistry is widely available in most cytology laboratories but has limited value in urine cytology as there is currently no specific marker to detect malignancy. Indeed, immunocytochemistry is generally unreliable in unfixed urine samples as cells show degenerative changes. Reliability can be improved, however, by collecting the urine in a suitable preservative. Case Study 8.5 demonstrates a rare occasion when immunocytochemistry was helpful in diagnosis.

There are newer techniques in development that are said to improve the sensitivity and specificity of detecting malignancy in urine samples. These include the NMP22 test and **fluorescent** *in situ* **hybridization (FISH)**, which show promise in improving on the sensitivity of cytology. FISH will be discussed in Chapter 13.

Fluorescent *in situ* hybridization (FISH)
A technique used to detect presence or absence of DNA sequences on chromosomes.

Cross reference
Chapter 13.

SUMMARY

- Cytological evaluation of urine can detect most high-grade urothelial cell carcinomas which may arise anywhere in the urinary tract.

- Criteria for diagnosing low-grade urothelial carcinoma lacks specificity and may lead to false positive diagnosis.

- Cytology may identify some infective agents; cytologists in particular should beware of cell changes due to polyomavirus infection, as these can mimic high-grade urothelial carcinoma.

- 'Casts' may be identified in cytological preparations and may be conservatively reported.

- Specimen collection procedures do affect the presentation of cells and cytologists should consider this when assessing a urine sample.

- Iatrogenic changes can affect cell presentation and cytologists should adjust their diagnostic criteria to avoid a false positive diagnosis.

FURTHER READING

- Bardales RH (2002) *Practical Urologic Cytopathology*. Oxford University Press, New York.

- Bibbo M and Wilbur D (2008) *Comprehensive Cytopathology*, third edition. Saunders, Philadelphia.

- Rathert P, Roth S, and Soloway MS (1991) *Urinary Cytology Manual and Atlas*. Springer-Verlag, New York.

- Rosenthal DL and Raab SS (2005) *Cytologic Detection of Urothelial Lesions (Essentials in Cytopathology)*. Springer, New York.

DISCUSSION QUESTIONS

1. Evaluate and compare different preparation methods used in urine cytology.
2. Urine cytology is of no value in diagnosis of low-grade urothelial carcinoma. Discuss this statement.
3. Evaluate possible causes of false positive diagnoses in urinary tract cytology.

Answers to the self-check questions, and tips for responding to the discussion questions, are provided in the book's Online Resource Centre.

 Visit www.oxfordtextbooks.co.uk/orc/shambayati/

9

Serous effusions

Behdad Shambayati

In this chapter you will learn about the cytopathology of serous cavities. This is an interesting and exciting area diagnostically, as the cytologist will encounter a variety of disease processes.

Although benign conditions are encountered in serous effusion cytology, the main role of cytology is to diagnose malignancy, as serous cavities are commonly involved in cancer. The reporting of malignant cells in serous fluids has important clinical implications: most patients with a malignant effusion will have advanced disease and will require appropriate management. Of utmost importance is the avoidance of a false positive cytological diagnosis, as this can lead to inappropriate follow-up tests or even treatment.

Unlike other fields of cytology, where the application of ancillary techniques is hampered by a lack of good quality material, cytology cell preservation in serous effusion is generally excellent. Serous fluid often contains abundant viable cells—it is a very good culture medium in which cells can thrive. Consequently, it has become generally accepted that, with the judicious use of immunocytochemistry, cytological assessment of serous effusions can provide a definitive diagnosis in the majority of cases. In recent years, immunocytochemistry has also enabled the cytologist to predict the primary site of metastatic tumours, and it can greatly help the clinician in patient management.

It is not possible to describe every disease entity that may be encountered in serous effusions in this short chapter. The reader is encouraged to consult the reading list at the end of the chapter.

Learning objectives

By the end of chapter you should be able to:

- Describe the anatomy and physiology of serous cavities.
- Describe different types of serous effusions and factors relating to their formation.
- Outline different sampling and preparation techniques.
- Describe the normal cells found in serous fluids.
- Describe cytological features of non-neoplastic effusions.
- List common sites of metastatic spread to serous cavities in men and women.
- Describe features of different types of malignant cells commonly found in serous effusions.

■ Summarize risk factors, presentation, and diagnosis of malignant mesothelioma.

■ Give examples on the uses of immunocytochemistry in serous effusion cytology.

9.1 Anatomy and physiology of serous membranes

There are three body cavities, the pleural, peritoneal, and pericardial cavities, and these are lined by a membrane called **mesothelium** (Figure 9.1).

The cells forming the mesothelium are called **mesothelial cells**. Immediately below these cells is a thin layer of fibrous connective tissue, which includes lymphatic vessels and capillaries. Mesothelium is composed of two layers: one layer covers the wall of the cavity and the other lines the surface of the organs. The layer covering the wall of the cavity is called the **parietal layer**, and the layer covering the organs is referred to as the **visceral layer**. The mesothelium produces **serous fluid**, a lubricating fluid, which allows movement of the organs in the body cavity such as the beating of the heart, movement of the bowel, and expansion and contraction of the lungs.

The mesothelium is given different names depending on the anatomic site in which it is found. The **pleural mesothelium** covers the lungs, the **peritoneal mesothelium** lines most of the abdominal organs and the **pericardium** surrounds the heart. Mesothelium also covers the male and female reproductive organs. In males, the **tunica vaginalis testis** covers the testicles. In females, the **tunica serosa uteri** covers parts of the uterus. In this chapter, we will limit our discussion to the cytopathology of effusions from the pleural, pericardial, and peritoneal cavities.

The serous cavities are sometimes referred to as 'potential' cavities. In health, serous cavities normally contain only a very small amount of lubricating fluid, but in disease states large quantities of fluids may be produced, which can fill these 'potential' cavities. The serous fluid present in the pleural, peritoneal, and pericardial cavities is in a constant state of formation and removal. The fluid is formed by filtration of plasma through the capillary endothelium and is determined by several factors, including **hydrostatic pressure** within the capillary lumen, **colloid oncotic pressure** (osmotic pressure caused by proteins in plasma), rate of lymphatic drainage, and permeability of capillaries. Imbalance in these factors may result in the accumulation of fluid; such an accumulation is termed an **effusion**, and in the peritoneal cavity is generally called **ascites**.

Hydrostatic pressure in blood vessels
This is due to height of the blood above, which exerts a force on the blood lower down the body.

Colloid oncotic pressure
This is a type of osmotic pressure due to proteins present in the blood.

SELF-CHECK 9.1

What is the function of serous fluid and how is it produced?

Hypoalbuminemia
A condition where levels of albumin in the blood serum are low.

Clinically, effusions are either classified as **exudates** or **transudates** according to the amount of protein they contain. Transudates have low protein and low **specific gravity** (density). The cellular content of transudates is generally low. Transudates are caused by an imbalance of hydrostatic and oncotic pressure and are associated with kidney, heart, or liver failure, or **hypoalbuminemia**.

Exudates have higher protein content and specific gravity. They are caused by the increased permeability of capillaries, which can occur when the serous membranes are damaged by a disease process. Metastatic malignancy and infections are common causes of exudates. The exact mechanism through which malignancy causes the formation of an effusion is not fully understood, although it is possible that invasion of the parietal layer by the tumour cells obstructs the lymphatics and reduces fluid removal (1). There are some benign causes of exudates: these include infections, **connective tissue diseases** such as systemic lupus erythematosus (SLE) and rheumatoid disease, and inflammatory conditions such as **pancreatitis**.

Pericardium

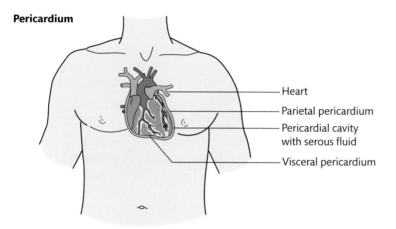

Heart
Parietal pericardium
Pericardial cavity with serous fluid
Visceral pericardium

Pleura

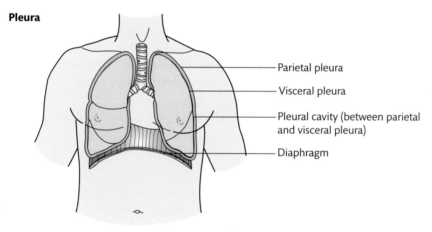

Parietal pleura
Visceral pleura
Pleural cavity (between parietal and visceral pleura)
Diaphragm

Peritoneum

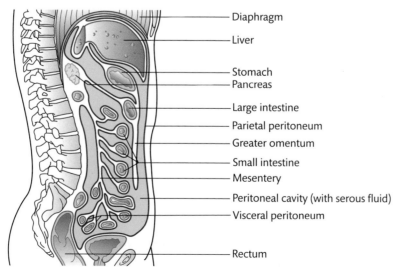

Diaphragm
Liver
Stomach
Pancreas
Large intestine
Parietal peritoneum
Greater omentum
Small intestine
Mesentery
Peritoneal cavity (with serous fluid)
Visceral peritoneum
Rectum

Connective tissue diseases

A group of diseases that primarily affect the connective tissues in the body. Many connective tissue diseases feature autoimmune activity, where the body's own immune system is involved in the inflammatory response against its own tissues. Systemic lupus erthematosus and rheumatoid disease are two examples of connective tissue disease.

Pancreatitis

Inflammation of the pancreas.

Cross reference

Section 9.4 for description of SLE and rheumatoid disease.

FIGURE 9.1

Schematic illustration of serous cavities.

What is the difference between an exudate and a transudate?

Key Points

An effusion develops when the amount of fluid entering the cavity exceeds the amount that is removed. This may be due to a combination of factors, including changes in hydrostatic pressure within the capillary lumen, colloid oncotic pressure, rate of lymphatic drainage, and permeability of capillaries.

9.2 Specimen collection and sample processing

Specimen collection

The aspiration of serous fluids is usually carried out under ultrasound guidance using local anaesthesia. It involves the insertion of a needle or cannula (tube) into the serous cavity. This procedure is termed **thoracocentesis** in the pleural space (Figure 9.2), **pericardiocentesis** in the pericardial space, and **paracentesis** in the peritoneal cavity.

The fluid is often collected into a bag or a large syringe. For cytological analysis at least 20 ml of fluid is required; optimally 40 ml will provide additional material for ancillary techniques. The specimen should be collected into a sterile container and sent to the laboratory without delay, although the sample may be stored in a fridge overnight without significant cellular deterioration.

Some serous effusions are rich in **fibrin**, which is a protein involved in the clotting of blood, and this may cause the sample to clot on standing. Clotted samples cannot be processed

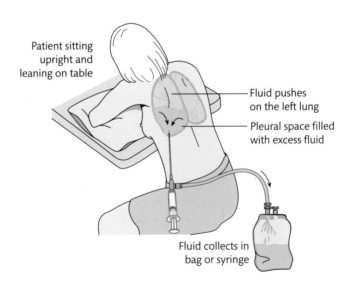

Patient sitting upright and leaning on table

Fluid pushes on the left lung

Pleural space filled with excess fluid

Fluid collects in bag or syringe

FIGURE 9.2
Illustration demonstrating thoracentesis.

using cytological preparation techniques, but can be fixed in formaldehyde and processed as a histological specimen. To prevent fluids clotting, anticoagulants may be added to the sample. **Sodium citrate** is a suitable cytological anticoagulant and is readily available. For a 20 ml universal container, 2 ml of 3.8% sodium citrate will inhibit clotting. Prefilled and adequately labelled containers with an expiry date, containing sodium citrate may be supplied to wards and clinics for convenience.

Sample processing

A variety of sample preparation methods are available. For a detailed description of methodology please refer to Chapter 2.

Cross reference

Chapter 2.

Normally, two alcohol-fixed slides are prepared for Papanicolaou staining and two air-dried slides for Diff Quick™ staining (Romanowsky type stain). Papanicolaou stain allows optimal assessment of nuclear details, whilst air-dried Diff Quick™ preparations are better for visualizing the cytoplasmic features. A good practice is to process half of the sample received, and use the remainder for making cell blocks, during which the cytological material is centrifuged and the cellular deposit is made into a histological cell block. This technique is particularly useful for immunocytochemistry, which is a method of analysing and identifying cell types based on the binding of antibodies to specific components of the cell.

As with all cytological preparatory techniques the aim is to concentrate cellular content and remove elements which do not contribute to the cytological assessment. Prior to processing, the macroscopic appearance and the volume of the sample should be noted and recorded. This has many benefits as it will allow the cytologist looking at the stained slide to relate the cytological picture to the macroscopic appearance of the fluid. Certain conditions produce serous effusions that have characteristic appearances (see the box below). For samples that do not appear to contain blood the cytocentrifuge technique is used. For blood filled samples, however, the **density gradient** method is employed (see density gradient technique on page 216).

BOX 9.1 *Macroscopic appearances of serous fluids*

Serous effusions can have varied appearances depending on their cellular and chemical content. Although the assessment of the colour of the fluid is a subjective exercise, it is a useful practice as it allows the person examining the final preparation to relate the cytological findings to macroscopic description. Figure 9.3 shows a range of fluids that may be received in a routine cytology laboratory.

Serous effusions that have low red blood cell content are generally yellow or straw coloured. The fluid becomes cloudier as the cell content increases. Figure 9.4 shows a range of appearances: the fluid on the right of the picture would be described as clear yellow, and the other two as cloudy yellow.

Red, pink, or orange colouring indicates the presence of red blood cells; the fluids in Figure 9.5a are bloodstained and require the use of density gradient technique to remove the red cells and concentrate the nucleated cells.

Infected fluids are often cloudy grey/green in colour. The sample in Figure 9.5b was from a patient who had a pleural **empyema** (pus in pleural cavity; see Case Study 9.1). This fluid would be described as **purulent** (containing pus).

Rarely, **chylous** fluids are received in the laboratory. These fluids have a milky white appearance due to high fat and lymphoid cell content as shown in Figure 9.5c.

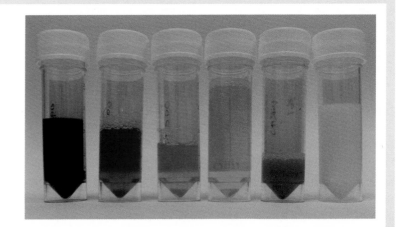

FIGURE 9.3
Serous fluids may exhibit a range of colours.

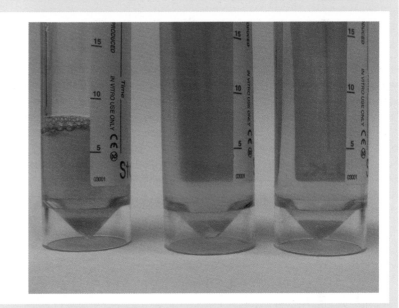

FIGURE 9.4
The universal bottle on the right is clear and low in cell content. The universal bottles in the middle and on the left have higher cellularity and appear cloudier.

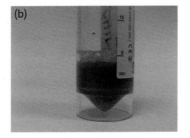

FIGURE 9.5
(a) Bloodstained serous fluids. (b) Purulent serous fluid. (c) Chylous serous fluid.

CASE STUDY 9.1

A 74-year-old male attended his GP complaining of fever, coughing green coloured sputum and suffering from chest pain. The GP examined the patient and based on his findings decided to refer him urgently to the A&E at his local hospital. He was treated with antibiotics for suspected (pneumococcal) pneumonia (the sputum culture confirmed this to be *Streptococcus pneumoniae*). After three days his symptoms did not improve and he developed **dyspnoea** (shortness of breath). A chest X-ray showed that a pleural effusion had developed. During thoracentesis, purulent pleural fluid was removed (see Figure 9.6a) and was sent for cytology and microbiology investigations.

Figure 9.6b is a cytological preparation which showed numerous polymorphs and cellular debris. The patient had a further thoracentesis and continued on antibiotic therapy. The effusion later resolved without further complication.

The collection of pus and debris in the pleural space is called an empyema. Empyema may occur as a result of bacterial pneumonia. Other causes include trauma and extension of infection from other sites. With appropriate timely antibiotic therapy, most empyemas resolve. However, some do not respond to therapy. The resulting inflammatory response may cause the fluid to thicken and the healing process can cause adherence of the visceral and parietal pleura, which may interfere with lung function. In these circumstances surgical intervention may be required to physically remove the adhesions and dead tissue.

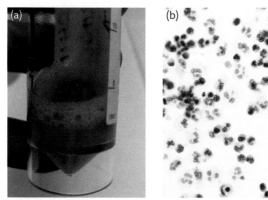
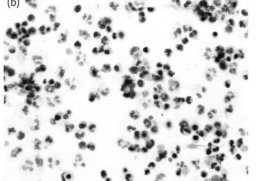

FIGURE 9.6
(a) Purulent serous fluid. (b) Highly cellular sample of serous fluid from a patient with empyema consists almost exclusively of polymorphs.

Centrifugation and direct preparation

This is the simplest and quickest method for preparing serous effusions. It involves spinning the sample in a centrifuge and making a preparation from the deposit. In most cases, this technique will yield a satisfactory preparation. When the sample is heavily bloodstained, however, the diagnostic cells may be diluted and obscured. In such cases, techniques that remove excess red blood cells are employed (see density gradient method).

Cytocentrifugation

This method involves spinning the cells directly onto the glass slides. It may be preceded by a centrifugation step to concentrate the cells. Cytocentrifugation produces a monolayer of cells, but is not suitable if the material is heavily bloodstained. This method is also not suitable if the sample is mucoid or gelatinous.

Density gradient method

This is an effective way of removing red blood cells from a bloodstained sample and concentrating the nucleated cells. In this method the fluid is carefully positioned on the density gradient solution and centrifuged. Differential migration during centrifugation results in the formation of visible cell layers containing different cell types. The bottom layer contains red blood cells; the layer immediately above the red blood cells contains the mesothelial and other large cells. There are many commercial density gradients on the market, including Lymphoprep™ and Histopaque®-1119.

Liquid based cytology

Liquid based cytology methods are applicable to serous effusion cytology, but pre-fixation does not allow the preparation of air-dried Diff Quick™ slides that enhance cytological assessment.

Cell blocks

Cell blocks made from centrifuged deposits can be processed as histological paraffin blocks. Varieties of commercial and in-house methods are available. The plasma-thrombin method is a way of making cell blocks in-house. This simple and inexpensive method involves combining centrifuged deposit of the fluid with commercially purchased blood plasma and **thrombin**. The resultant clot is fixed in formaldehyde then processed in the conventional manner and embedded in paraffin.

Thrombin
A protein involved in the coagulation process (clotting) in blood.

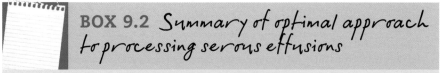

BOX 9.2 Summary of optimal approach to processing serous effusions

- Request at least 20 ml of freshly collected fluid.
- Describe the volume, colour and consistency of the sample.
- Use density gradient to remove red blood cells if material is bloodstained.
- Prepare both air-dried and alcohol fixed slides for MGG and Papanicolaou staining.
- Prepare cell blocks for immunocytochemistry.

9.3 **Reporting terminology**

Most laboratories in the UK describe the cellular content of the specimen followed by a concluding statement, which divides the effusion into the following categories:

- Negative for malignancy
- Suspicious for malignancy (should be used very rarely)
- Malignant

With careful use of immunocytochemistry, most fluid samples could be reported as either negative or malignant. The term 'atypical' should be avoided as it does not contribute to patient management. The 'suspicious' category is only used when diagnostic cells are too few in number to allow for a confident report of malignancy. In this situation, a further specimen is requested. We must avoid making false positive diagnoses, as finding malignant cells in serous effusions generally implies poor **prognosis** (outcome) and may result in unnecessary treatment and emotional stress for the patient. When malignancy is present, however, immunocytochemistry can be used to suggest the origin of the malignant cells. This will be discussed later in this chapter.

9.4 **The cells of effusions**

Mesothelial cells, macrophages, and lymphocytes are present in almost all fluid samples in varying proportions. Red blood cells may also be present, either as a component or as a contaminant during the sampling procedure.

Mesothelial cells

Mesothelial cells, when injured or stimulated, divide freely in serous fluids, and it is common to see variation in size and nuclear features. Mesothelial cells are usually present as discrete single cells, or in small clusters. Clustering appears to be seen more in ascitic fluids, but generally the presence of tightly cohesive clusters in benign effusions is unusual and should alert the cytologist to assess the sample further to exclude malignancy. Malignant mesothelioma, which will be discussed later in this chapter, sometimes presents as morphologically normal mesothelial cells in clusters.

Key Points

Presence of tightly cohesive clusters of mesothelial cells is an unusual finding and should alert the cytologist to examine the sample carefully to exclude malignancy.

Mesothelial cells are round in shape, and stain deep green with Papanicolaou and blue to purple/blue with MGG staining (see Figure 9.7).

This staining characteristic is due to the presence of **intermediate filaments** (cytoskeletal proteins) in the cytoplasm. A clear zone can sometimes be seen at the edge of cytoplasmic borders under high magnification. This is due to the presence of long **microvilli** (microscopic cellular membrane protrusions) on their cytoplasmic surface. This important morphological feature can be seen under the electron microscope. Indeed, prior to the invention of immunocytochemistry, the observation of this morphological feature was the main diagnostic criterion for the identification of mesothelial cells.

Mitotic figures (aggregation of chromosomes) are common in benign fluids (Figure 9.8a). Mesothelial cells in a group (Figure 9.8b) are separated by a clear narrow zone which is referred to as a 'window'.

Mesothelial cells generally have one nucleus, but binucleation and multinucleation can sometimes be seen. The nucleus is round, positioned centrally or just off centre, and the chromatin pattern is often described as being finely stippled (comprising small dots), but this can vary. The nuclei often have a single nucleolus. Mesothelial cells that are actively dividing in response to injury or stimulus are often called **reactive mesothelial cells**. Reactive mesothelial cells (Figure 9.9) show a spectrum of changes that can resemble malignancy. These changes may

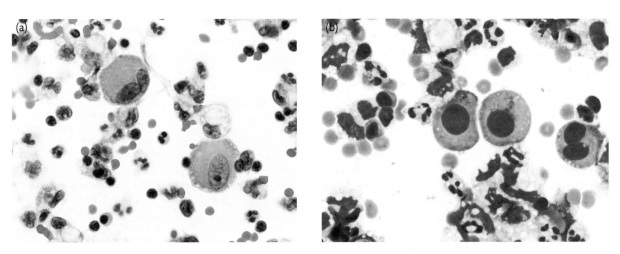

FIGURE 9.7
Mesothelial cells. (a) Many characteristic features of mesothelial cells can be seen in this photomicrograph: dense cytoplasm, clear zone at the cytoplasmic borders, eccentrically placed nuclei, and single nucleolus. (b) Another example of mesothelial cells.

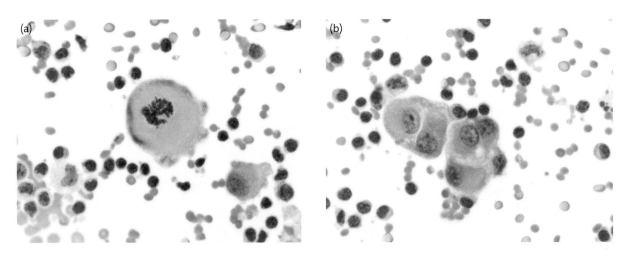

FIGURE 9.8
Mitotic figures. (a) Mesothelial cell in mitosis. (b) Mesothelial cell in a group separated by a 'window'.

BOX 9.3 Cytological features of mesothelial cells

- Usually isolated cells
- Round cells
- Dense staining cytoplasm
- Round central nucleus
- Stippled chromatin
- Single nucleolus
- Stain positive with calretinin, CK 5/6, and thrombomodulin
- Benign mesothelial cells stain with desmin
- Generally, negative staining with EMA

Calretinin

A calcium binding protein. Antibodies to this protein are reactive with cells of mesothelial origin. This antibody stains both the nucleus and the cytoplasm of the cell.

Thrombomodulin

A membrane protein. Antibodies to this protein are immunoreactive with mesothelial cells and some adenocarcinomas.

Cytokeratin 5/6

This is an antibody to cytokeratins 5 and 6. This antibody stains mesothelium and squamous carcinoma, but is usually negative in adenocarcinoma.

Desmin

An antibody to muscle tissue useful in serous effusion cytology as it reacts with benign mesothelial cells.

Epithelial membrane antigen (EMA)

A glycoprotein that is present in a wide variety of epithelial cells. It is useful in fluid cytology as antibodies to this protein are immunoreactive with most adenocarcinomas and mesotheliomas, but unreactive in benign mesothelial cells.

Vacuole

An enclosed compartment within cells which may contain fluids or occasionally solid matter.

include a coarsened chromatin pattern and enlarged nucleoli. In these cases, careful analysis of cytomorphology, together with clinical details and immunocytochemistry, should resolve the diagnostic dilemma.

Immunocytochemistry can be used to identify mesothelial cells, which usually stain positive with the immunocytochemical markers **calretinin** (Figure 9.10a), **thrombomodulin**, and **CK 5/6**. (We discuss these markers in more detail later in the chapter.) Benign mesothelial cells also stain positive with **desmin** (Figure 9.10b). However, they generally do not stain with **epithelial membrane antigen (EMA)** (Figure 9.10c) or, if they do, it will be weak staining.

Macrophages

Macrophages are phagocytic cells (which engulf waste matter) that are present in almost every serous effusion. These large cells can have a varied appearance. Their cytoplasm may contain **vacuoles** (Figure 9.11).

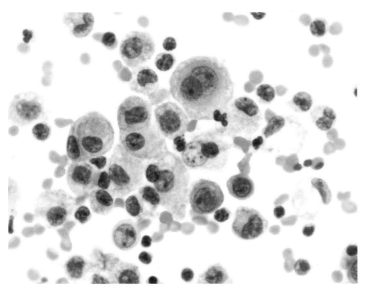

FIGURE 9.9

Reactive mesothelial cells. This population of mesothelial cells are relatively enlarged in size and have large, hyperchromatic, irregular nuclei.

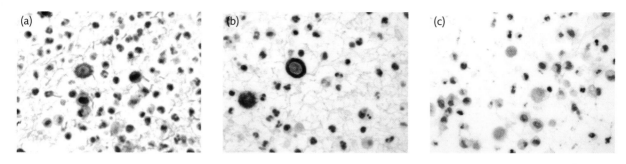

FIGURE 9.10

Staining of mesothelial cells. (a) Calretinin staining in mesothelial cell. Note that the staining is both nuclear and cytoplasmic. (b) Desmin staining of mesothelial cells. (c) EMA staining is either absent or very weak in benign mesothelial cells.

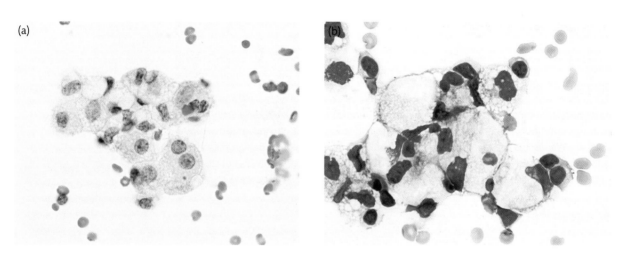

FIGURE 9.11

Two images depicting macrophages showing cytoplasmic vacuolation.

Indeed, degeneration may cause the formation of a large vacuole that fully displaces the nucleus to the side, though this happens only rarely. The visual appearance of the macrophage after such nuclear displacement is referred to as a **signet ring** (Figure 9.12).

Cross reference

Section 9.6.

Occasionally, adenocarcinoma may also present as signet rings, where the nucleus is displaced to one side by a globule of mucus, but there are usually sufficient abnormal nuclear features to identify this.

Cross reference

Section 9.5.

The nucleus of a macrophage can have a varied shape; while it mostly appears round, it can sometimes be indented or kidney shaped. Multinucleation is not commonly seen, except in **rheumatoid disease**, where this feature is part of the diagnostic picture, as discussed later in the chapter.

CD68

A glycoprotein expressed by macrophages.

Macrophages have different staining characteristics to mesothelial cells. This is particularly evident in MGG preparations where cytoplasmic staining is very pale. Macrophages may often adhere together and form loose clusters, but the staining characteristic is helpful in differentiating them from mesothelial cells (Figure 9.13). Macrophages stain positive with immunocytochemical marker **CD68**. Mesothelial cells do not stain with CD68.

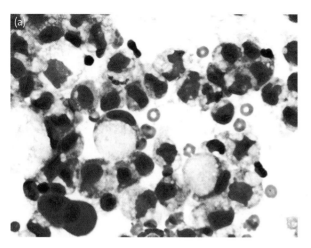
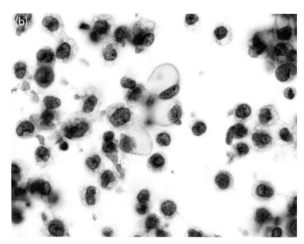

FIGURE 9.12
Two images depicting macrophages with displaced nuclei.

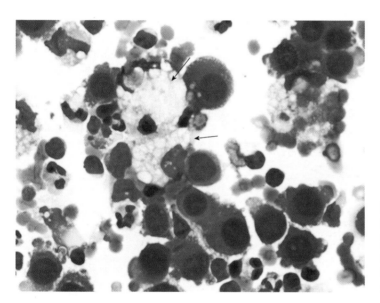

FIGURE 9.13
Macrophages and mesothelial cells. Macrophages (arrows) have translucent clear cytoplasm with small irregularly shaped nuclei. Mesothelial cells, in contrast, are deeply stained with large round nuclei.

BOX 9.4 *Cytological features of macrophages*

- Varied shape
- Granular or vacuolated cytoplasm
- Pale staining cytoplasm in MGG stain
- Nuclei often bean shaped
- Stain positive with CD68

How do mesothelial cells and macrophages differ in their morphology?

Leucocytes

Cross reference

We discussed morphology of leucocytes in Chapter 3, Section 3.6.

Leucocytes (or white blood cells) are present in almost every effusion. Lymphocytes are a common finding, though their number will vary; in certain conditions lymphocytes predominate, as we discuss later on. Neutrophils and eosinophils can also be present in varying numbers. Their morphology is similar to that found in blood films.

It is generally not possible to deduce the cause of most non-specific effusions. A peritoneal effusion due to **liver cirrhosis** is morphologically similar to a peritoneal effusion due to **renal failure**. However, there are instances in which the cytological picture may suggest the possible cause.

Liver cirrhosis

Scarring of liver which will eventually lead to loss of liver function. There are many causes for cirrhosis, including alcoholism and infection with hepatitis B and C virus.

Renal failure

A condition where the kidneys are not functioning adequately.

Rheumatoid effusion

Cross reference

Chapter 10, Section 10.6.

Rheumatoid arthritis is an autoimmune disease that causes chronic **granulomatous inflammation** of the joints. Rheumatoid arthritis can also cause inflammation in other organs. Most rheumatoid effusions occur in the pleural cavity, but occasionally a few develop in the pericardium. Rheumatoid inflammation of the pleura is called **rheumatoid pleuritis**, which causes chest pain with deep breathing or coughing. The effusion is often one sided, but is occasionally bilateral.

Granulomatous inflammation

An inflammatory response characterized by accumulation of macrophages. Macrophages with an epithelia-like appearance are called epithelioid. Macrophages may also fuse together to form giant cells.

Rheumatoid pleuritis

A pleural effusion that occurs in some patients as a result of systemic manifestations of rheumatoid arthritis.

An effusion caused by rheumatoid disease has distinctive features. Macroscopically, the fluid is often green/grey in colour and appears purulent. Microscopically, the slide background is dominated by necrotic granular debris (Figure 9.14a), which can have a varied staining pattern

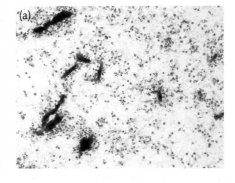
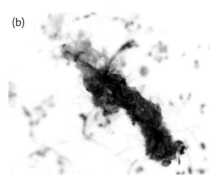

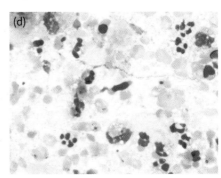

FIGURE 9.14
Rheumatoid effusion.
(a) Polymorphs in background of necrotic material.
(b) Clumped strand-like material. (c) Multinucleated giant cells. (d) Elongated macrophages.

BOX 9.5 *Cytological features of rheumatoid effusions*

- Multinucleated macrophages
- Elongated macrophages
- Granular debris
- Cholesterol crystals

with Papanicolaou stain (Figure 9.14b), but stains blue with MGG. The background may also include polymorphs and macrophages.

Rheumatoid effusions are generally long-standing and cholesterol crystals are often produced. Cholesterol crystals dissolve during processing but their outline can be seen in MGG preparations. Mesothelial cells are often absent in rheumatoid effusions: the most characteristic cells are multinucleated macrophages (Figure 9.14c) and elongated macrophage forms (Figure 9.14d).

Systemic lupus erythematosus

Systemic lupus erythematosus, or SLE, is an autoimmune disease of the connective tissues that can affect various organs, including the serous cavities. Pleural cavities are most often involved; the pericardium is involved occasionally but the peritoneum only rarely. The diagnosis of SLE is usually already established clinically. The cytologist can therefore actively search and look for **LE cells**, which are seen in some cases of SLE. The LE cell is a macrophage or neutrophil which has ingested nuclear matter from another cell (Figure 9.15). LE cells are usually scanty and are difficult to find without prior knowledge of clinical history.

LE cell

A macrophage or neutrophil which has ingested nuclear matter from another cell. It is seen in SLE.

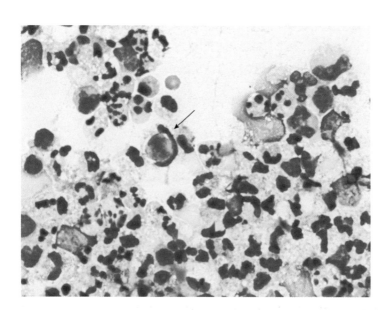

FIGURE 9.15
Systemic lupus erythematosus. The LE cell (arrow) is a macrophage or neutrophil which has ingested nuclear matter from another cell.

Purulent effusions

Bacterial infections may cause acute pleuritis, peritonitis, and pericarditis. These are all medical emergencies that require prompt treatment.

Macroscopically, the fluid is green/grey. The microscopic picture consists of numerous neutrophils, degenerate debris, and occasional macrophages. Mesothelial cells are usually absent (see Case Study 9.1).

Lymphocytic effusions

Tuberculous pleuritis

A pleural infection caused by *Mycobacterium tuberculosis*.

Chronic lymphocytic leukaemia

A type of slow-developing leukaemia affecting the lymphocytes.

Small lymphocytic lymphoma

A B cell lymphoma which is similar to chronic lymphocytic leukaemia.

Lymphocytic effusions are not specific to a particular disease. This pattern can be seen in patients with **tuberculous pleuritis**, post-pneumonia, and in association with carcinoma (Figure 9.16).

In patients with tuberculosis, the sample consists almost entirely of small lymphocytes. Mesothelial cell are often absent.

Effusions that consist of small lymphocytes are morphologically very similar to those due to **chronic lymphocytic leukaemia (CLL)** and **small lymphocytic lymphoma**. These are malignancies of B cell lymphocytes and immunocytochemistry can be helpful in diagnosis.

Eosinophilic effusions

We commonly see eosinophilic pleural effusions in patients who have had repeat thoracentesis. Eosinophils have bi-lobed nuclei and could be mistaken for neutrophils in Papanicolaou stained preparations as their cytoplasm stains green and the characteristic cytoplasmic granules are not easily noted. In MGG stained slides, however, the pink-red granules are easily seen (Figure 9.17).

SELF-CHECK 9.4

What typical cytological features are seen in rheumatoid effusions?

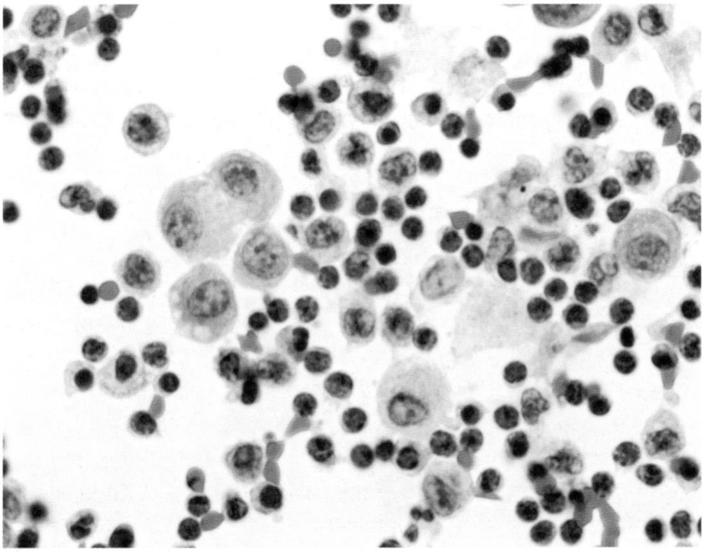

FIGURE 9.16
Lymphocyte rich effusion.

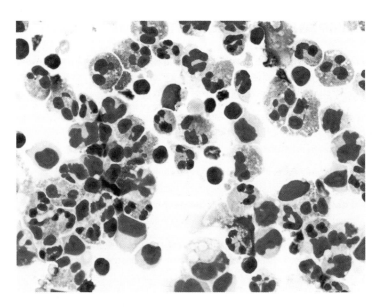

FIGURE 9.17
Eosinophilic pleural effusion.

9.5 Neoplastic effusions

Malignant cells in effusions are due either to malignant mesothelioma or more commonly to a metastatic malignancy. Most tumours can spread to serous cavities, though some have a greater tendency to this than others. The relative frequency of different tumours differs according to age, gender, and the serous cavity. For example, in children, lymphoma is the most common tumour involving the serous cavities, followed by other childhood tumours. In men, pleural involvement is common in lung carcinoma, followed by gastrointestinal tumours. In women, pleural involvement is most common in breast carcinoma, followed by lung. Table 9.1 gives a summary of the common pattern of metastasis.

In most cases the site of the primary malignancy is already known, but occasionally the detection of malignant cells in a serous effusion is the first manifestation of metastatic malignant disease of unknown primary origin (also called an **occult malignancy**). The most common occult malignancy in men and women to present with pleural effusion is lung cancer. Peritoneal involvement of an occult malignancy is usually due to gastrointestinal malignancy in men, and ovarian carcinoma in women. In these cases, careful attention to morphology and use of immunocytochemistry will allow a primary site to be established, which helps to narrow **differential diagnosis**. In some cases, however, a primary site is never found and the patient is managed **palliatively** (by reducing the severity of disease symptoms).

Differential diagnosis
The processes of distinguishing between different diseases.

TABLE 9.1 Common sites of metastatic spread to serous cavities in men and women.

	Pleural	Peritoneal
Men	Lung Gastrointestinal tract Lymphoma	Gastrointestinal tract Lymphoma
Women	Breast Lung Gastrointestinal tract Lymphoma	Ovary Breast Gastrointestinal tract Lymphoma

Identification of malignant cells in serous effusion

In some cases, the nuclear changes are so grossly abnormal that diagnosis can be made very easily, but as you will see later, some malignant cells have very subtle appearances. This makes serous fluid cytology very challenging, even for the experienced cytologist.

A good rule of thumb for identifying malignant cells in serous effusions is first to identify normal mesothelial cells, then look for another population of cells that is morphologically different. This second population is often referred to as the 'foreign population' (Figure 9.18a).

Cross reference

See Chapter 10, Section 10.7 for discussion on small cell anaplastic carcinoma.

In this photograph the predominant cell types are the mesothelial cells and lymphocytes. The second or foreign population of cells that stand out from this background are the hyperchromatic cells, which form tight clusters. These clusters of cells have abnormal nuclear morphology and were reported as malignant cells showing features of small cell anaplastic carcinoma of the lung.

This method of cell analysis works when malignant cells are present in small numbers, thus allowing their recognition against a background of normal cells. In some cases, however, malignant cells may be numerous and form the predominant cell type (Figure 9.18b). In this scenario we have to look for the usual cytological features, such as nuclear to cytoplasmic ratio, nuclear hyperchromasia, size of nucleoli, and in some cases rely on immunocytochemistry to resolve the diagnostic dilemma.

In Figure 9.18b, there is only a single population of cells and diagnosis of malignancy must be made on nuclear morphology.

Key Points

Malignant cells may appear as a second 'foreign population' amongst benign mesothelial cells.

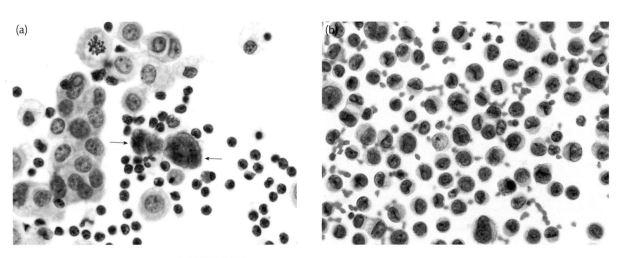

(a)   (b)

FIGURE 9.18

Identifying malignant cells in serous effusions. (a) Two cell population. Mesothelial and small lymphocytes predominate in this field; the malignant cells (arrows) are visibly different and form the second population. (b) Single population of malignant cells (confirmed breast carcinoma). In this field mesothelial cells are absent and diagnosis of malignancy must be made on abnormal nuclear features.

We will first look at the varied appearances of metastatic malignancy in serous cavities, as they constitute the majority of malignant effusions, and examine the cytology of mesothelioma later in the chapter.

Adenocarcinoma

Adenocarcinomas are the most common malignancy found in effusions. Adenocarcinoma cells may present as clusters of **papillary** (nipple-like protrusion or projection) or **acinar** (gland-like) forms, or rarely as isolated discrete cells.

Key Points

Adenocarcinoma is the most common metastatic malignancy encountered in effusions.

The most common presentations of adenocarcinoma cells are multilayered clusters (Figure 9.19a) and spherical (Figure 9.19b) or papillary-like structures (Figure 9.19c) sometimes associated with **psammoma bodies** (Figure 9.19d). The nuclei may show irregular nuclear contours, **pleomorphism**, marked nuclear enlargement, hyperchromasia, and the presence of large nucleoli. Furthermore, the cytoplasm may include vacuoles (Figure 9.19e). These vacuoles are usually degenerative in nature, but could be due to the production of mucins by the tumour cells. Adenocarcinoma presenting as signet rings (Figure 9.19f) usually originates from the stomach, but lung and pancreatic adenocarcinoma may also present this way.

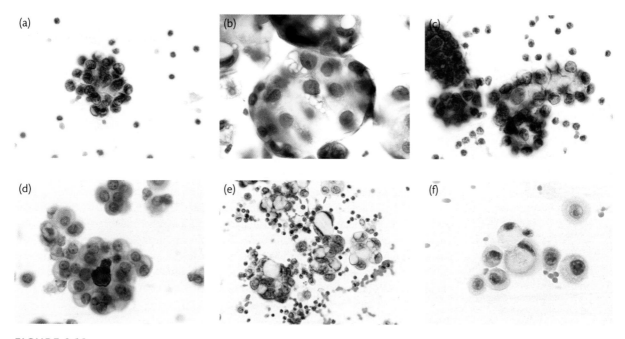

FIGURE 9.19

Adenocarcinoma. (a) Cluster of malignant cells. (b) Confirmed breast carcinoma showing spherical clusters of malignant cells. (c) Confirmed ovarian carcinoma showing papillary and acinar clusters of malignant cells. (d) Confirmed ovarian carcinoma showing psammoma body. (e) Confirmed ovarian carcinoma showing vacuolated malignant cells. (f) Confirmed lung carcinoma showing signet ring formation.

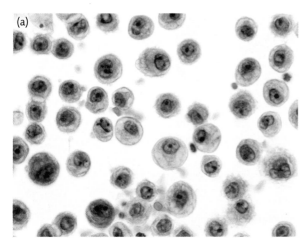
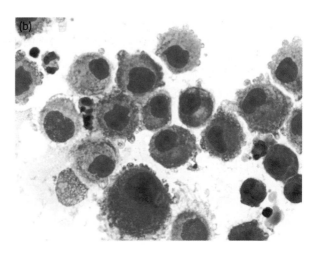

FIGURE 9.20

Adenocarcinoma (confirmed lung carcinoma). (a) This single population of malignant cells can be difficult to recognize. Careful examination reveals irregular nuclear outlines, variation in nuclear size, and enlarged nucleoli. (b) Same case as in (a). MGG stain is useful as it exaggerates cytological features. The abnormal nuclear outline is more obvious with this stain.

BOX 9.6 *Cytological features of adenocarcinoma*

- **Large clusters**
- **Cytoplasmic vacuolation**
- **Isolated free cells forms**
- **Stains positive with epithelial markers Ber EP4, CEA, EMA, etc.**

Ber EP4

An antibody that reacts to all epithelial cells except the superficial layer of squamous epithelia cells, hepatocytes, and parietal cells (acid secreting cells of stomach), it is useful in serous effusion cytology as it does not label mesothelial cells and is rarely reactive with mesotheliomas.

Carcinoembryonic antigen (CEA)

A glycoprotein expressed by foetal cells, and only expressed in small amounts by normal adult epithelial cells, its expression is greatly increased in many adenocarcinomas, including those of colonic, pancreatic, and lung origin. It is useful in effusion cytology as it is not expressed by mesothelial cells or mesothelioma.

Adenocarcinoma may also present cytologically as isolated discrete cells. This is a common cause of false negative diagnosis, where they are mistaken for reactive mesothelial cells. These tumour cells have little or no tendency to form aggregates (Figures 9.20a and b). This pattern is sometimes seen in metastatic breast and lung carcinoma, but other adenocarcinomas can also produce this picture. At low power, tumour cells may resemble mesothelial cells, but a high-power examination will reveal an increased nuclear-to-cytoplasmic ratio, irregular nuclear outlines and prominent nucleoli. Adenocarcinoma cells usually stain positive with epithelial markers **Ber EP4** and **carcinoembryonic antigen (CEA)** and EMA.

Squamous carcinoma

Although squamous carcinoma is common, squamous carcinomas are rarely seen in serous cavities. When a squamous carcinoma is found it is usually due to carcinoma of lung, cervix, and larynx (Figure 9.21a). These cancers metastasize late in the disease process, so the primary tumour is usually known to the cytologist. The presentation of malignant cells is variable. Most often, the typical keratinization is absent, but the cytoplasm is usually dense and stains deep

> **BOX 9.7** *Cytological features of squamous carcinoma*
>
> - Large clusters or single cells
> - Dense staining cytoplasm, sometimes keratinized
> - Hyperchromatic nuclei with irregular outlines
> - Occasionally showing nuclear pyknosis
> - Co-expression of p63 and CK 5/6

(a)

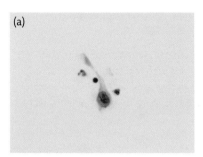

(b)

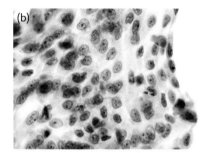

(c)

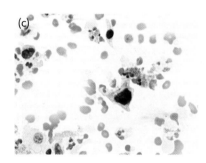

FIGURE 9.21
Squamous carcinoma. (a) Confirmed carcinoma of larynx. This highly keratinized tadpole cell was a rare finding in this sample, which consisted mostly of polymorphs and lymphocytes. (b) Confirmed squamous carcinoma of lung. This sample included sheets and large clusters of malignant cells. On cytological features alone it would not be possible to suggest the cell type. (c) Confirmed squamous carcinoma of lung. The malignant cells in the sample have dense staining cytoplasm, with angular pyknotic nuclei.

green in Papanicolaou stain (Figure 9.21b). The cells of squamous carcinoma may be arranged either as clusters or isolated cells. The nuclear abnormality includes hyperchromasia, irregular nuclear outlines and nuclear pyknosis (Figure 9.21c).

In poorly differentiated squamous carcinoma where keratinization is absent, immunocytochemistry is helpful in reaching the diagnosis. **p63** and CK 5/6 antibodies are used in identification of poorly differentiated squamous carcinoma, particularly if these antibodies are expressed together.

> **p63**
> A tumour suppressor gene product, p63 is expressed by squamous carcinomas.

Small cell anaplastic carcinoma

Small cell anaplastic carcinoma (small cell carcinoma) may be seen in pleural and pericardial cavities, but rarely in peritoneal cavities. This tumour usually arises in the lung, but occasionally originates from other organs including the bladder, cervix, or oesophagus. This highly malignant tumour metastasizes widely. Indeed, the discovery of malignant cells in serous effusion is sometimes the first sign of lung cancer. Small cell carcinoma will be discussed in more detail in Chapter 10; please read this chapter for background information.

> **Cross reference**
> Chapter 10, Sections 10.7 and 10.8.

Small cell anaplastic carcinomas have morphologically distinct appearances in serous effusion preparations (see Figure 9.22a). The cells are small in size (usually twice the size of lymphocytes),

CD56

Antibody to CD56 reacts with many normal and neoplastic tissues. It is useful in identification of pulmonary small cell carcinoma.

Chromogranin

A glycoprotein located on neurosecretory granules in neuroendocrine cells. Antibodies to this glycoprotein are useful in identification of tumours of neuroendocrine origin.

Synaptophysin

A glycoprotein present in neuroendocrine cells. Antibodies to this protein are useful in identifying tumours of neuroendocrine origin.

and are seen as discrete single cells, but most often in clusters. Look at Figures 9.22b and c, which demonstrate some of the features of small cell anaplastic carcinoma. The cells have very high nuclear-to-cytoplasmic ratio: the nucleus often fills most of the cell, with only a scanty rim of cytoplasm. The nuclear chromatin is finely granular and hyperchromatic. Nucleoli are not a typical feature of small cell carcinoma, but are seen in some cases.

The feature which typifies this tumour is the tendency of the cells to mould around each other, giving a mosaic-like effect due to the small amount of cytoplasm between them. This is better seen in air-dried MGG preparations than in wet-fixed Papanicolaou stained slides. These features are so characteristic that a definitive diagnosis can be made on conventional preparations. However, immunocytochemistry is often used to support the diagnosis. Small cell carcinoma is of neuroendocrine origin and can be stained with neuroendocrine markers, including **CD56** (Figure 9.23), **chromogranin**, and **synaptophysin**, amongst others. The cells of small cell carcinoma can occasionally look similar to those of lymphomas (see next section), but lymphoid cells tend to appear as discrete single cells, whereas clusters are common in small cell anaplastic carcinoma.

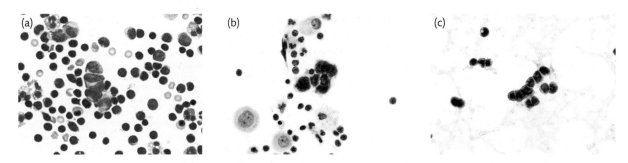

(a)   (b)   (c)

FIGURE 9.22

Small cell carcinoma of lung (pleural fluid). (a) Malignant cells have scanty cytoplasm and appear moulded together. (b) Malignant cells have coarse granular chromatin, and are crescent shaped and angulated. (c) This appearance is sometimes referred to as 'Indian file'.

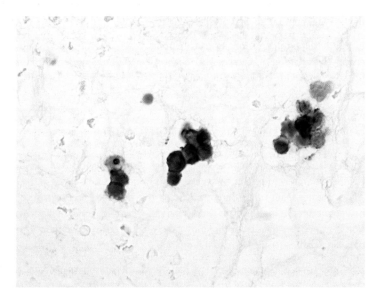

FIGURE 9.23

Small cell carcinoma of the lung. Staining for CD56.

> **BOX 9.8** Cytological features of small cell carcinoma
>
> - Small discrete cells and clusters
> - Hyperchromatic nuclei with high nuclear-to-cytoplasmic ratio
> - Nuclear moulding giving a mosaic appearance
> - Stains positive with CD56 and other neuroendocrine markers

Lymphoma

Lymphoma is the cancer of the lymphoid cells originating mostly in lymph nodes, and is broadly divided into **Hodgkin's** and **non-Hodgkin's lymphoma (NHL)**. Hodgkin's lymphoma is named after Dr Thomas Hodgkin, who in 1832 first described what was initially called Hodgkin's disease. Other types of lymphoma were subsequently described, called non-Hodgkin's lymphoma. They are broadly divided into those of B cell and T cell origin. Hodgkin's and non-Hodgkin's lymphomas are further subdivided based on various features. Six types of Hodgkin's lymphoma and at least 61 types of non-Hodgkin's lymphoma have been described by various international medical committees, including the World Health Organization.

Malignant effusions due to lymphomas are uncommon and usually there is a prior biopsy-proven diagnosis. In a small number of cases, however, it may be the first manifestation of disease. Lymphomas can be seen in pleural, peritoneal, and pericardial effusions.

Classification of non-Hodgkin's lymphoma on cytological preparation is not easy and is usually limited to the description of the size of the cell as either large or small. The cytological features of lymphoid cells are better appreciated in MGG stains, where air-drying exaggerates cell features. Cytological preparations are usually highly cellular and composed of discrete cells (Figure 9.24a). The cells of large cell lymphoma are easy to recognize due to the large cell size (which could be the same size as a mesothelial cell). The cells have a high nuclear-to-cytoplasmic ratio and a nucleus which often fills the cell (Figures 9.24b and c). The nuclear outline could be round or irregular, with coarse nuclear chromatin. Nucleoli may also be present.

Cross reference
Chapter 13, Section 13.4.

Small cell lymphomas are much harder to recognize as they can look very similar to normal small mature lymphocytes (Figure 9.25). Flow cytometry or immunocytochemical studies may be helpful in these circumstances.

Cells of lymphoid origin stain positive with immunocytochemical marker **CD45**.

Hodgkin's lymphoma is characterized by the identification of **Reed-Sternberg cells**—large, multi- or binucleated cells with very prominent nucleoli. Reed-Sternberg cells stain positive with immunocytochemical markers **CD30** and **CD15**, which can further help in their identification.

CD45
A protein present on the surface of lymphoid cells. CD45 is also known as leukocyte common antigen.

CD30
This is a membrane protein expressed on Hodgkin's Reed-Sternberg cells.

CD15
This is found on neutrophils. It is expressed in Hodgkin's disease, some T cell lymphomas, and leukaemias.

SELF-CHECK 9.5

Small cell anaplastic carcinoma can rarely be confused with malignant lymphoma. Which immunocytochemical antibodies are helpful in resolving this differential diagnosis?

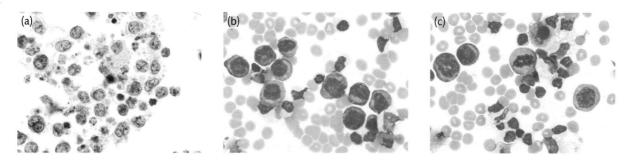

FIGURE 9.24

Malignant lymphoma (large cell type). (a) The lymphoid nature of the cells is not so readily recognizable, but the cells are occurring as single dissociated cells with coarse chromatin and contain one or two nucleoli. (b) Same case as in (a). MGG stain is particularly useful in assessment of lymphoid cells and their lymphoid nature is more easily recognizable. (c) Mitotic figure.

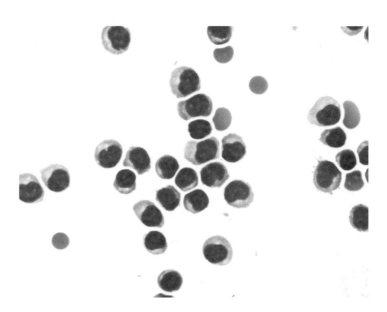

FIGURE 9.25

Malignant lymphoma (small cell type). This was fluid from a patient who had low-grade B cell non-Hodgkin's lymphoma. These monotonous lymphoid cells are small in size (compare with red cells for comparison), and close examination shows slight indentation in some of the nuclei.

BOX 9.9 Cytological features of lymphoma

- Discrete cells
- High nuclear-to-cytoplasmic ratio
- Irregular nuclear outlines
- Prominent nucleoli
- Reed-Sternberg cells (Hodgkin's lymphoma)
- Stains positive with lymphoid markers

Other metastatic cancers

Essentially, any cancer may involve the serosal surfaces. While it is not possible to recognize the origin of these cytologically, the use of immunocytochemistry, adequate clinical history, and the comparison of morphology to the previous biopsy may make it possible to suggest or confirm the primary origin.

Malignant melanoma is a cancer that develops from melanocytes. Most malignant melanomas arise in the skin, but small numbers arise in other tissues. These rare **extracutaneous** cancers include the eye, vagina, and urogenital sites amongst others. Effusions due to melanoma are rare, and cytological appearances are quite variable. The cells are usually dissociated, round to oval in shape, and may resemble mesothelial cells. The nuclei may have prominent nucleoli. Melanin may be present in some cases; this appears as brown staining pigment in Papanicolaou stain or black to navy in MGG. This pigment can be demonstrated with the Masson-Fontana silver method, which stains melanin as black granules. Immunocytochemistry is the method of choice, however, as melanin is only present in a minority of cases. Useful markers include **Melan A** and **HMB45**.

Sarcomas, **germ cell tumours**, and urothelial carcinomas (transitional cell carcinomas) are some examples of other tumours that are rarely seen in serous effusions. Sarcomas are tumours of connective tissues and rarely metastasize to serosal surfaces. Even when this does occur, it is usually later in the course of the disease.

Germ cell tumours originate from the germ cells, the cells that develop in the embryo and become the cells that make up the reproductive system in males and females. Like sarcomas, germ cell tumours rarely metastasize to serosal surfaces.

Urothelial carcinoma is a common malignancy, but it is rarely associated with serous effusions. In the rare cases that it metastasizes to serous cavities it is in patients with advanced cancers.

Melan A
Antibodies to this gene product are useful in identification of metastatic melanoma.

HMB45
A monoclonal antibody that reacts with melanomas and occasional carcinomas.

Malignant mesothelioma

Malignant mesothelioma is the primary malignant tumour of the mesothelium and arises most commonly in the pleura. It is seen less commonly in the peritoneum, and only rarely in the pericardium and tunica vaginalis of the testis. Mesothelioma is associated with **asbestos** exposure. The word asbestos comes from the Greek word meaning 'indestructible', as from very early on its flame retardant properties were recognized. Asbestos is a naturally occurring fibrous mineral that is mined in many parts of the world and has been in use for thousands of years. Indeed, there are records of its use by ancient Persians, Greeks, and Romans. There are three main types of asbestos: **crocidolite** (blue asbestos), **amosite** (brown asbestos), and **chrysotile** (white asbestos). It is thought that blue and brown asbestos are more hazardous forms of asbestos due to the structure of their fibres. Asbestos use increased during the Industrial Revolution, when it was used initially in heavy industries for insulating boilers and pipes. Later, however, its unique property of being resistant to heat, acid, and alkali made it an ideal building material. Asbestos was also used in some common household goods such as oven gloves, mats, and flooring. It is thought that at the height of its use up to 3,000 products contained asbestos.

From ancient Roman times the hazards of asbestos were recorded: Roman naturalist Gaius Plinius Secundus, better known as Pliny the Elder, wrote that the slaves working in asbestos quarry mines were dying young and many had lung ailments. The first link between asbestos and lung disease in British literature dates back to 1924: W E Cooke, a pathologist, linked inhalation of asbestos dust to **lung fibrosis** (2). There were some 200 publications between the

1920s and 1970s describing various links between asbestos exposure and respiratory disease, but it was only in 1985 that the import of blue and brown asbestos was banned in the UK, with the white asbestos ban following in 1999. In November 2006 the Health and Safety Executive published the Control of Asbestos Regulations 2006. This regulation reinforces previous regulations and also continues to ban the second-hand use of asbestos products such as asbestos cement sheets and asbestos boards and tiles, including panels which have been covered with paint, or textured plaster containing asbestos. Most developed countries have a similar ban, but in developing countries asbestos is still used as a building material.

Incidence and mortality statistics

Mesothelioma is a rare disease: in 2005 there were 2,164 new cases (2.7 per 100,000 **age standardized**). This figure comprises 1,827 cases in males and 337 in females. Most of these cases were in people who had worked with or had been exposed to asbestos, but there are a number of cases for which no asbestos exposure could be accounted for. There is currently no effective therapy for mesothelioma and hence mortality rates are very high (median survival ten months); in 2006, there were 1,996 deaths (2.3 per 100,000 age-standardized). The expected number of deaths amongst males is predicted to increase to a peak of 2,038 around the year 2016 (3).

Risk factors

The main risk factor for mesothelioma is exposure to asbestos. Most people with malignant mesothelioma have worked or lived in places where they inhaled asbestos. It may take up to 40 years after exposure to asbestos for malignant mesothelioma to develop. Family members of asbestos workers have also been shown to have an increased risk of developing mesothelioma as a consequence of the inhalation of asbestos dust brought home on the clothing of asbestos workers. Other minerals such as **erionite**, which have similar fibre structure to asbestos, can be linked to the development of mesothelioma. In the Cappadocia area of Turkey, for example, where this mineral is present in the soil, a study showed high incidence of mesothelioma due to erionite exposure (4).

Simian virus 40 (SV40) is present in some human malignant mesotheliomas. SV40 virus, a DNA virus found in both monkeys and humans, has the potential to cause tumours in monkeys. Available evidence appears sufficient to link SV40, either alone or in conjunction with asbestos, to malignant mesotheliomas (5). However, a lack of epidemiological evidence makes it difficult to consider the full impact of SV40. The exact mode of human infection by SV40 is not known, but SV40 virus has been identified in the injectable form of polio vaccine which was in production from 1955 to 1962. The vaccine contamination may have occurred during the production of vaccine which utilized monkey kidney cells.

SELF-CHECK 9.6

List risk factors for malignant mesothelioma.

Presentation

The most common symptom is breathlessness due to build-up of fluid and chest pain. Some patients present with more general symptoms such as weight loss and loss of appetite.

Investigation of patients with suspected mesothelioma

Diagnosis of mesothelioma is difficult to make as symptoms of the disease are not unique to it. The diagnosis is most often made after careful assessment of clinical and radiological findings, in addition to a confirmatory cytology and/or tissue biopsy result. It is important to establish a correct diagnosis early, because in the UK mesothelioma is recognized by the Department of Work and Pensions (DWP) as an industrial disease. Consequently, people diagnosed with mesothelioma and their families can get financial assistance by claiming Industrial Injuries Disablement Benefit, through the War Pensions Scheme, or by suing their previous employer for damages.

A history of asbestos exposure may increase clinical suspicion, but some patients with mesothelioma have no known knowledge of having been exposed to asbestos. Patients generally have a chest X-ray, which may highlight the build-up of fluid, and CT scanning may show the presence of **pleural plaques**, the dense areas that are seen on the inner surface of the ribcage and diaphragm, which indicate previous exposure to asbestos. The fluid can be tapped and sent for cytological examination using a syringe, or a small sample of tissue can be taken using a special needle for histological assessment.

Video-assisted thoracoscopic surgery (VATS) is the preferred method for obtaining tissue for diagnosis of mesothelioma, but in some patients this procedure may be not be suitable and a needle biopsy of the pleura may be the recommended method of obtaining material for histological diagnosis.

Treatment

There is currently no cure for mesothelioma. However, **pleurectomy/decortication** may be performed very rarely for very early disease. In this procedure, an attempt is made to remove all the pleural membrane which shows evidence of tumour. This surgery has had varying success and is only suitable for a small number of patients used in combination with chemotherapy and radiation. The usual treatment for mesothelioma is palliative chemotherapy and radiotherapy.

Histology of mesothelioma

Histologically, these tumours show either **epithelial**, **sarcomatoid**, or **mixed** pattern. The epithelial form, as the name suggests, has an epithelial growth pattern; the sacromatoid type appears as elongated spindle-shaped cells that are irregularly shaped and often overlap one another; and the mixed pattern has elements of both epithelial and sarcomatoid cells. The epithelial form is the most common type.

Cytology of malignant mesothelioma

The diagnosis of mesothelioma in cytological preparations is challenging even for experienced cytologists, as in some cases malignant mesothelial cells may look very similar to reactive mesothelial cells. This similarity often leads to a false negative diagnosis. You may recall earlier in this chapter (Section 9.5) we mentioned that looking for a second population or a foreign population of cells was a useful tip in identifying malignant cells. Unfortunately this rule of thumb is not applicable to the diagnosis of mesothelioma as the cells show a spectrum of subtle changes from benign reactive mesothelial cells to malignant.

Video-assisted thoracoscopic surgery (VATS)

This a form of minimally invasive surgery, commonly known as 'keyhole' surgery. While the patient is under general anaesthesia, usually up to three incisions are made in the chest wall. The surgeon introduces a video camera via an endoscope (long thin hollow tube) into one of the incisions to visualize the chest cavity. Various other instruments can be introduced into the chest via the other incisions. VATS can be used to obtain diagnostic biopsies from the lung, pleura, or diaphragm, and also perform therapeutic procedures such as lung resections.

Pleurectomy/decortication

Surgical removal of the pleura for treatment of mesothelioma.

The next diagnostic challenge after the identification of malignant cells is to establish their mesothelial origin, as drawing a distinction between malignant mesothelioma and metastatic malignancy is important for therapeutic purposes. We routinely employ immunocytochemistry to resolve these diagnostic dilemmas, and we will look at this later in this chapter.

As mentioned earlier, there are three different histological subtypes. The sacromatoid type is generally difficult to diagnose cytologically as the tumour does not shed its cells readily into the effusion. The discussion below is therefore limited to the recognition of epithelial and mixed pattern varieties.

The epithelial type can have a varied appearance; the most common presentation is of numerous cell clusters (Figure 9.26a), which may show branching. These clusters may

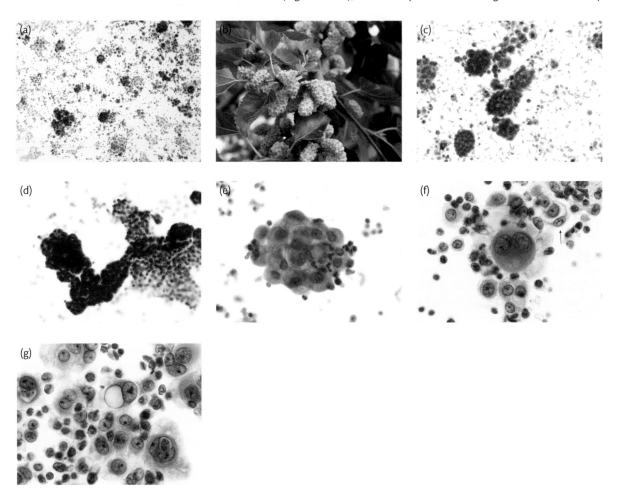

FIGURE 9.26
Malignant mesothelioma. (a) This low-power photomicrograph shows a highly cellular specimen and clustering of cells. (b) White mulberry (photograph courtesy of Behnam Shambayati). (c) These three-dimensional clusters of mesothelial cells with knobbly contour are sometime described as 'mulberry-shaped'. (d) Malignant mesothelioma showing many large papillary aggregates. (e) This cluster shows minimal cytological atypia, and nuclear to cytoplasmic ratio is normal. (f) Malignant mesothelioma showing large binucleated cell with dense cytoplasm. Also note cell engulfment (arrow). (g) Dense staining mesothelial cells showing binucleation, prominent nucleoli, and cell engulfment.

be composed of as few as 20 to several hundred cells and have rounded mulberry-like external contours (Figures 9.26b–d). These cells may sometimes show deceptively 'normal' cytology; an increased nuclear-to-cytoplasmic ratio, which is the cornerstone of cytological diagnosis in malignancy, is often not seen, and the nuclei may show minimal chromatin abnormality (Figure 9.26e). In these cases careful consideration of clinical history and use of immunocytochemistry may be required to suggest a diagnosis. In most cases, though, malignant cells are larger than normal mesothelial cells, with prominent nucleoli and dense staining cytoplasm (Figures 9.26f and g). In Papanicolaou stained preparations, this staining characteristic is similar to the metaplastic squamous cells seen in cervical cytology.

Malignant mesothelial cells may appear to engulf other cells (Figure 9.27). This **cell engulfment**, where one cell wraps around another cell, is not unique to mesothelioma and reactive mesothelial cells also engulf each other. In mesothelioma, however, this is a more frequent finding.

Another feature which is sometimes seen in Papanicolaou stained smears is the presence of small cells which stain orange (Figure 9.28) and resemble degenerate keratinized cells, similar to those seen in cervical samples.

Mesothelioma may sometimes present as a pure population of isolated cells with very little clustering; this is a common cause of false negative cytology, where these single malignant cells are interpreted as reactive mesothelial cells.

As mentioned earlier, immunocytochemistry is a useful tool available to the cytologists in resolving some of these diagnostic dilemmas. There is currently no true specific marker of mesothelioma or mesothelial cells. There are mesothelial markers that are highly selective in staining mesothelial cells, but occasionally show positive staining in adenocarcinoma cells. The general approach to immunocytochemistry and difficult cases, including that of mesothelioma versus adenocarcinoma, is to use a panel. A selection of markers positive for adenocarcinoma, supplemented by a range of mesothelial markers, should guide the cytologist to a definitive diagnosis. In the next section the use of immunocytochemistry panels will be explained in more detail.

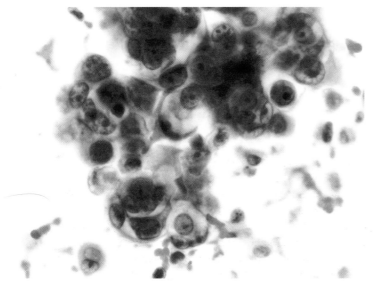

FIGURE 9.27
Malignant mesothelioma showing cell engulfment.

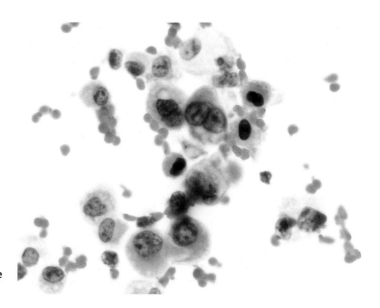

FIGURE 9.28
Malignant mesothelioma showing squamous-like cells in a background of mesothelial cells.

BOX 9.10 *Cytological features of malignant mesothelioma*

- Large clusters with mulberry-like features
- Cell enlargement
- Dense cytoplasm
- Prominent nucleoli
- Cell engulfment
- Small orange staining cells
- Positive immunocytochemical staining with mesothelial markers and negative staining with adenocarcinoma markers

Immunocytochemical markers in fluid cytology

Immunocytochemistry has become an important complementary method to cytological diagnosis. A wide range of antibodies can be applied to cell blocks, and the results can be relied upon because the cells are generally well preserved and are present in large numbers.

Immunocytochemistry is useful in cases where it is difficult to distinguish between a population of malignant cells and reactive mesothelial cells. However, it is important to note that there is currently no antibody that could be used as an absolute marker of malignancy in serous effusions. Consequently, immunocytochemistry can only be used to *support* a cytological diagnosis.

In many laboratories, positive staining with epithelial membrane antigen (EMA) is used to support a diagnosis of malignancy. EMA is a glycoprotein that is present in a wide variety of epithelial cells. EMA stains most adenocarcinomas and malignant mesotheliomas. Reactive mesothelial cells usually do not stain at all, or stain very weakly.

Desmin, an antibody to muscle tissue, can be used to differentiate between reactive mesothelial cells, mesothelioma, and adenocarcinoma. Desmin is generally expressed only in benign mesothelial cells; malignant mesothelioma and metastatic adenocarcinoma are unreactive and therefore do not stain.

SELF-CHECK 9.7

Which antibodies are useful in helping to differentiate reactive mesothelial cells from malignant cells in serous effusions?

Another scenario where immunocytochemistry is useful is in differentiating malignant mesothelioma from metastatic carcinoma. In the past, the 'markers' which were used to distinguish mesothelioma from adenocarcinoma were 'negative markers'—those expressed in adenocarcinomas, but not in mesotheliomas. These markers included antibodies to carcinoembryonic antigen (CEA) and BerEP4. Recently, however, 'positive markers' expressed by mesothelial cells and mesotheliomas have become available. These markers, which are normally 'positive' in mesotheliomas and 'negative' in adenocarcinomas, include calretinin, thrombomodulin, cytokeratin 5/6 and **Wilm's tumour gene product (WT-1)**.

None of the antibodies described so far are 100% specific for mesothelial tissue or epithelial tissues and there is always a possibility of cross reactivity. For this reason it is important to use a panel of positive and negative markers when assessing suspected mesothelioma.

Immunocytochemistry may also be helpful in predicting the primary site of metastatic carcinoma. In the past, the prediction of the site of origin of a tumour was dependent on correlating clinical information and morphological features. There are several 'site specific' antibodies that are useful in this situation, including **thyroid transcription factor 1 (TTF1)**, **prostatic specific antigen (PSA)**, and **prostatic acid phosphotase (PAP)** amongst others. Thyroid transcription factor 1 is a nuclear transcription factor that is expressed in normal lungs and

Wilm's tumour gene product (WT-1)

Antibodies to WT1 protein, the product of WT1 gene, react with mesothelium and are unreactive with a majority of adenocarcinomas, with the exception of some ovarian carcinomas.

Thyroid transcription factor 1 (TTF1)

Nuclear immunoreactivity with TTF1 antibody is useful in identifying adenocarcinomas of pulmonary origin. It is expressed in normal lung and in thyroid, and in the cancers arising from these organs.

Prostatic specific antigen (PSA)

This is a protein produced by the cells of prostate gland. Antibodies to this protein are useful in identification of metastatic adenocarcinoma of prostate.

Prostatic acid phosphotase (PAP)

This antibody to PAP is useful in identification of metastatic adenocarcinoma of the prostate.

CASE STUDY 9.2

A 55-year-old man was referred to the A&E department by his GP for evaluation of severe abdominal distension, vomiting, and nausea. There was no significant past medical history.

After initial physical examination he was sent for CT scanning which confirmed the presence of significant ascites and moderate thickening of the stomach wall. Paracentesis was performed and 1.5 litres of fluid was drained, and the sample was sent for cytological analysis.

The cytological appearances of the ascitic fluid are shown in Figures 9.29a–d. The markedly cellular fluid included numerous dispersed cells which had round nuclei. Many of the cells had a single nucleolus (Figures 9.29a–c). Occasional mitotic figures were also seen (Figure 9.29d). The background cells included a mixture

of inflammatory cells. In summary, the cells appeared mesothelial in origin and their presentation favoured a benign effusion. Immunocytochemistry was also performed on a cell block. The majority of the cells stained strongly positive for desmin (Figure 9.29e) and calretinin (Figure 9.29f) and were unreactive with EMA (Figure 9.29g), further supporting their benign nature.

Endoscopy was performed to evaluate the cause of vomiting and nausea. This revealed a gastric ulcer, which was then biopsied. The biopsy showed adenocarcinoma with features consistent with gastric origin.

Cytology is often requested in the early stages of ascitic fluid assessment. Malignant cells are seen in ascitic fluids when tumour cells are in the lining of the peritoneum, but there are many cases in which cytology is negative.

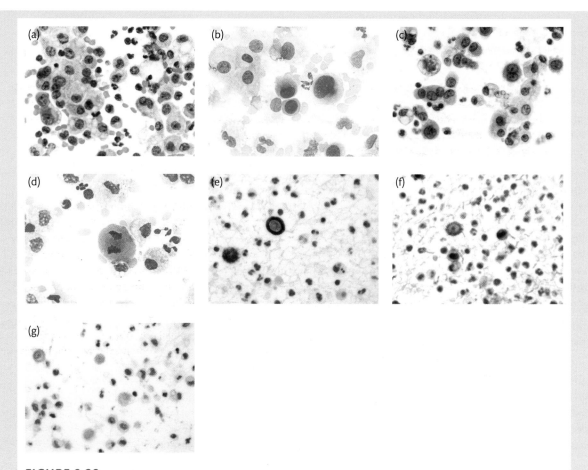

FIGURE 9.29
Reactive mesothelial cells. (a) Reactive mesothelial cells showing increase in nuclear to cytoplasmic ratio, coarse chromatin, and hyperchromasia, (b) and (c) show further examples of reactive mesothelial cells, while (d) shows mesothelial cell in mitosis. (e) Benign mesothelial cells showing positive staining with desmin. (f) Meosthelial cells showing staining with calretinin. The staining is both nuclear and cytoplasmic. (g) Benign mesothelial cells either show very weak staining or are generally unreactive with EMA.

Oestrogen receptor (ER)

This is expressed by some breast carcinomas. Oestrogen receptors are required for oestrogen stimulated growth and proliferation of breast cancers. Immunocytochemical expression of ER in malignant cells in serous effusion is useful in identification of metastatic breast carcinoma.

thyroid and in the cancers arising from these organs. Positive staining of malignant cells in a serous effusion is highly likely to indicate a lung primary as thyroid carcinomas rarely metastasize to the serosal surfaces. PSA is a protein produced by the cells of prostate gland and PAP is an enzyme produced by the prostate gland. Antibodies to these are useful when metastatic cancer from the prostate is suspected.

Supportive evidence of metastatic breast carcinoma can be obtained by nuclear staining with **oestrogen receptor (ER)**. Oestrogen receptors are required for oestrogen stimulated growth and proliferation of breast cancer. They are expressed to some degree in 50–80% of breast tumours. A strong ER staining favours a breast carcinoma over other sites. However, other malignant tumours, including ovary, endometrium, and uterine cervix, also express ER.

CASE STUDY 9.3

A 75-year-old male presented with shortness of breath. A CT scan showed the presence of a **peripleural** mass (mass around the pleura) and small pleural effusion. The fluid was tapped for diagnostic reasons and sent for cytological analysis. It appeared cloudy yellow in colour and viscous in consistency (Figure 9.30).

The cytological appearances are shown in Figures 9.31a–d. The fluid was cellular and contained both dissociated single cells and cell clusters (Figure 9.31a). The majority of the cells had dense cytoplasm with peripheral vacuoles (Figure 9.31b). Many of the cells were binucleated, with prominent nucleoli. Occasional cell engulfment (Figure 9.31c) and 'windows' between cells were also seen. Occasional grossly enlarged cells were also present (Figure 9.31d).

In summary, the cytological features suggested mesothelial origin, and nuclear abnormality favoured a diagnosis of malignant mesothelioma. The differential diagnosis considered was florid reactive mesothelial hyperplasia, which may result in significant pleomorphism. The diagnosis was substantiated by performing immunocytochemistry. The cells stained strongly with calretinin (Figure 9.32a), cytokeratin 5/6 (Figure 9.32b), thrombomodulin (Figure 9.32c), and EMA (Figure 9.32d), and negative with BerEP4, CEA, and desmin.

Calretinin, cytokeratin 5/6, and thrombomodulin are markers for mesothelial cells. BerEP4 and CEA are glandular markers for stain adenocarcinomas. Desmin and EMA are very useful markers in distinguishing benign from malignant mesothelium. EMA stains neoplastic cells in serous fluids and desmin appears to be expressed only in reactive mesothelium. This staining pattern supported the morphological diagnosis of malignant mesothelioma. The highly viscous nature of the fluid is also worth noting; this is due to increased levels of hyaluronic acid which is produced in mesothelioma.

FIGURE 9.30
Viscous pleural effusion.

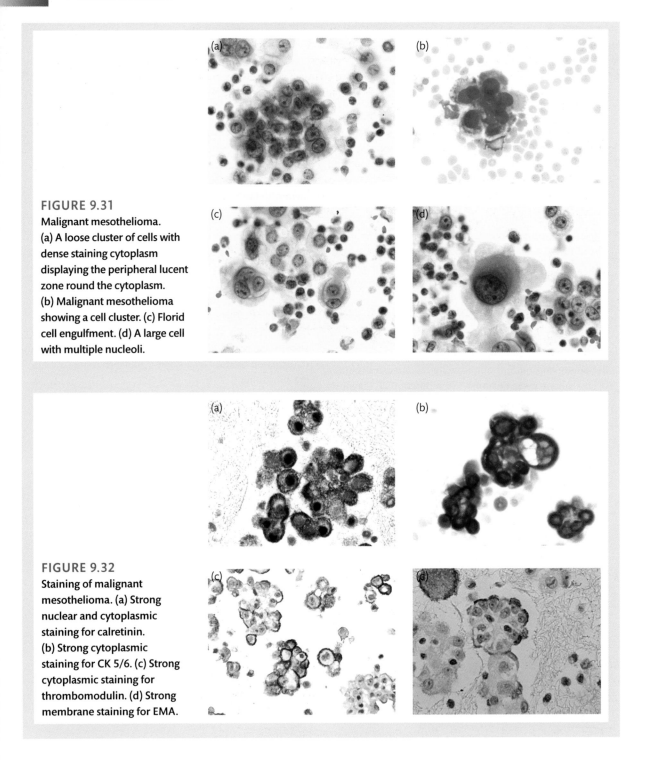

FIGURE 9.31

Malignant mesothelioma.
(a) A loose cluster of cells with dense staining cytoplasm displaying the peripheral lucent zone round the cytoplasm.
(b) Malignant mesothelioma showing a cell cluster. (c) Florid cell engulfment. (d) A large cell with multiple nucleoli.

FIGURE 9.32

Staining of malignant mesothelioma. (a) Strong nuclear and cytoplasmic staining for calretinin.
(b) Strong cytoplasmic staining for CK 5/6. (c) Strong cytoplasmic staining for thrombomodulin. (d) Strong membrane staining for EMA.

Another approach to identifying the origin of metastatic tumours is the examination of the differential expression of cytokeratins, **cytokeratin 7** and **cytokeratin 20**, in the malignant cells. Epithelia in the human body usually express specific cytokeratins according to the type of epithelium. The differential expression of cytokeratin 7 and 20 may help in determining the origin of metastatic carcinomas. Table 9.2 is a summary of differential expression of common tumours which may be encountered in serous effusions.

Cytokeratin 7 (CK7)

This is present in many carcinomas, including the majority of lung, breast, ovarian, and endometrial adenocarcinomas.

Cytokeratin 20 (CK20)

This is present in many carcinomas, including gastrointestinal and transitional cell carcinomas.

Key Points

Immunocytochemistry can be used to confirm malignancy in equivocal cases. It is also useful in distinguishing adenocarcinoma from malignant mesothelioma, and in helping the cytologist to predict the primary site of a metastatic malignancy.

Case Studies 9.4 and 9.5 illustrate the use of immunochemistry in resolving common diagnostic scenarios that are encountered routinely.

TABLE 9.2 Expression of cytokeratin 7 and cytokeratin 20 in common tumours.

Cytokeratin 7 positive/cytokeratin 20 negative
Majority of non-small cell lung carcinomas
Breast carcinoma
Ovarian carcinoma
Endometrial carcinoma

Cytokeratin 7 negative/cytokeratin 20 positive
Colorectal carcinoma

Cytokeratin 7 positive/cytokeratin 20 positive
Pancreatic carcinoma
Transitional cell carcinoma

Cytokeratin 7 negative/cytokeratin 20 negative
Small cell carcinoma
Squamous cell carcinoma
Prostatic carcinoma

CASE STUDY 9.4

A 60-year-old male presented with left-sided pleural effusion. He had a previous history of colonic carcinoma. The cytological features are shown in Figures 9.33a and b.

The preparations included numerous clusters of large pleomorphic cells, with hyperchromatic nuclei containing coarse granular chromatin and nucleoli (Figure 9.33a). Some of the clusters included large vacuoles (Figure 9.33b). These morphological features were consistent with metastatic carcinoma. Immunocytochemical stains were performed on the cell block to determine the origin of the adenocarcinoma cells. The adenocarcinoma cells stained positively for cytokeratin 7 (Figure 9.33c), but negative for cytokeratin 20.

This staining pattern is suggestive of a lung primary rather than a colonic primary. The cells also stained strongly with TTF1 (Figure 9.33d), thus further establishing the lung as the primary site. This case demonstrates the usefulness of immunocytochemistry in establishing the primary site of a malignant effusion in a patient with history of other neoplasms.

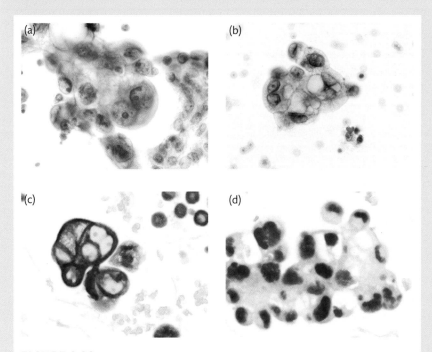

FIGURE 9.33
Adenocarcinoma. (a) Cluster of highly abnormal cells, which are easy to identify as malignant. (b) Cluster of highly vacuolated malignant cells. (c) Strong cytoplasmic staining for CK 7. (d) Strong nuclear and cytoplasmic staining for TTF1.

CASE STUDY 9.5

A 52-year-old woman with previous history of breast cancer presented with large pleural effusion. The effusion was quite cellular and at low power examination resembled reactive mesothelial cells (Figure 9.34a). At higher power the atypical nature of the cells became evident; the fluid contained many dissociated/single cells with a high nuclear cytoplasmic ratio, hyperchromatic nuclei with chromatin clumping, and irregular nuclear outlines (Figure 9.34b). Occasional nuclei also contained large nucleoli.

The above features are those of a metastatic adenocarcinoma with dissociated cell pattern. Immunocytochemistry was performed and tumour cells stained positive with EMA

(Figure 9.34c), cytokeratin 7, BerEp4 (Figure 9.34d), ER (Figure 9.34e), and negative with cytokeratin 20, and TTF1. Positive staining with EMA confirms the malignant nature of the cells and BerEp4 confirms the epithelial origin. Strong nuclear staining with ER favours breast carcinoma and negative staining with TTF1 excludes lung primary. Differential expression of cytokeratin 7 and cytokeratin 20 (cytokeratin 7 positive, cytokeratin 20 negative) also supports a breast primary and excludes colonic carcinoma.

The small dissociated cells can easily be overlooked as reactive mesothelial cells. However, careful attention to cytomorphology and careful use of immunocytochemistry and correlation with clinical history can help in resolving this diagnostic dilemma.

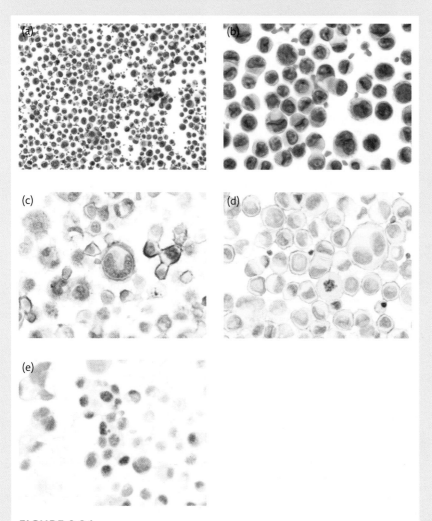

FIGURE 9.34

Adenocarcinoma. (a) Malignant cells dispersed as isolated malignant cells. (b) Close examination shows that many of the malignant cells are vacuolated. (c) Strong cytoplasmic staining for EMA. (d) Strong membrane staining for BerEP4. (e) Strong nuclear staining for ER.

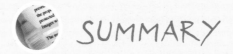

SUMMARY

- The main role of serous fluid cytology is to diagnose malignancy, as serous cavities are commonly involved in cancer.

- Correlation between cytological findings and clinical history is important in reaching a correct diagnosis.

- It is generally not possible to deduce the cause of most benign effusions from the cytological features; however, in rheumatoid disease and systemic lupus erythematosus there are typical cytological features which help in indicating the specific cause.

- Malignant cells in serous fluids are more commonly due to metastatic malignancy than to mesothelioma (primary malignant tumour of the mesothelium).

- The relative frequency of involvement by metastatic malignancy differs according to age, sex, and serous cavity.

- In a proportion of cases, the effusion may be the presenting feature of malignant disease. In these cases immunocytochemistry may be helpful in providing a definitive diagnosis of primary tumour.

- Most malignant mesotheliomas initially present with serous effusion.

- Cells from a malignant mesothelioma show diverse cytological features. Once the diagnosis of malignancy is established immunocytochemistry may be helpful in confirming the mesothelial nature of the cells.

FURTHER READING

- **Cibas ES and Ducatman BS (2009)** *Cytology: Diagnostic Principles and Clinical Correlates*, third edition. Sanders, Philadelphia.

- **Gray W (2002)** *Diagnostic Cytopathology*. Churchill Livingstone, Edinburgh.

- **Shidham VB and Atkinson BF (2007)** *Cytopathologic Diagnosis of Serous Fluids*. Saunders, Philadelphia.

DISCUSSION QUESTIONS

1. Discuss methods for the removal of red blood cells from serous effusions.
2. Discuss identification of malignancy of unknown origin in serous effusions.
3. Compare and contrast cytological features of adenocarcinoma and malignant mesothelioma. How can immunocytochemistry complement cytological findings?

Answers to the self-check questions, and tips for responding to the discussion questions, are provided in the book's Online Resource Centre.

 Visit www.oxfordtextbooks.co.uk/orc/shambayati/

10

Lower respiratory tract cytology

Behdad Shambayati

In this chapter, you will learn about cytology of the lower respiratory tract. This branch of cytology is not new: its use dates back as early as 1845. However, it was not until the 1960s that respiratory cytopathology was developed into a reliable diagnostic test (1). In the 1980s, with technical advances in radiological techniques, it became possible to image small lesions, and with improvements in sampling devices the use of cytology as a diagnostic test increased. It is encouraging to note that with advances in imaging and development of molecular diagnostics, morphologic examination of respiratory material still contributes greatly to patient management.

Although the respiratory tract is subject to many benign disease processes, these generally do not produce cellular changes, which would allow a specific diagnosis. The main role of cytology is in diagnosis of lung cancer.

Learning objectives

- Describe the anatomy of the respiratory tract.
- Describe the normal cells found in respiratory samples.
- Outline different sampling and preparation techniques.
- Describe common infections identified in respiratory tract cytology.
- Summarize incidence, mortality, risk factors, presentation, and treatment of lung cancer.
- Describe cytological features of different types of lung cancer.
- Give examples of the uses for immunocytochemistry in respiratory cytology.

10.1 Anatomy and normal cytology of the respiratory tract

The primary function of the respiratory system is to supply blood with oxygen and remove carbon dioxide. Figure 10.1 shows the simplified anatomy of the respiratory tract. The respiratory system is divided into two sections: the upper respiratory tract, which includes the nasal cavity, pharynx, and larynx, and the lower respiratory tract, which is the major focus of diagnostic cytopathology, and includes the trachea, bronchi, and lungs. The main function of the upper respiratory tract is transportation of air to the lungs, warming and humidifying the air, removing dust particles, and acting as a first line of defence against microorganisms. The lower respiratory tract is also involved as a defence mechanism and the terminal part is responsible for gaseous exchanges.

Air enters the respiratory system through the mouth and the nose. The air then passes through the nasal cavity and pharynx, through the larynx and the trachea. The trachea splits into two smaller tubes, left and right main bronchi. The lungs differ anatomically in that the left lung has two lobes (Figure 10.2), but the right lung has three lobes. The right bronchus divides into three lobar bronchi, to supply the upper, middle, and lower lobe of the right lung, and the

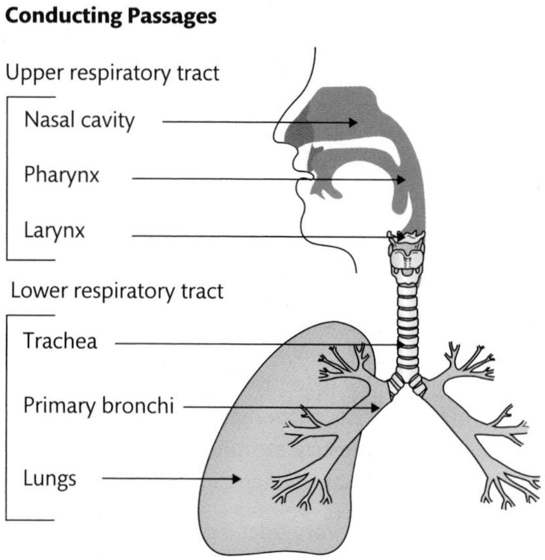

FIGURE 10.1
Line drawing demonstrating main components of upper and lower respiratory tract.

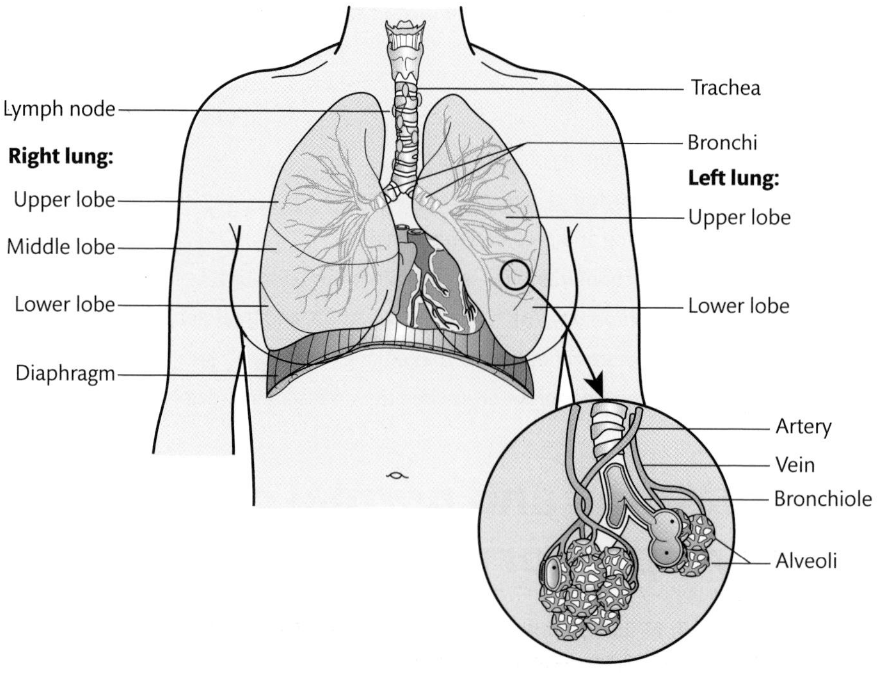

FIGURE 10.2
Line drawing showing components of the respiratory tract.

left bronchus branches to supply the upper and the lower lobe of the left lung. The bronchial tubes divide further into many smaller tubes, which connect to tiny sacs called alveoli. The alveoli are the basic functional component of the lungs and are surrounded by capillaries. Oxygen present in inhaled air passes into the alveoli and diffuses through the capillaries into the arterial blood. Meanwhile, carbon dioxide carried in the blood from the veins is released into the alveoli. Carbon dioxide follows the same path out of the lungs during exhalation.

SELF-CHECK 10.1

What is the function of the upper and lower respiratory tract?

Two main types of epithelium line the respiratory tract: non-keratinizing squamous epithelium and **respiratory epithelium** (a type of columnar epithelium). Non-keratinizing squamous epithelium covers areas that are liable to frictional forces such as those caused by swallowing of food or by touch. These areas include the front part of the nasal cavity, central and lower portions of pharynx, and parts of the larynx. Non-keratinizing squamous epithelium cells commonly exfoliate. They can be recognized in respiratory samples as **superficial and intermediate squamous** cells (Figure 10.3a). They are similar in all aspects to the superficial and intermediate cells of the female genital tract that you learned about in Chapter 4. Respiratory epithelium lines the major part of the nasal cavity, parts of the larynx, trachea, and bronchi. Respiratory epithelium is referred to as **pseudostratified epithelium** (Figure 10.3b), as it appears to be multilayered, even though it consists of a single layer of epithelial cells. The impression of multilayering is due to the position of nuclei in different levels of the cells in the epithelium. The respiratory epithelium consists of ciliated columnar cells (Figure 10.4) interspersed with mucus secreting goblet cells (Figure 10.4b).

Cross reference
Chapter 4.

The ciliated cells have basal, round to oval shaped nuclei with finely granular chromatin. The luminal part (facing the lumen) is covered by cilia; these are numerous and at their point of attachment form a terminal plate (Figure 10.5).

The main function of ciliated cells is to move the bronchial secretions toward the pharynx. The mucus secreting goblet cells are fewer in number. As the name implies they are shaped like

(a)
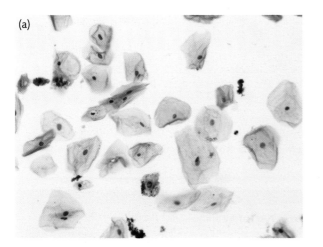
(b)

FIGURE 10.3

Epithelial cells of the respiratory tract. (a) Normal, superficial, and intermediate squamous cells. (b) Normal respiratory epithelium composed mainly of ciliated cells and occasional goblet cells (bronchial biopsy, medium power H&E).

(a)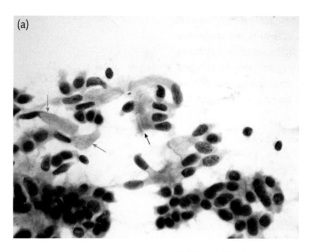

(b)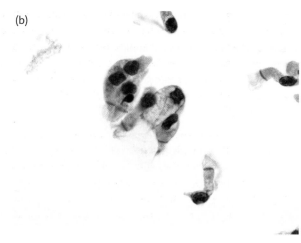

FIGURE 10.4

(a) Ciliated bronchial epithelial cells and occasional goblet cells (red arrows). Bronchial epithelial cells have oval nuclei. Cilia are attached at terminal plate (black arrow).
(b) Ciliated bronchial epithelial cells and single goblet cell.

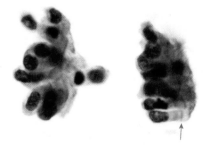

FIGURE 10.5

Ciliated bronchial epithelial cells. Note the terminal plate (arrow).

Neuroendocrine cells

These receive neural input and can release various messenger molecules into the blood. In the lung, they may act as receptors for oxygen levels.

Clara cells

These cells produce various proteins which protect the bronchiolar epithelium. They also have the potential to differentiate into ciliated cells.

a drinking goblet due to the distension of the cytoplasm by mucus (Figure 10.6). Goblet cells also have basally placed nuclei.

SELF-CHECK 10.2

What do you understand by the term 'respiratory epithelium'?

Other cells present in the respiratory epithelium, which are not identified easily in cytological samples, are **basal cells** or **reserve cells** (the precursor of goblet and columnar cells), **neuroendocrine** cells, and non-ciliated columnar cells called **Clara cells**, which line the terminal bronchiole.

The cells lining the alveoli are known as type I and type II pneumocytes. Type I pneumocytes are large, thin cells stretched across a large surface and cover the areas where gaseous exchange takes place. Type II pneumocytes produce **surfactant**, which has a dual function of preventing alveolar collapse during the breathing cycle and protecting the lungs from injuries and infections caused by foreign bodies and pathogens. **Type II pneumocytes** also act as reserve cells or **progenitor** cells and can differentiate into **type I pneumocytes** when they need to be replaced.

Alveolar macrophages (also known as histiocytes) are macrophages found in the pulmonary alveolus. They are **phagocytes**; their role is engulfing, digesting, and removal of inhaled particles and pathogens. They also stimulate lymphocytes and other immune cells to respond to foreign matter. Alveolar macrophages vary in appearance depending on their activity. The cytoplasm may be vacuolated and evidence of inhaled particles (usually carbon) may be seen as brown granules in Papanicolaou stain (Figure 10.7) or black in Romanowsky stains, such

Surfactant
A lipoprotein secreted by alveolar cells, its function is to reduce the surface tension of fluids in the lung.

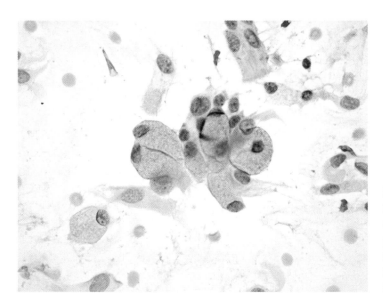

FIGURE 10.6
Bronchial brushings showing groups of goblet cells. Their cytoplasm is distended due to presence of mucin. It is generally unusual to see so many goblet cells together, except in hyperplasia (see Figure 10.4a). This patient had a history of asthma.

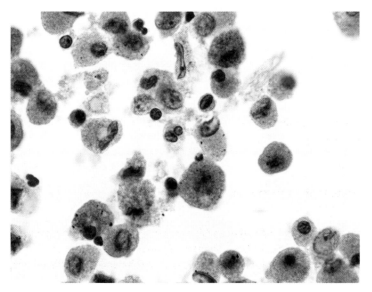

FIGURE 10.7
Alveolar macrophages. Note variation in cell and nuclear size. The cytoplasm is foamy and includes ingested carbon particles.

BOX 10.1 *Summary of different cell types in respiratory tract*

Upper respiratory tract

- Ciliated columnar cells
- Squamous cells

Trachea and bronchi

- Ciliated columnar cells
- Goblet cells
- Basal cells
- Neuroendocrine cells

Terminal bronchioles

- Non-ciliated columnar cells (Clara cells)
- Alveoli
- Type I and II pneumocytes
- Alveolar macrophages

as May-Grünwald-Giemsa (MGG) and Diff-Quik™. They may have a single round or bean-shaped nucleus, with granular chromatin. Binucleation and multinucleation occur, particularly in response to foreign matter.

Occasional white blood cells such as neutrophils and lymphocytes may be seen in most respiratory samples, but a noticeable increase in their numbers is indicative of disease.

Other cellular and non-cellular components of the respiratory system

There are a variety of non-cellular elements that can be seen in respiratory tract samples. These include materials inhaled, produced by the host, or introduced in the laboratory as contaminants. Their recognition is important to avoid misinterpretation.

Curschmann's spirals, named after German physician Heinrich Curschmann (1846–1910) who first described them, are strands of mucus that are formed in the lumen of small bronchi. They are occasionally seen in respiratory tract samples. It was thought that their presence was diagnostic of conditions such as asthma; we now know they can be seen in any condition where there is an increase in production of mucus. Curschmann's spirals vary in structure, but they have a coiled appearance, with a dark central part and translucent periphery (Figure 10.8).

Inspissated mucus (Figure 10.9) refers to thickened mucus; these look different to Curschmann's spiral in that they have no structure and stain dark blue in Papanicolaou stain.

Charcot-Leyden crystals, first described by French physician Jean-Martin Charcot (1825–1893), and German physician Ernst Viktor von Leyden (1832–1910), are orangeophilic needle-shaped

(a)

(b)

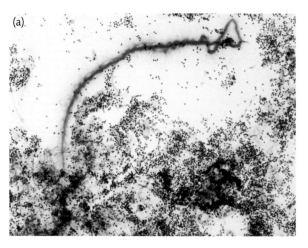
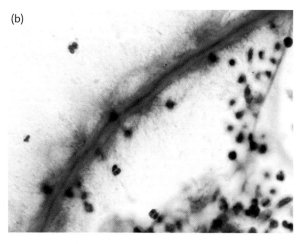

FIGURE 10.8

Curschmann's spiral. (a) Low-power view of a Curschmann's spiral amongst polymorphs.
(b) High-power view of a Curschmann's spiral. It consists of a central core and feathery
outer layer.

FIGURE 10.9

Darkly stained strands of inspissated mucus.

structures derived from degenerating eosinophils. They are seen in patients with allergic disorders, such as asthma or eosinophilic pneumonia (Figure 10.10 and Case Study 10.1).

Ferruginous (containing particles of iron) or **asbestos bodies** are formed when filamentous
particles become coated in protein and iron. This is often due to inhalation of asbestos (see
Chapter 9 for more information on asbestos) where small asbestos fibres are enveloped by
macrophages (Figure 10.11). Ferruginous bodies are often called asbestos bodies.

Undigested food particles such as fragments of plant tissue and meat fibres are sometimes
observed in sputum samples. The plant tissue has a characteristic rectangular shape (Figure
10.12a). Meat fibres (Figure 10.12b) can be recognized by the presence of cytoplasmic cross
striation.

Cross reference
Case Study 10.1.

Cross reference
Chapter 9, Section 9.5.

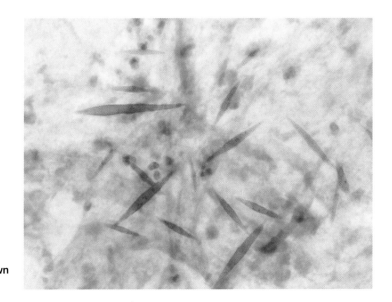

FIGURE 10.10
Charcot-Leyden crystals. These needle-shaped, orange-pink crystals are formed from breakdown products of eosinophils.

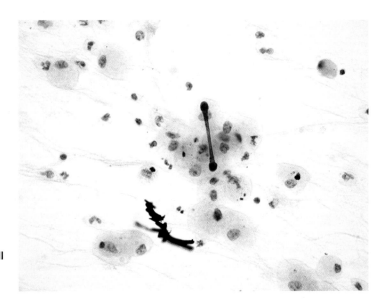

FIGURE 10.11
Ferruginous or asbestos bodies are dumbbell shaped structures of mineral fibres coated by protein.

FIGURE 10.12
(a) Vegetable cells. The intense staining, prominent cell borders, and refractile cellulose wall are features of vegetable matter and helpful in distinguishing these from malignant cells. (b) Meat fibres showing cross striation of muscle tissue.

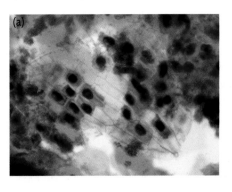

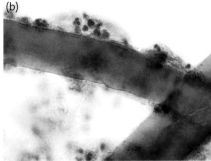

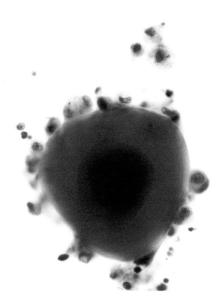

FIGURE 10.13
Corpora amylacea showing concentric radial striation.

Corpora amylacea

These glass-like structures (Figure 10.13) are thought to be derived from degenerating cells and are rarely seen in respiratory tract samples. Corpora amylacea have no known significance in disease; they are also seen in samples from the prostate and brain.

Cross reference
Chapter 8, Section 8.5.

10.2 **Specimen collection and sample processing**

Cytological diagnosis is in general dependent on optimally obtained and prepared material. The choice of collection method is largely dependent on the status of the patient; whether they are fit enough to tolerate a procedure, the location of the lesion, and the differential diagnosis. Although cells obtained by different techniques have similar morphologic characteristics, there are important differences, caused by cell preservation and different specimen processing regimes. There are several techniques in common use for obtaining cellular material for the diagnosis of lung cancer. The oldest method is sputum cytology, which depends on the spontaneous exfoliation of cells. There are four bronchoscopic techniques: bronchial washing, bronchial brushing, bronchoalveolar lavage, and transbronchial needle aspiration (TBNA). Fine needle aspiration (FNA) complements other diagnostic methods and is used in obtaining material through the chest. These procedures will be described below.

Sputum

Sputum is a mixture of mucus and cells that is **expectorated** (coughed up) from the respiratory tract. The patient is asked to collect material that is produced from an early morning deep cough into a specimen container. This sampling procedure could be repeated over several days to increase the sensitivity of the test.

Sputum cytology is one of the easiest methods of investigating the respiratory tract as sample collection is simple and causes no discomfort to the patient. In the recent past, sputum cytology was commonly used as a front-line investigation when lung cancer was suspected. With the advent of bronchoscopy and FNA, its use has declined. Today, sputum cytology is reserved for symptomatic patients who may be too ill to tolerate other procedures.

Sputum samples can be processed in several ways (see Chapter 2). All respiratory tract material is considered potentially infective and sample preparation should take place with appropriate health and safety consideration.

The simplest method is to pour the sputum sample into a Petri dish and examine the sample for tissue fragments or blood, and make smear preparations from these areas. The smears are alcohol fixed and stained with Papanicolaou stain. Sputum could also be prepared using one of the liquid based cytology (LBC) methods. These all involve a pre-treatment step to dissolve the mucus with a subsequent concentration of specimen.

Bronchial brushings

The first recorded use of the bronchoscope was in 1897, by Gustav Killian (1860–1921), a German laryngologist, to remove a piece of impacted pork bone from the main bronchus of a man (2). This was a rigid bronchoscope, consisting of a straight piece of metal inserted through the mouth. Rigid bronchoscopy is used mainly by thoracic surgeons these days; it is useful for removal of foreign objects from the airways or when a larger biopsy sample is needed, also used for **disobliteration therapy**—like cryotherapy or laser, and also for **stenting** large airways. Rigid bronchoscopy is a very uncomfortable procedure and hence is performed under general anesthetic.

Disobliteration therapy
This is the opening of closed structures due to disease.

Stenting
A stent is a tube, commonly made of wire, mesh, or plastics placed in open structures in the body to keep them open. In lung cancer, a stent can sometimes be used to keep an airway open that is narrowed or blocked due to a tumour.

The flexible bronchoscope was developed in the 1960s and has since been refined. Using fibre optic technology, it allows visualization of greater parts of the respiratory tract (Figure 10.14).

A channel in the bronchoscope allows small instruments to be inserted to provide diagnostic samples. Such instruments include fine plastic brushes or fine biopsy forceps. The flexible bronchoscope is inserted through the nose or mouth. The procedure is well tolerated by patients and is carried out using mild sedation and local anaesthesia.

Bronchial brushings can be taken from the surface of a tumour visualized with the bronchoscope. The brush can then gently be rolled onto the surface of the glass slide and air-dried for Diff-Quik™ staining, and rinsed in LBC preservative to allow the production of further slides and cell blocks for immunocytochemistry (see Chapter 2).

Bronchial brushing is a valuable procedure for the bronchoscopist, as it allows the sampling of areas that are prone to bleeding if biopsied. The fine brush can also be passed through narrowed bronchial tubes to sample malignant lesions which cannot be visualized and are out of reach of biopsy forceps. It is recommended that a cytologist is present during the bronchoscopy procedure, to microscopically assess the adequacy of the sample by rapidly staining and examining air-dried samples. This practice reduces unnecessary repeat sampling and reduces the risk of bleeding to the patient.

Bronchial washings

This procedure involves instilling several 20 ml aliquots of isotonic saline through a bronchoscope and with the aid of suction aspirating the material into a trap. This fluid will include mucus and cells which have been exfoliated. The fluid is centrifuged and the concentrate is

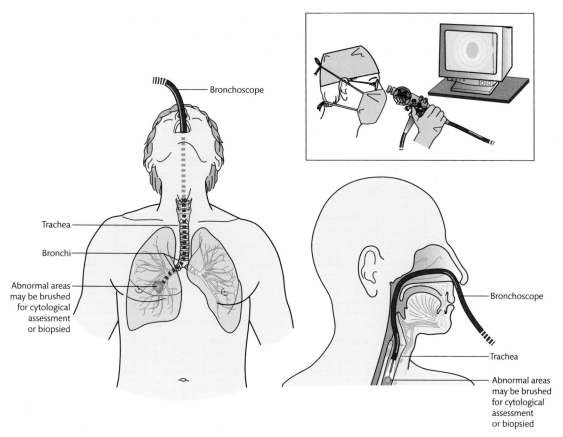

FIGURE 10.14
Diagrammatic representation of bronchoscopy procedure.

used to make smears, or is processed using LBC methodology. A Megafunnel cytocentrifuge may also be prepared from the cell concentrate. As with any cytological sample it is important to prepare bronchial washings without delay as cells suspended in saline undergo degenerative changes.

Bronchoalveolar lavage

Bronchoalveolar lavage (BAL) involves wedging the bronchoscope in a sub-segmental bronchus and instilling 20 ml aliquots of isotonic saline up to a total of around 120 ml; with the aid of gentle suction lavage fluid is gathered into a collection trap. The aliquot collected first is generally discarded to avoid contamination of the sample with upper respiratory tract material. This procedure is repeated until 40–60 ml of fluid is collected in the trap (approximately 40–70% of total saline instillate). The technique is useful in diagnosis of opportunistic infections in immunocompromised patients (see later sections in this chapter) and investigation of **interstitial lung disease**; these are groups of inflammatory lung diseases affecting the **interstitium** (the tissue and space around the air sacs of the lungs). A cytocentrifuge preparation is the ideal method for BAL samples. This is one of the few instances whereby LBC is not recommended as it removes white blood cells which form part of the **differential cell count**. See reading list for further discussion on interstitial lung diseases.

Interstitial lung disease
This is the term used to describe various types of inflammatory diseases that affect the tissue around the air sacs. These diseases generally cause progressive scarring of the lung tissue which eventually interferes with oxygen absorption.

Differential cell count
Technique of counting cell types in a sample and expressing these as percentages of the total number of cells in the sample.

Transbronchial fine needle aspiration

Transbronchial fine needle aspiration (TBNA) is a technique that allows sampling of lesions via the bronchoscope that lie beneath the bronchial surface. This is a relatively new technique and its use was first described in 1981 (3).

The most common application of TBNA is in the diagnosis of lung cancer and sampling of mediastinal lymph nodes. The **mediastinum** is the area between the lungs that contains the heart, aorta, oesophagus, trachea, and lymph nodes. Metastatic spread of cancer cells to mediastinal lymph nodes is one of the indicators used in **staging** (how far the tumour has spread) of lung cancer, which is an important determinant of treatment and prognosis. The lesion is aspirated via a needle attached to a fine tube, which is passed down the bronchoscope.

This technique has recently been further refined by the use of ultrasound to guide the bronchoscopist to sample smaller nodes and those in more distant stations in the mediastinum, traditionally the role of the thoracic surgeon performing mediastinoscopy. This technique is called **endobronchial ultrasound guided transbronchial needle aspiration (EBUS-TBNA)**.

The technique can even be extended to be used within the oesophagus to sample other **lymph node stations**.

Lymph node stations
This is the term used to describe the anatomical position of lymph nodes for lung cancer staging. These are numbered from 1 to 14.

Although TBNA is well tolerated by patients, it is important for a cytologist to be present during the procedure and microscopically assess the quality of the sample; this will avoid unnecessary repeat sampling which can lead to bleeding and complications. For a TBNA sample to be considered satisfactory, in the absence of malignancy, it should contain lymphocytes. The aspirated material is gently spread onto the surface of the glass slide and air-dried, for Diff-Quik™ staining. The remainder of the sample should be rinsed in LBC preservative for processing in the laboratory. Cell blocks prepared from needle washings will allow application of further ancillary testing including immunocytochemistry.

Transthoracic fine needle aspirations

Transthoracic fine needle aspiration involves insertion of a fine needle through the thoracic wall to obtain material for cytological examination. This technique is used primarily in patients with lung lesions that are peripherally positioned, as shown with cross-sectional imaging, or who have had a negative bronchoscopy. A radiologist under computed tomography (CT) guidance carries out the technique. The procedure has a low complication rate: the most common complications of transthoracic FNA are **pneumothorax** (air in pleural cavity) and haemorrhage. Pneumothorax, which causes the lungs to collapse, can cause chest pain and make breathing difficult. Small pneumothoraces resolve without any medical intervention, as the trapped air is gradually absorbed. A large pneumothorax may require further intervention to remove the air by insertion of a chest drain under local anaesthesia. Pneumothorax may develop because of repeated aspiration; it is therefore useful for a cytologist to attend the procedure and offer on-site analysis of sample adequacy, and appropriately advise the radiologist, thus avoiding unnecessary repeat sampling. After preparation of a direct smear for rapid staining with Diff-Quik™ stain, the needle is rinsed in LBC preservative, which can be used for making further cytological preparations in the laboratory or producing a cell block.

Biopsy imprints

This technique allows microscopical assessment of biopsies, thus minimizing the number of biopsies needed and reducing the risk of complications that can occur during these procedures.

The biopsy is placed on the glass slide and gently rolled off into the histological fixative. Exfoliated cells can be stained with Diff-Quik™ and examined under a microscope by a trained cytologist.

CASE STUDY 10.1

A 52-year-old man attended his GP complaining of a persistent cough and several episodes of coughing up blood. The GP referred him for a chest X-ray, which showed a cavitating lesion with radiological features suggestive of possible TB or cancer. A CT scan was arranged, which showed changes which were more likely due to an inflammatory process. He underwent bronchoscopy, which showed normal carina and bronchial tree; the source of bleeding could not be seen. Bronchial washings were sent to cytology. Cytological preparations showed numerous eosinophils, and many eosinophilic needle-shaped crystalline structures (Figure 10.15), consistent with Charcot-Leyden crystals. No abnormal cells were seen.

The patient was discussed at respiratory MDT and as his condition appeared to be inflammatory rather than neoplastic in nature, he was prescribed **corticosteroids** and offered a repeat CT scan four months later. He was later seen in clinic and his symptoms appeared to have improved.

Charcot-Leyden crystals represent breakdown products of eosinophils. They can be seen in asthma, parasitic infections, and **eosinophilic pneumonia**. Eosinophilic pneumonia encompasses a group of diseases that occur when eosinophils gather in the lung and can cause damage to the alveoli. They generally respond well to treatment with steroids.

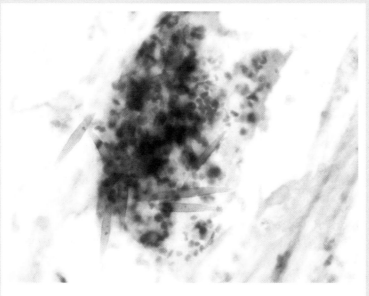

FIGURE 10.15
Charcot-Leyden crystals.

Routine staining methods

As in other branches of cytology, Papanicolaou stain is widely used as the routine stain, because it allows the cytologist to scrutinize the nuclear features. Romanowsky staining, such as May-Grünwald-Giemsa (MGG) or one of the rapid methods such as Diff-Quik™ (see below) are useful complements to Papanicolaou stain, as they allow better visualization of cytoplasmic features and enhance cell size.

Ancillary techniques—immunocytochemical markers

In the past few years, immunocytochemistry has become widely available and has greatly improved the specificity of cytological diagnosis. There are many antibodies that are useful in respiratory cytology and these will be mentioned in Section 10.8.

SELF-CHECK 10.3

List some of the advantages and disadvantages of different sampling methods in respiratory cytology.

10.3 **Reporting terminology**

Most laboratories write a descriptive report, detailing cytological features and cell types, followed by a concluding statement, dividing the report into the following four categories:

- Unsatisfactory or non-diagnostic
- Negative for malignancy (infections noted)
- Suspicious for malignancy (should be used sparingly)
- Malignant cells present—cell type suggested.

Although there are several types of lung cancer, these are broadly divided into two main categories, this being important for subsequent treatment and management of the tumour: small cell carcinoma and non-small cell carcinoma. These will be discussed further in Section 10.7, but are briefly mentioned here for clarity. Treatment regimes for these two subtypes of lung cancer are very different and a malignant cytology report must include indication of the subtype of lung cancer when possible.

10.4 **Assessment of sample adequacy**

An adequate cytological specimen must include representative material confirming the origin of the sample or sufficient number of abnormal cells to allow for confident diagnosis of malignancy. It is important for the cytologist to recognize these scenarios and request a repeat test if the sample is unsatisfactory for assessment. On-site assessment of adequacy is an important service that should be offered by all cytology laboratories and sufficient resources should be made available from the onset. Obtaining a representative sample at first visit is not only important to the patient, but contributes to the efficient management and overall diagnostic pathway of the patient, which is particularly significant given the current pressures to meet cancer turnaround times. The attending cytologist reviews rapidly stained direct smears and can request repeat sampling when material is deemed inadequate for diagnosis.

For a sputum sample to be considered adequate it must contain alveolar macrophages; this would indicate that the sputum sample is from deep within the lung and has sampled alveolar spaces. Presence of columnar cells without alveolar macrophages is not an indication of adequacy, as columnar cells may be originating from the nasal passages or upper respiratory tract. Samples that contain only squamous cells would suggest origin from the mouth or upper respiratory tract and are considered unsatisfactory. Some contaminants may render the sample unsatisfactory as they obscure cellular detail. Food particles such as vegetable matter and meat fibres may mask large areas of the sample and cover diagnostic material.

Bronchial brushings and washings are generally quite cellular and cytological findings should be correlated to the clinical details when considering adequacy of the sample. Bronchoalveolar lavages should consist predominantly of macrophages and be virtually free from respiratory epithelium and squamous cells, if the first collected aliquot has been discarded. Fine needle aspiration samples from solid tissue should contain interpretable cells corresponding to the site aspirated.

10.5 Inflammatory and non-neoplastic cytology

Respiratory epithelium, like other epithelia in the body, responds to injury. This could manifest itself in several ways, including metaplasia, inflammatory changes, and **hyperplasia**. Recognizing these changes is vital, as these reactive changes could be misinterpreted as neoplastic and cause unnecessary harm to the patient.

Most respiratory samples will include some level of inflammatory changes. Inflammatory changes may affect both squamous and respiratory epithelial cells. Inflammatory processes may cause keratinization and nuclear degeneration in squamous cells. Inflammatory changes in respiratory epithelial cells may manifest in different ways: ciliated columnar cells may increase in size, and multinucleation (Figure 10.16a) is a common feature. Degenerating respiratory epithelial cells may exhibit different staining reactions from those normally expected; the cytoplasm may stain orange rather than the usual green with Papanicolaou stain (Figure 10.16b).

Hyperplasia
This is the physiological response to stimuli. It may lead to increase in cell size and number.

(a)

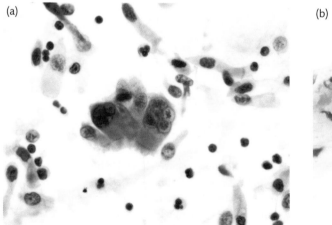

(b)

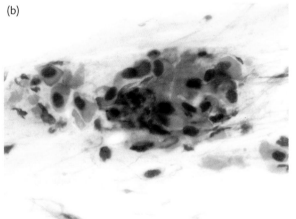

FIGURE 10.16

(a) Multinucleated bronchial epithelial cells. Multinucleation is a common inflammatory response. (b) Degenerative change in respiratory epithelial cells. These cells have taken up the eosin component of the Papanicolaou stain.

Creola bodies

These are clusters of ciliated epithelial cells sometimes seen in sputum samples of patients with asthma.

Large papillary groups of bronchial cells called **Creola bodies** (Figure 10.17) are sometimes seen in sputum samples from asthmatic patients. In asthma, bronchi periodically narrow, and produce more mucus than usual, causing breathing difficulties and coughing. This chronic inflammatory cycle may lead to dislodging of large groups of cells.

Bronchial, goblet, and reserve cells may undergo hyperplasia. Hyperplasia is the proliferation of cells under normal physiological conditions. The nuclear and cytoplasmic changes that occur in hyperplasia can sometimes make the cells appear abnormal, and care must be exercised not to interpret these benign changes as malignant. Bronchial cell hyperplasia can occasionally show very bizarre nuclear changes and form clumps: the clue to their benign nature is the presence of cilia (Figure 10.18a). Hyperplastic goblet cells may exfoliate in sheets and their cytoplasm may appear distended because of increased mucin production (Figure 10.18b). Reserve cells,

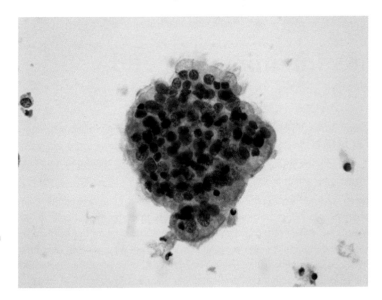

FIGURE 10.17
Creola body. An example of a Creola body which was found in a sputum sample of an asthmatic patient. The cell group has a distinct perimeter due to presence of cilia.

(a)

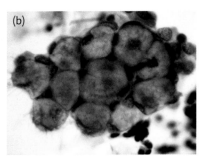
(b)

(c)

FIGURE 10.18
The nuclear and cytoplasmic changes that occur in hyperplasia. (a) High-power view of hyperplastic bronchial epithelial cells. This three-dimensional cluster of epithelial cells shows minor variation in nuclear size and chromatin pattern. Their smooth border is due to presence of cilia. These cell groups are a classic cause of false positive diagnosis. (b) High-power view of hyperplastic goblet cells. The cytoplasm is greatly distended due to presence of mucin. (c) Reserve cell hyperplasia. Reserve cells are rarely seen except in reactive states. They are small in size (compare the size with adjacent columnar cell), have hyperchromatic nuclei, and may show evidence of nuclear moulding. It is important not to confuse them with small cell carcinoma (see Figure 10.6).

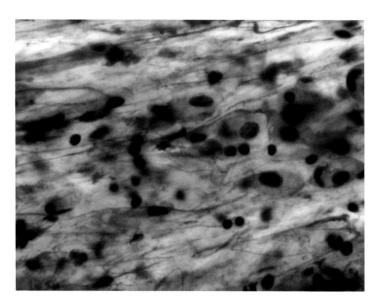

FIGURE 10.19
Atypical metaplasia. These dissociated squamous cells found in a sputum sample of a lifelong smoker show keratinization and have hyperchromatic nuclei, which are regular in size.

which are normally very inconspicuous, form clusters of crowded cells (Figure 10.18c) in reserve cell hyperplasia.

Squamous metaplasia is a common response of respiratory epithelium to injury and initially it looks very similar morphologically to that which occurs in cervical epithelia. As it progresses it can undergo keratinization and is termed **atypical metaplasia** (Figure 10.19).

Cross reference
Chapter 4, Section 4.3.

Atypical metaplasia
This is recognized as dissociated squamous cells, which show increase in nuclear/cytoplasmic ratio and sometimes keratinization.

SELF-CHECK 10.4

Why is the recognition of hyperplastic bronchial epithelium important?

10.6 Infections

Although detection of infections is not the primary role of respiratory cytology, occasionally infective agents can be identified. These include cytological changes due to viral infections, identifying morphologically distinct fungi, and occasionally parasites (see Case Study 10.2). In this section, the most commonly encountered infections are described. For a more comprehensive description, please consult the textbooks listed in the further reading section at the end of this chapter.

Bacterial infections

Bacterial infections may cause **pneumonia** (lung infection) and this can manifest itself as an **exudate** (protein rich fluid) containing numerous polymorphs (Figure 10.20a).

The filamentous bacteria actinomyces (Figure 10.20b) is commonly seen in sputum samples as a contaminant from the back of the throat.

Diagnosis of infection by *Mycobacterium tuberculosis* (TB) is best achieved by culture in the microbiology laboratory; occasionally cytology samples may include evidence of **granulomatous inflammation** that accompanies this infection. Granuloma is a collection of macrophages, called **epithelioid histiocytes** as they have a resemblance to epithelial cells (Figure 10.21).

(a)
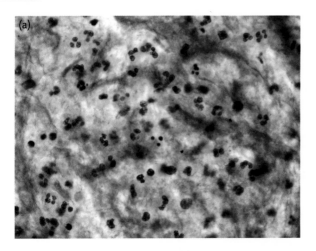

(b)

FIGURE 10.20

Bacterial infection. (a) Inflammatory exudate. This sputum sample from a patient with pneumonia shows a mixture of inflammatory cells in necrotic debris. (b) Actinomyces. These filamentous branching bacteria are often found in sputum samples and are of no significance.

(a)
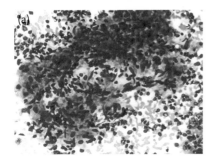

(b)
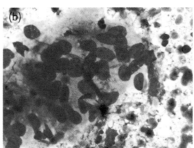

(c)
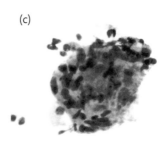

FIGURE 10.21

Granuloma. (a) Low power, TBNA. Aggregate of epithelioid histiocytes intermixed with lymphoid cells from a patient with confirmed TB. (b) TBNA. Epithelioid histiocytes with elongated nuclei. (c) Bronchial washing. Epithelioid histiocytes are elongated macrophages which have pale staining cytoplasm. The cytoplasm does not contain any ingested matter such as carbon.

Sarcoidosis

This is an inflammatory condition that causes formation of granulomas in different organs, including lungs, lymph nodes, and skin.

Ground-glass nuclei

In cytology this term is used to describe nuclear features that appear due to margination of chromatin, which gives the appearance of 'ground glass'. Often used to describe nuclear features due to herpes virus infection.

The sample may also include multinucleated giant cells (Figure 10.22a) in a degenerative background (Figure 10.22b).

These findings are not specific for TB as epithelioid cells may also be seen in other granulomatous inflammatory conditions, such as **sarcoidosis** or fungal infections. Mycobacterium is demonstrated with **Ziehl-Neelson** (ZN) stain. The organisms stain deep red colour, but require careful microscopy, as they may be sparse in numbers (Figure 10.23).

Herpes simplex virus

Morphologic changes of herpes infected cells can be recognized in cytological preparations from the respiratory tract. These are similar to the changes that you read about in Chapter 4, namely: multinucleation, nuclear moulding, and **ground glass nuclei**, which occur due to

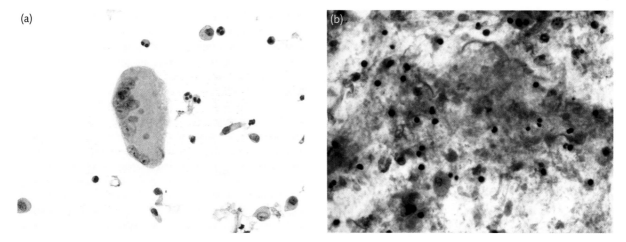

FIGURE 10.22

(a) This bronchial washing sample from a patient with confirmed TB included numerous giant cells. The nuclei are typically distributed at one pole of the cell and cytoplasm is devoid of any ingested carbon. (b) Necrotic debris in a BAL sample from a patient with confirmed TB infection.

FIGURE 10.23

Ziehl-Neelson (ZN) stain. The slender, slightly curved bacilli are stained magenta pink (arrows).

peripheral margination of chromatin (Figure 10.24). Herpes simplex most commonly affects **immunocompromised** patients (see below) where it infects the trachea and bronchi.

Cross reference
Chapter 4, Section 4.7.

Pneumocystis jiroveci

Pneumocystis jiroveci, also known as **Pneumocystis carinii**, was previously classified as a protozoan, as it resembles a parasite morphologically, but based on nucleic acid and biochemical analysis it is now re-classified as a fungus. It is widespread in nature, and is present in the soil and air, but it causes disease in immunocompromised individuals. Immunocompromised patients, also called **immunodeficient**, have reduced or absent ability to fight infections.

Immunodeficiency is either acquired (secondary), or genetic (primary). Most cases of immunodeficiency are acquired, either due to specific diseases, such as HIV (**human immunodeficiency virus**), which causes AIDS (**acquired immune deficiency syndrome**), or due to medications which suppress the immune system, as seen in patients receiving organ transplants.

Infection of *Pneumocystis jiroveci* occurs by inhalation, causing pneumonia; patients present with dry cough, fever, and shortness of breath. The organism can be seen in BAL samples and occasionally in deep cough sputum samples and bronchial washings. In Papanicolaou stained preparations individual organisms are not visible; instead a cluster of organisms is seen embedded in a proteinaceous material which appears as green foamy alveolar casts (Figure 10.25a).

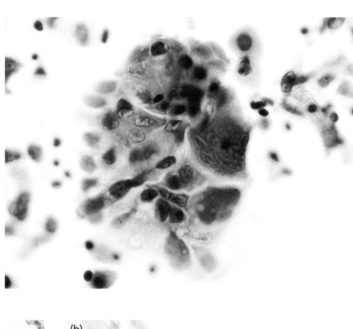

FIGURE 10.24
Herpes infection. Multinucleated bronchial epithelial cells with characteristic nuclear moulding.

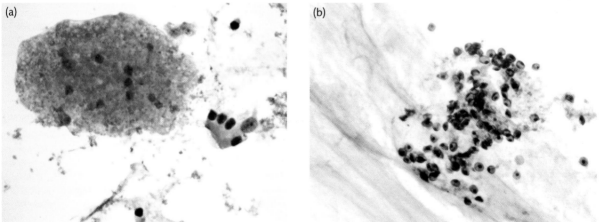

(a)

(b)

FIGURE 10.25
Pneumocystis jiroveci (bronchial washings). **(a) This is an example of a typical proteinaceous alveolar cast with foamy appearance. (b) Visualization with Grocott's stain. The cysts are stained black; within each cyst a dot-like structure is also stained.**

The individual cysts can be demonstrated using silver stain; **Grocott's methenamine silver** shows the cysts as small circular structures with a central dot-like structure (Figure 10.25b). Other methods suitable for identification include immunofluorescence or polymerase chain reaction (PCR).

Aspergillosis

Aspergillus is a filamentous fungus which is present in the soil and air, and commonly contaminates food. It does not normally cause disease, as most people are naturally immune to it and resist infection. When disease does occur it may take several forms, from an allergic reaction to its spores, where cytological samples will include numerous eosinophils and Charcot-Leyden crystals, to a severe life-threatening condition, where invasion of lung tissue occurs, leading to necrosis of lung tissue.

The presence of aspergillus in respiratory tract samples may be due to contamination from the mouth or due to genuine infection. Sputum samples are often contaminated with mouth flora: the presence of aspergillus in BAL samples is highly suggestive of an infective process. Aspergillus is recognized in Papanicolaou stained cytological preparations as **septate hyphae** (divided into discrete structures) which branch at 45° angles (Figure 10.26).

 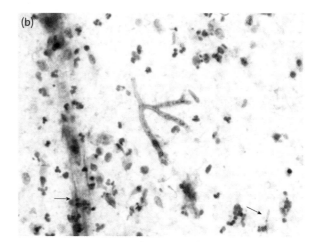

FIGURE 10.26

Aspergillus. (a) Tangled mass of Aspergillus is shown in this photomicrograph. The hyphae branch at approximately 45|. (b) Aspergillus (sputum). In this patient there was a severe allergic response and the sample included numerous eosinophils and Charcot-Leyden crystals (arrows).

Parasitic infections

In the developed world, parasitic infections involving the respiratory tract are rare. These include infections caused by protozoa and parasitic worms. Please consult the reading list for more description on this subject.

Case study 10.2 describes an infection by a parasitic worm.

CASE STUDY 10.2

A 66-year-old man who had recently emigrated from Vietnam to the UK presented with fever, shortness of breath, and **haemoptysis** (coughing up blood) to the A&E department. He was admitted for observation. When giving his medical history he also mentioned that he had suffered from diarrhoea for a long time. During his stay in hospital, sputum samples were sent for cytological assessment and stool samples sent to microbiology. Cytological preparations were negative for malignant cells, but included thread-like organisms that were recognized as larvae of *Strongyloides stercoralis* (Figure 10.27). Examination of stool sample confirmed the presence of larvae. He was treated with **ivermectin and thiabendazole**, which are anti-parasitic medications. The patient's general condition improved after a week, he was discharged, and followed up in outpatient clinic.

Strongyloides stercoralis is a **nematode** (a roundworm belonging to the phylum Nematoda) that can infect humans. It is a rare infection in developed countries, but is prevalent in societies where soil is contaminated with faecal matter.

The larvae exist in two forms, the **filariform** (threadlike) larvae, and the free-living **rhabditiform** (first larval stage) which lives in the soil. The infection occurs when larvae penetrate the skin and enter the circulation. The larvae eventually enter the lung and move up the upper respiratory tract until they are swallowed. In the small intestine, the larvae mature and produce ova which mature into rhabditiform larva that are passed into the faeces.

Strongyloides generally does not cause major symptoms in healthy individuals; however, in immunocompromised patients where widespread multi-organ infection can occur, mortality is as high as 80%.

The patient in this case did not have any major contributory underlying condition.

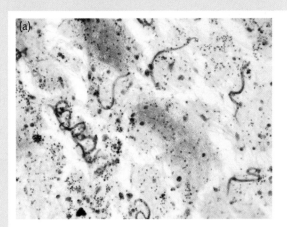
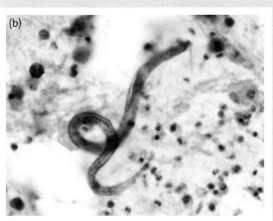

FIGURE 10.27
Two sputum samples testing positive for Strongyloides. Numerous filariform larvae are seen in sample (a).

10.7 Lung cancer

Lung cancer is the most common cancer in the world, with 1.3 million cases diagnosed every year (4). In the UK, it is the second most common cancer after prostate cancer in men, and in women it is the third most common after breast cancer and bowel cancer. Lung cancer is the most common fatal malignancy in both men and women. The poor survival rate is mainly due to its late diagnosis, when the cancer is already at the stage when curative treatment is not possible and many patients are elderly, with existing medical conditions that make them unfit for radical treatment. Most lung cancers are caused by smoking.

Incidence, mortality, and survival statistics

Table 10.1 summarizes some available statistics on incidence and mortality.

Incidence

Most cases of lung cancer are diagnosed in people over the age of 60, and rarely diagnosed in people under 40 years of age. The highest incidence is between the ages of 75 and 84 years. Lung cancer incidence rate has decreased overall between 1997 and 2006, due to cessation of smoking mostly among men; where the rate in men has decreased by 21%, the incidence rate in women has changed little (5–8). In the UK there is wide geographical variation in lung cancer rates: the highest incidence is seen in Scotland and northern England and the lowest incidence in Wales and southern England (9), mirroring smoking habits.

Mortality

Lung cancer is the most common cause of death due to cancer in both men and women (5–8, 10). Within the UK, variations in mortality follow the smoking habit, with the highest recorded mortality in Scotland. In the UK, male cancer mortality has decreased steadily between 1982 and 2007. In women, during the same period the rates increased until the late 1990s; they have since levelled off. The prediction for the future is that death rates will generally fall. However, in Scotland the death rate in women will continue to rise slightly up to 2010–2014 (11). The different patterns of lung cancer death rates in men and women could be explained by previous smoking behaviour; men started smoking earlier than women, and smoked more heavily; it is now apparent that more men gave up smoking than women.

Survival

Survival outcome for lung cancer is very low as most cancers are discovered late, when the tumour has spread to other areas, or it is anatomically situated in an area that cannot be surgically removed. Radical treatment is also not feasible in some patients as they may have an underlying medical condition. After diagnosis, 20% of patients are alive after one year; this falls to 5% at five years. Survival rates for different subtypes of lung cancer are discussed later in this chapter.

SELF-CHECK 10.5

What could be the possible explanation for the drop in lung cancer death rates in men, whilst the rate in women is increasing?

TABLE 10.1 Incidence, mortality, and survival statistics for lung cancer.

	Incidence (2006)		Mortality (2007)	
	Male	Female	Male	Female
Total	22,381	16,646	19,637	14,872
Age standardized rate (per 100,000 population)	60.8	37.1	51.5	31.3

Screening for lung cancer

The health agenda in many countries is to detect cancer early before it has spread to other organs and improve the survival rate. Lung cancer screening was initially trialled using chest X-rays and sputum cytology. Although slightly more cancers were detected, no improvement in overall survival was seen (12).

Helical or spiral CT
This is three-dimensional CT scanner that can produce high-resolution images with lower X-ray dose to the patient.

Low-dose **helical CT** is a potential imaging method with promise for screening, due to its lower radiation dosage and thus ability to be repeated more frequently. There have been several trials in the USA and Japan, with promising results. However, the numbers needed to be screened to pick up even a small number of cancers is enormous and thus extremely costly. Before screening is implemented it must be proven that detecting lung cancer earlier improves survival rates. With any screening programme, the risk and benefits must also be fully evaluated to minimize false positive results which put patients through unnecessary investigations, operations, and psychological stress.

Risk factors

Cigarette smoking is the main cause of lung cancer. Risk of developing lung cancer is related to both the number of cigarettes smoked and duration of smoking habit. A person smoking ten cigarettes a day has eight times the chance of dying from lung cancer and a person smoking 25 cigarettes a day has 25 times the risk compared to a non-smoker (13). Second-hand smoke or passive smoking (breathing the smoke of others) increases the risk of developing lung cancer. Life-long non-smokers exposed to second-hand smoke at home or in their occupation have their risk of lung cancer raised about a quarter, while heavy exposure at work doubles the risk (14). Tobacco smoke contains 55 carcinogens that have been evaluated by the International Agency for Research on Cancer (IARC), which showed that there was sufficient evidence for carcinogenicity in either laboratory animals or humans (15).

Radon gas, a naturally occurring radioactive gas released during decay of uranium 238, is considered the second most important factor after cigarette smoking. Extensive epidemiologic evidence from studies of underground miners and complementary animal data has documented that radon causes lung cancer in smokers and non-smokers (16). Radon is present in the air we breath at very low levels, but at high concentration radon becomes a risk factor to the general population; this typically occurs in well-insulated, tightly sealed buildings. Ground floors or basements, because of their closeness to the ground, typically have the highest radon levels (17). In certain parts of the UK, particularly in Devon and Cornwall, high levels of radon have been recorded. Radon gas may be responsible for 9% of lung cancers in Europe (17). Other risk factors include occupational exposure to asbestos and polycyclic hydrocarbons (18).

BOX 10.2 *Summary of risk factors that contribute to development of lung cancer*

- The single most important risk factor is smoking.
- The risk increases with quantity of cigarettes smoked, duration of smoking, and starting age.
- Stopping smoking results in a reduction in risk of developing cancer.
- Other factors include exposure to radon gas, polycyclic hydrocarbons, and asbestos.

Presentation

For a comprehensive description of investigation and treatment of lung cancer consult the guidance issued by National Institute for Clinical Excellence in its publication no. CG24 and available on its website (19).

Patients commonly present with haemoptysis, chronic cough, **dyspnoea**, chest pain, excessive fatigue, and **cachexia**. Weight loss is often quite severe in spite of good nutrition, and may be due to effects of hormones produced in lung cancer.

Patients may also present with enlarged lymph nodes, particularly in the neck region, or **finger clubbing** (see Case Study 10.3). In some cases, patients do not present with chest symptoms, but symptoms due to metastatic disease such as pain, **pathological bone fractures**, or neurological symptoms.

SELF-CHECK 10.6

What are the symptoms of lung cancer?

Investigation

All patients with suspected lung cancer will usually under go a **contrast-enhanced chest CT**; this is a technique where a contrast media is injected intravenously to increase the sensitivity of CT scanning. The scan may also include the liver and adrenals, which will be important when **staging** the cancer (staging will be discussed later in the chapter).

Patients with central lesions who are well enough to tolerate the procedure will undergo bronchoscopy, when bronchial washings, brushings, or biopsies may be taken.

Sputum cytology, which was the one of the main diagnostic modalities in lung cancer diagnosis, is rarely used these days, except in patients who have centrally placed masses or are unable to tolerate, or do not wish to undergo, a bronchoscopy or other invasive tests. Patients with lesions located in the periphery of the lung will undergo a percutaneous transthoracic fine needle aspiration or biopsy. In some patients, the diagnosis is obtained by sampling the enlarged mediastinal lymph nodes using the TBNA technique; a small number of patients will require surgery to obtain a diagnostic sample.

A **positron emission tomography (PET)** scan is required to stage the cancer fully if radical treatment is a possible option.

Types of lung cancer

Lung cancer is divided into two main types, based on histological appearance: **non-small cell lung cancer** (NSCLC) (often called non-small cell carcinoma) and **small cell lung cancer** (SCLC), commonly called small cell carcinoma. Treatments for SCLC and NSCLC are very different, and it is essential to have an unequivocal cytological or histological diagnosis.

Non-small carcinoma is the most common type; approximately 80% of lung cancers are of this type. Non-small cell carcinoma is subdivided into three main subtypes as approaches to diagnosis, staging, prognosis, and treatment are similar:

- Squamous carcinoma
- Adenocarcinoma
- Large cell carcinoma

Dyspnoea
Shortness of breath.

Cachexia
This is severe weight loss and muscle mass associated with disease. Cachexia is seen in cancers, AIDS, and other chronic progressive diseases.

Finger clubbing
This is the change in the shape of fingertips and nails that is seen in some patients with lung cancer and heart disease. The fingertips appear larger and nails curve more than usual. It is thought to be due to fluid accumulation in the fingertips.

Pathological fracture
These occur when a disease process weakens the bones significantly so that fractures occur during normal activities such as walking. Many disease processes, including infections, generalized bone diseases such as osteoporosis, and metastatic tumours can cause pathological fractures. Malignant tumours that metastasize to the bone include carcinomas of lung, breast, prostate, thyroid, and kidney, amongst others.

Positron emission tomography (PET)
A PET scanner (a machine similar to a CT scanner in appearance) detects **positrons** (positively charged electrons) emitted by a radionuclide in the organ or tissue being examined. PET scanning measures metabolic activity; in this procedure a radionuclide such as **fluorodeoxyglucose (FDG)** is administered into a vein. After a short waiting period, this chemical accumulates in biologically active tissues. Most commonly these days, the PET scanner is combined with CT scanning, which allows for simultaneous visualization of metabolically active areas in relation to anatomical site. PET scanning is commonly used when staging the tumour.

Squamous carcinoma (Figure 10.28a) makes up 20–30% of lung cancer, and is associated with smoking. Tumours most often arise in the **hilum** of the lung (area where the bronchus, blood vessels, nerves, and lymphatics enter or leave the lung). Squamous carcinoma and adenocarcinoma have a premalignant precursor stage that may undergo morphological changes from hyperplasia to metaplasia and through to intraepithelial neoplasia. These early changes have been termed **bronchial intraepithelial neoplasia** or early central airways lung cancer.

Adenocarcinoma (Figure 10.28b) constitutes 30–40% of lung cancers. It usually occurs in the periphery of the lung as a discrete mass arising from the proximal airways. Adenocarcinomas arise from glandular cells, such as the mucous goblet cells, Clara cells, and type II pneumocytes. The incidence of adenocarcinoma is rising: adenocarcinoma is the most common lung cancer in ex-smokers. Another less common subtype, called **bronchoalveolar cell carcinoma (BAC)** (Figure 10.28c), occurs as a diffuse tumour arising from the terminal air spaces. Bronchoalveolar carcinoma has a typical radiological appearance, known as ground-glass opacities, that helps identify this tumour type.

Large cell carcinomas (10–15% of lung cancers) are aggressive, fast growing, central tumours and, by definition, at light microscopy show no squamous or glandular differentiation; therefore the diagnosis is made by exclusion (Figure 10.28d).

Small cell lung cancer and bronchial carcinoid

Small cell carcinoma or small cell anaplastic carcinoma (15–20% of lung cancers) was previously known as 'oat' cell carcinoma, as the small nuclei resemble oat grains (Figure 10.29).

This tumour usually occurs centrally and metastasizes very early to the hilar lymph nodes, brain, and liver. Sometimes a patient presents with metastases and the primary may be too small and difficult to find. The majority of patients have a strong smoking history.

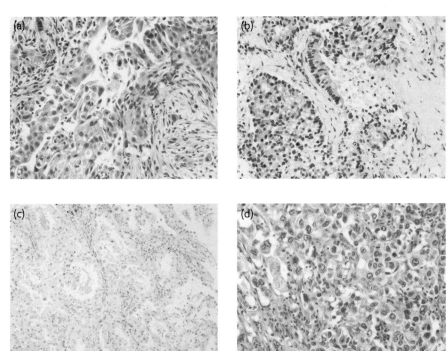

FIGURE 10.28
Types of lung cancer.
(a) Squamous carcinoma (bronchial biopsy, medium-power H&E).
(b) Adenocarcinoma (bronchial biopsy, medium-power H&E).
(c) Bronchoalveolar carcinoma (bronchial biopsy, medium-power H&E).
(d) Large cell undifferentiated carcinoma (bronchial biopsy, medium-power H&E).

Small cell carcinoma belongs to the group of tumours known as **bronchial neuroendocrine tumours** as they arise from neuroendocrine cells in the bronchus. Neuroendocrine cells are specialist cells that release hormones as result of neural stimulus and are found throughout the body. Pulmonary neuroendocrine cells may be involved in regulation of oxygen levels. This is likely done by detecting decreased oxygen or increased carbon dioxide levels and sending chemical messages to help the lung adjust to these changes. It is noted that people living in high altitudes, where oxygen levels are lower, have higher numbers of neuroendocrine cells in their lungs. Similar granules are seen in **bronchial carcinoid** tumours, which, in contrast to small cell carcinomas, are slow growing tumours.

Two types of bronchial carcinoids are recognized; the first and most common is referred to as **typical carcinoid** (Figure 10.30), arising mostly as a central tumour. This is a low-grade tumour, capable of local invasion, but rarely metastasizes. The second type is referred to as **atypical carcinoid**, which has the potential to metastasize. Bronchial carcinoid is not associated with smoking and may occur at any age.

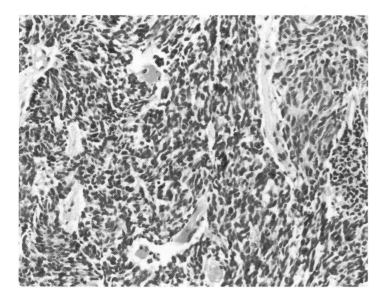

FIGURE 10.29
Small cell anaplastic carcinoma (bronchial biopsy, medium-power H&E).

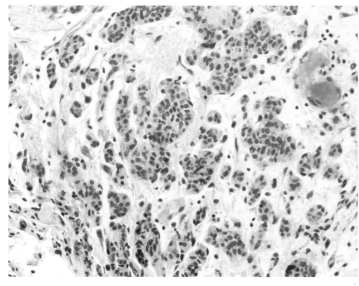

FIGURE 10.30
Bronchial carcinoid (bronchial biopsy, medium-power H&E).

Other primary and secondary tumours

Other primary tumours such as bronchial lymphomas, sarcomas, and adenoid cystic carcinomas, which are rare, may also occur; these will not be discussed here, please see the further reading list at the end of this chapter for more information.

The lungs are a common site for metastases from many cancers, including those from breast, gastrointestinal tract, and the kidneys. The patient may present with similar symptoms as a patient with primary lung cancer, but as the treatment is different, it is vital to make the correct diagnosis.

Staging

As with all cancers, the stage of the tumour has important **prognostic** (outlook) significance. To stage the tumour means to determine how large the tumour is and how far it has spread. Non-small cell carcinomas are staged 1–4 (see box staging of lung cancer). Stages 1 and 2 are classified as early disease, stage 3A as locally advanced, and stages 3B and 4 as advanced. Lung cancer presents late: up to 50% of patients present with advanced disease and only 20% have cancer that has not spread by the time of diagnosis.

BOX 10.3 *Key facts on main types of lung cancer*

- Squamous cell carcinoma, comprise 20–30% of lung cancers and are often found in the centre of the lung (hilum), closely associated with cigarette smoking.
- Adenocarcinoma, comprise 30–40%, are often found in the outer areas of the lung.
- Large cell carcinoma, comprise 10–15%, can occur in any parts of lung.
- Small cell carcinoma, comprise 15–20% of lung cancers, usually arise in the hilum, strongly associated with smoking.

BOX 10.4 *Staging of lung cancer*

The staging system described below is very simplified. For further explanation consult the further reading list and the references at the end of the chapter or on the Web.

Generally, there are two staging systems: the TNM and number system. The TNM takes into account three factors: the size of the tumour (T), lymph node involvement (N), and whether the tumour has spread anywhere else in the body (M).

The TNM staging is used widely in the world to stage lung cancer and has been revised seven times since its publication in 1997 (20).

The number system uses information from the TNM staging and divides the cancer into stages 1 to 4. For example, in stage 1 the cancer is limited to the lung and in stage 4 the cancer has spread to other parts of body, such as liver or the brain. Stages 2, 3A, and 3B are in-between stages.

Small cell lung cancer may be staged using the number system, but many centres stage the cancers as **limited stage disease** when cancer is only in the chest and **extensive disease** when the cancer has spread outside the chest.

Treatment

Treatment options depend on factors such as type of lung cancer (non-small cell versus small cell), stage of the cancer, position of the tumour, and patient's fitness to tolerate the therapy. Treatments for small cell carcinoma and non-small cell carcinoma are very different. Surgery is only applicable to non-small cell carcinoma when detected early (usually stages 1 and 2). The surgeon may remove a small part of the lung (**wedge resection**), one of the lobes (**lobectomy**), or the entire lung (**pneumonectomy**). Radical chemo-radiotherapy regimes can be offered to those patients with locally advanced disease or local disease who are deemed inoperable, with good results. More commonly, chemotherapy and radiotherapy are given to patients with a palliative intent when the cancer is metastatic.

Small cell carcinoma spreads very quickly throughout the body and patients do not benefit from surgery. Chemotherapy is the main treatment for this type of cancer. Combination chemotherapy and radiation is given to patients with extensive disease, to prolong their life and relieve their symptoms (**palliative**).

Bronchial carcinoids are usually treated by surgery. Typical carcinoids with low malignant potential are treated with conservative surgery. Carcinoid tumours which are suspected to have a malignant potential may require more extensive surgery.

Prognosis

Lung cancer has one of the worst survival outcomes of any cancer. The outcome is dependent on stage and type of lung cancer. Overall, for all types of lung cancers irrespective of stage at the time of diagnosis, only 25% of patients will live for one year and 8% will live for five years.

For non-small cell lung cancer the five-year survival varies from 58 to 73% for stage 1A to only 2–13% for stage 4.

For small cell lung cancer the five-year survival statistic is worse, with only 38% of stage 1A patients surviving for five years, and 1% for stage 4.

Bronchial carcinoids have excellent prognosis; for a typical carcinoid the five-year survival rate is 94%. This rate is reduced to 71% when regional lymph nodes are involved.

SELF-CHECK 10.7

Why is staging of lung cancer important?

10.8 **Cytology of lung cancer**

Squamous cell carcinoma

Cytological diagnosis of squamous carcinoma relies on identification of squamous cells exhibiting malignant features. In well-differentiated squamous carcinoma, cytological samples may include loosely cohesive and dispersed cells, showing abnormal nuclear features such as hyperchromasia, irregular chromatin pattern, increased nucleocytoplasmic ratio and very abnormal nuclear shape (Figure 10.31a).

The nuclei can occasionally appear pyknotic (Figure 10.31b); nucleoli are not commonly seen in squamous carcinoma. In Papanicolaou stained preparations the cytoplasm of keratinized cells is dense green (Figure 10.31c) or intense yellow-orange in colour (Figures 10.31d and e), and pale blue in MGG stained slides (Figures 10.31f and g). Squamous cells can take bizarre shapes, including tadpole or fibre-like (Figures 10.31h and i).

Many squamous carcinomas undergo central necrosis and this can be seen in cytological samples as necrotic and inflammatory debris. Moderately and poorly differentiated squamous

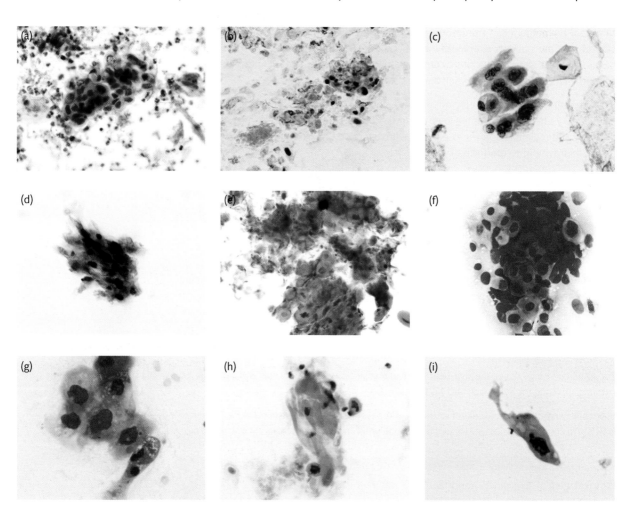

FIGURE 10.31

Squamous cell carcinoma. (a) This cluster of cells shows most features of squamous cell carcinoma: hyperchromasia, irregular chromatin pattern, increased nucleocytoplasmic ratio and very abnormal nuclear shape, and keratinization. (b) Squamous cell carcinoma showing pyknotic nuclei. (c) Squamous cell carcinoma showing dense green cytoplasmic staining of malignant cells. Elsewhere in this sample the typical orangeophilic keratinized cells were present. (d) and (e) show further images of squamous cell carcinoma; note the highly keratinized cells with pyknotic nuclei in (e). (f) and (g) show squamous cell carcinoma with MGG stain. (h) and (i) show 'tadpole' cells; note the smaller malignant cells with pyknotic nuclei in (h).

carcinomas do not show keratinization and cells may not exhibit any features that would allow identification of their squamous origin (Figure 10.32). Moderately and poorly differentiated tumours are often reported as non-small cell carcinoma.

A common diagnostic pitfall, which leads to a false positive diagnosis, is misinterpretation of keratinized metaplastic cells or degenerate bronchial epithelial cells. To minimize the risk of a false positive diagnosis it is important to avoid making a diagnosis on a small number of cells or poorly preserved cells.

False negative diagnoses can occur when necrotic or inflammatory debris dominates the sample and obscures malignant cells (which may be few in number) in the background.

Currently, treatment regimes for squamous carcinoma and adenocarcinoma are similar, but with development of novel drugs, the physician will benefit from more accurate typing of tumours, in particular non-small cell carcinoma. Immunocytochemistry can be helpful in suggesting a differential diagnosis as most squamous carcinomas react with **CK 5/6** and **p63** (Figure 10.33) antibodies.

> **Monoclonal antibody to cytokeratins 5 and 6 (CK 5/6)**
> This is useful in identification of squamous carcinoma and cells of mesothelial origin.
>
> **Nuclear immunoreactivity with p63**
> This is seen in squamous cell carcinomas and in urothelial carcinomas. Adenocarcinomas generally do not express p63.

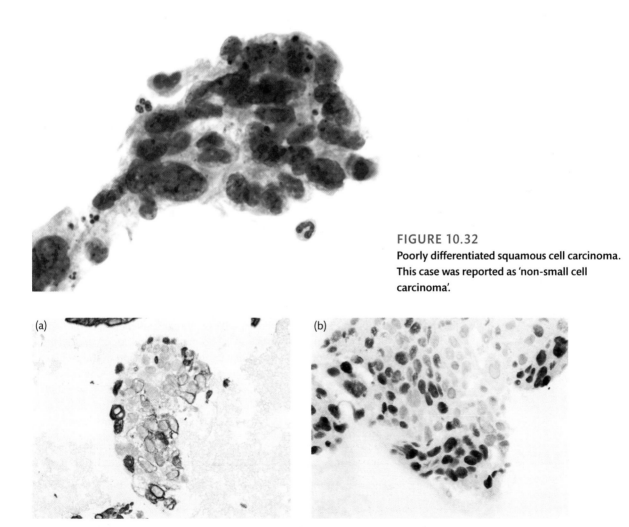

FIGURE 10.32
Poorly differentiated squamous cell carcinoma. This case was reported as 'non-small cell carcinoma'.

(a)
(b)

FIGURE 10.33
Squamous cell carcinoma (cell block). (a) CK 5/6 antibody. The staining is cytoplasmic. (b) p63 antibody. The staining is nuclear.

BOX 10.5 *Cytomorphology of squamous carcinoma*

- Numerous abnormal, dissociated squamous cells
- Intense yellow-orange staining of squamous cells
- Bizarre shaped cells, including tadpole and fibre-like. Background of necrotic or inflammatory debris

CASE STUDY 10.3

A 67-year-old man visited his GP with a six-month history of worsening shortness of breath (SOB). The GP requested an urgent chest X-ray that showed abnormal changes. The patient was referred urgently to the respiratory clinic. On questioning, he mentioned that he had also developed a productive cough and over the past month had noticed blood in his sputum. He was a smoker and had smoked 20 cigarettes a day for 40 years. He also complained of weight loss and tiredness. On examining him the consultant noted that his fingers were clubbed (the depositing of extra fluid or tissue under the fingernails). Chest X-rays showed a mass in the left upper lobe; CT images confirmed the presence of the mass and extensive lymph node involvement, as well as the presence of liver metastases.

Bronchoscopy was performed and confirmed the presence of a tumour. Bronchial brushings were taken and sent for cytological analysis. Figure 10.34 shows a heavily blood-stained smear with hyperchromatic cells which have an abnormal chromatin pattern. The cytoplasm of some of the abnormal cells stained orange with Papanicolaou stain, confirming the presence of keratin. The sample was reported as showing malignant cells, from a non-small cell carcinoma, with features suggestive of squamous carcinoma.

The patient was treated with local radiotherapy to control his symptoms.

The majority of people with lung cancer present with some symptoms. The symptoms may be due to the direct effects of the primary tumour, the effects of metastatic tumours in other parts of the body, or to disturbances of hormones, blood, or other systems caused by the cancer. Shortness of breath may be due to obstruction of the respiratory tract, or to collection of pleural fluid around the lung, reducing the lung volume. Haemoptysis may have many causes, but should be taken seriously and investigated. Patients with cancer can have a range of non-specific symptoms such as weight loss or tiredness. This weight loss is referred to as cachexia; it is thought to be caused by effects of a signalling molecule's **tumour necrosis factor-alpha (TNFα)**. Finger clubbing or digital clubbing is the term used to describe changes in the shape of nails and fingernails; this is seen in patients with advanced heart disease or lung cancer. The exact cause of clubbing is not known but is thought to be caused by fluid collecting in the soft tissues at the ends of the fingers. At later stages extra bone may form on the finger joints.

Non-small cell carcinoma could be resected surgically depending on the site, but this patient's tumour was too advanced to be considered for curative treatment and he was therefore offered **palliative** treatment (to reduce the severity of the symptoms of disease).

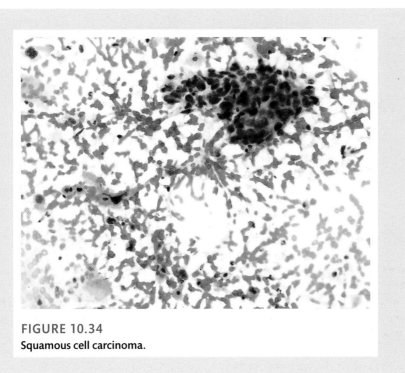

FIGURE 10.34
Squamous cell carcinoma.

Adenocarcinoma

It was mentioned earlier that there are two subtypes of adenocarcinoma: a more common bronchogenic type that occurs as a solid mass, arising from the proximal airways, and a diffuse type called bronchoalveolar carcinoma, arising from the terminal airways. Cytological distinction of the two subtypes is not easy and some believe that it is not possible, but there are some clues, which, together with clinical correlation, can suggest a differential diagnosis.

Bronchogenic carcinoma often presents as cell aggregates or three-dimensional clusters. Occasionally, acinar formation (Figure 10.35a) and mucin vacuoles (Figure 10.35b) can be seen, suggesting the glandular origin of the cells. The cytoplasm is more delicate in nature, often transparent or foamy. Nuclear abnormality may not be as pronounced as seen in squamous carcinoma (Figure 10.35c); hyperchromasia may occur, but sometimes the nuclei are **normochromatic** (normal staining). Chromatin pattern could vary from fine to coarse granularity. Nuclei are often enlarged and show **anisonucleosis** (variation of nuclear size) within a cluster. Nucleoli are often seen in adenocarcinoma (Figure 10.35d).

In poorly differentiated adenocarcinoma, more pronounced nuclear abnormalities may be seen: cells may appear in large clusters or in a dispersed pattern and it may be impossible on light microscopy alone to suggest a glandular origin. Moderately and poorly differentiated adenocarcinoma is often reported as non-small cell carcinoma (Figure 10.36).

False positive diagnosis can occur when reactive changes or repair in bronchial epithelial cells is interpreted as malignant. A useful clue to the benign nature of reactive bronchial epithelial cell is the presence of cilia; cilia are lost when malignant transformation occurs in respiratory epithelium and detecting cilia is frequently used to support a benign diagnosis.

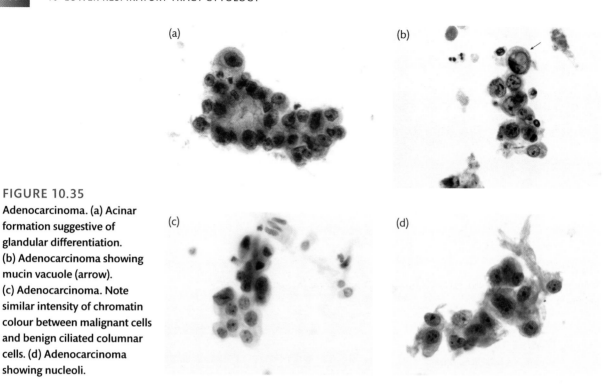

FIGURE 10.35

Adenocarcinoma. (a) Acinar formation suggestive of glandular differentiation. (b) Adenocarcinoma showing mucin vacuole (arrow). (c) Adenocarcinoma. Note similar intensity of chromatin colour between malignant cells and benign ciliated columnar cells. (d) Adenocarcinoma showing nucleoli.

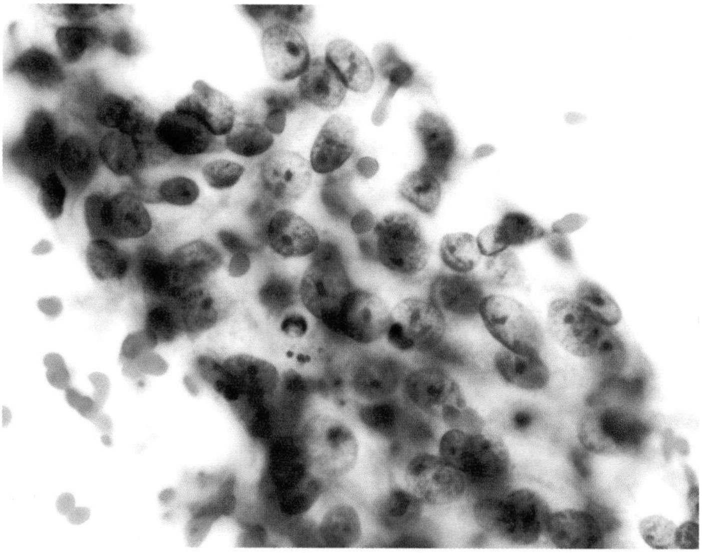

FIGURE 10.36

Adenocarcinoma. This case was reported as 'non-small cell carcinoma'.

Most lung adenocarcinomas react with **thyroid transcription factor 1 (TTF1)** and **CK7** antibodies (Figure 10.37), and are negative for **CK20**.

The other rarer subtype of adenocarcinoma, bronchoalveolar carcinoma, may present as small groups of regular sized glandular cells forming three-dimensional aggregates (Figure 10.38). Nuclear **pleomorphism** (variability in the size and shape) is minimal and at first look they may be indistinguishable from reactive bronchial epithelial cells. Multiple aggregates of varying size are more often seen in malignancy than in reactive bronchial epithelial cells.

(a)
(b)

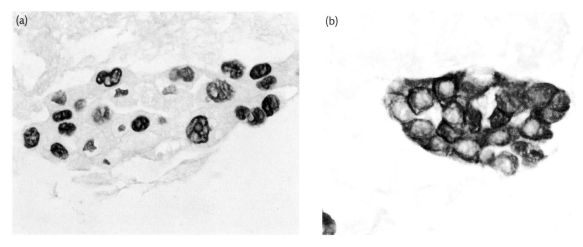

FIGURE 10.37
Adenocarinoma (cell block). (a) Reaction with TTF1 antibody. (b) Reaction with CK7 antibody.

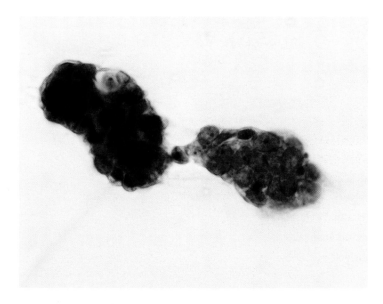

FIGURE 10.38
Bronchoalveolar carcinoma. These three-dimensional clusters of relatively uniform cells were scattered throughout the sample. The radiologic appearances were also suggestive of bronchoalveolar carcinoma.

Critical evaluation of nuclear detail will reveal variation amongst neighbouring cells, including varied chromatin pattern or presence of nucleoli. These changes and the absence of cilia are helpful features in confirming the neoplastic nature of the clusters.

The other exfoliation pattern seen in this tumour is a single cell pattern (Figure 10.39a), which looks remarkably similar to alveolar macrophages at first, but careful evaluation will reveal nuclear enlargement and anisonucleosis. Occasionally malignant cells may be seen inter-mixed amongst macrophages; in this setting comparison of nuclear features, such as presence of nucleoli, nuclear enlargement, and increased nuclear size, should be helpful in diagnosis (Figure 10.39b). Immunocytochemistry is not always helpful in bronchoalveolar carcinoma as some subtypes stain positive with CK20 and negative for TTF1. In these circumstances, clinico-pathological correlation is required to reach the correct diagnosis.

Cross reference
Chapter 9.

(a)

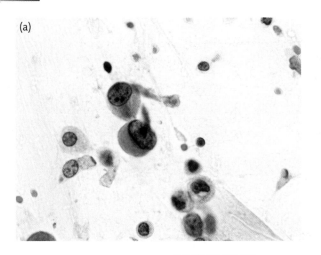

(b)

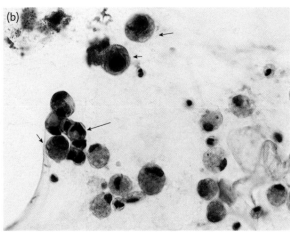

FIGURE 10.39

Bronchoalveolar carcinoma. (a) Single cell exfoliation pattern. (b) Malignant cells (arrows) are seen here intermixed with alveolar macrophages. Some of the macrophages include carbon particles.

BOX 10.6 *Cytomorphology of adenocarcinoma*

Bronchogenic type

- Three-dimensional clusters
- Presence of acinar structures and mucin vacuoles
- Delicate cytoplasm
- Nuclear enlargement
- Anisonucleosis within a cluster
- Fine to coarse chromatin pattern
- Presence of nucleoli

Bronchoalveolar carcinoma type

- Small groups of regular sized glandular cells
- Minimal nuclear abnormality
- Macrophage-like single cell pattern

Large cell anaplastic carcinoma

Large cell anaplastic carcinoma is an undifferentiated cancer, composed of large sized tumour cells that do not show any squamous or glandular features. It is quite likely that poorly differentiated squamous or adenocarcinomas are often reported as large cell anaplastic carcinoma.

In cytological samples, this undifferentiated tumour is often reported as non-small cell carcinoma, large cell type. Cytological recognition of cells from this undifferentiated tumour is very easy as cells show a myriad of malignant features, such as high nucleocytoplasmic ratio, hyperchromasia, anisonucleosis, and prominent nucleoli. Cells generally appear as single cells or as loosely disorganized pleomorphic cells (Figure 10.40).

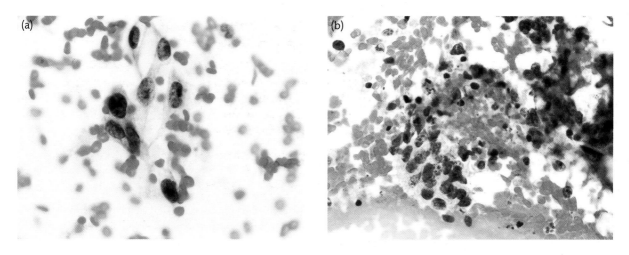

FIGURE 10.40
(a) Undifferentiated malignant cells. Reported as 'non-small cell carcinoma'. (b) Large cell undifferentiated carcinoma. This is from the same case as 10.28d.

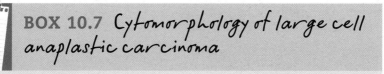

BOX 10.7 Cytomorphology of large cell anaplastic carcinoma

- Loosely cohesive cell clusters or single cells
- High nuclear/cytoplasmic ratio
- Hyperchromasia, anisonucleosis
- Prominent nucleoli

Small cell carcinoma

The cytomorphology of small cell carcinoma varies slightly with the method of specimen collection, as cellular preservation and collecting media affect the cellular presentation. Once these minor variations are considered, cytological presentation of small cell carcinoma is highly specific on light microscopy and when complemented with appropriate immunocyto-chemistry greatly enhances cytological accuracy.

In directly prepared sputum samples, the exfoliated cells are generally small, similar size to lymphocytes, and appear as loosely dissociated cells or loose elongated aggregates within the streaks of mucus (Figure 10.41a). The nuclei may be very hyperchromatic and frequently degenerate and pyknotic. The cells show nuclear pleomorphism within the aggregate, together with nuclear moulding, which is one of the main criteria for recognizing small cell anaplastic carcinoma (Figure 10.41b). Nucleoli are not easily seen.

Bronchial washing samples are collected either in physiological saline or in LBC collection fluid. The cells are better preserved than in sputum samples, but form tight clusters.

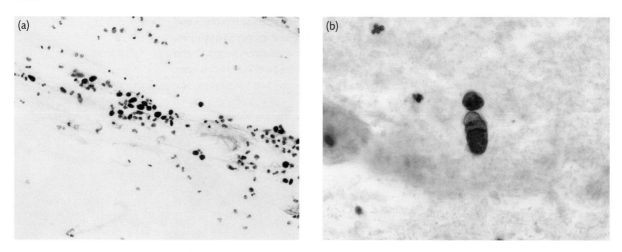

FIGURE 10.41

Small cell anaplastic carcinoma (sputum, direct preparation). (a) Small hyperchromatic dissociated 'oat grain' like nuclei. There is no evidence of nuclear moulding. Elsewhere in the specimen more substantial diagnostic features were present. (b) Same case as (a). These cells have scanty cytoplasm and show nuclear moulding.

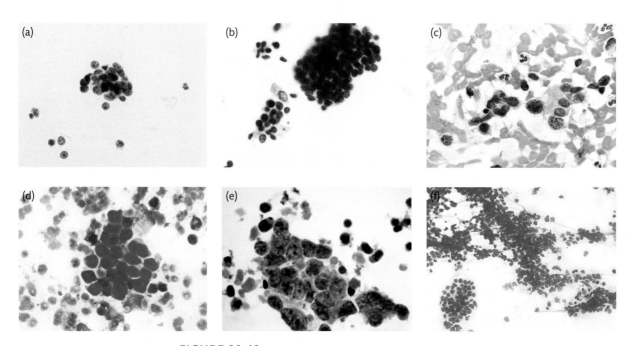

FIGURE 10.42

Small cell anaplastic carcinoma. (a) Tight cluster of malignant cells. The cytoplasm is scanty, and the nuclei have coarse chromatin pattern and nuclear moulding is present (bronchial washing, LBC). (b) Small cell anaplastic carcinoma (bronchial washing, LBC). (c) Bronchial brushing, direct preparation. (d) Bronchial brushing, direct preparation, MGG stain.
(e) Small cell anaplastic carcinoma (FNA). Small nucleoli are evident in some of the cells.
(f) FNA, direct preparation, MGG. This photograph demonstrates 'crush' artefact that occurs during spreading of fragile tumour cells onto the slide.

Examining these clusters at high power will reveal small nuclei with coarse chromatin and evidence of nuclear moulding (Figures 10.42a and b). In bronchial brushings and FNA samples, the better preserved cells appear larger in size, and display variation in nuclear chromasia and chromatin pattern (Figure 10.42c). In MGG preparations which require air-dried preparations, cells appear larger and nuclear moulding is exaggerated, which greatly facilitates cytological typing of this tumour, and this feature makes MGG the ideal stain to use when carrying out on-site assessment of bronchial brushings (Figure 10.42d). Nucleoli that are difficult to see in sputum and bronchial washings are often seen in bronchial brushings (Figure 10.42e). Cells of small cell anaplastic carcinoma are very fragile and break up easily during preparation. This 'crush' artefact produced during direct spreading appears as fine strands of nuclear material, a feature that is not often seen in any other malignancy (Figure 10.42f). Small cell anaplastic carcinoma undergoes necrosis and recognition of necrotic material is an important feature when distinguishing small cell carcinoma from bronchial carcinoid.

Immunocytochemistry is also helpful in differentiating between small cell and non-small cell lung cancer; small cell carcinomas are reactive with **CD56** (Figure 10.43a), **NSE (neuron specific enolase)** (Figure 10.43b), and other neuroendocrine markers, whereas non-small cell carcinomas are generally unreactive.

Neuron specific enolase (NSE)

This is expressed by many tumours, including those of neuroendocrine origin.

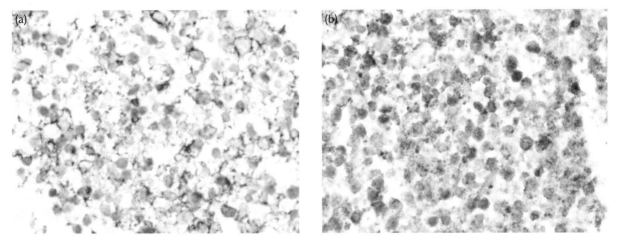

FIGURE 10.43
Small cell anaplastic carcinoma (cell block). (a) Reaction with CD56 antibody. (b) Reaction with NSE antibody.

BOX 10.8 Cytomorphology of small cell anaplastic carcinoma

Directly prepared sputum

- Loosely dissociated cells
- Loose elongated aggregates within the streaks of mucus
- Hyperchromatic, degenerate and pyknotic nuclei
- Nuclear pleomorphism within the aggregate, together with nuclear moulding

Bronchial washings

- Tight hyperchromatic clusters
- Coarse chromatin within clusters
- Evidence of nuclear moulding

Bronchial brushings and FNA

- Obvious nuclear moulding in MGG preparations
- 'Crush' artefact
- Necrotic material

CASE STUDY 10.4

This is a case study of a patient diagnosed with lung cancer and takes you through the clinical journey.

A 61-year-old male attended his GP complaining of increasing cough and shortness of breath for the previous three weeks. He had lost one stone (about 6 kg) in weight in the same period. He also mentioned that he had coughed up considerable amounts of blood on two occasions. He was an ex-smoker and had given up smoking ten years previously, but prior to that had smoked 20 cigarettes a day for 30 years.

The GP arranged for a chest X-ray and referred him to the respiratory physician. Chest X-ray showed a right sided mass and the GP told the patient there was a clinical suspicion of cancer. A bronchoscopy was arranged, which was attended by a cytologist for on-site assessment of sample adequacy. At bronchoscopy, the lesion appeared fragile and prone to bleeding. The consultant physician took a single brush sample to minimize the chance of bleeding. A direct smear was made in the theatre, stained with Diff-Quik™ (Figure 10.44)

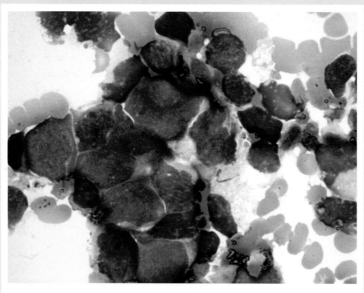

FIGURE 10.44
Small cell anaplastic carcinoma (high-power MGG stain).

and examined under the microscope. The brush was also rinsed in CytoLyt solution. Once back in the laboratory a ThinPrep® slide was made and stained with Papanicolaou stain. The remainder of the sample was made into a cell block for immunocytochemistry.

Direct and ThinPrep® slides showed numerous single and loose clusters of cells with scanty cytoplasm. The cells had hyperchromatic nuclei and nuclear moulding was seen within the clusters. Cytological features were those of small cell anaplastic carcinoma. A cell block was also processed and stained with CD56 and NSE, further confirming the neuroendocrine origin of the cells.

A CT scan was arranged, and showed extensive involvement of mediastinal lymph nodes and presence of liver metastases. The patient was discussed at MDT and was referred to oncology for palliative chemotherapy. He was given four cycles of **cisplatin** and **etoposide** (chemotherapy drugs) every three weeks. A chest X-ray taken after the third cycle showed that the chest lesion had resolved completely, and CT showed partial response in his liver. He was followed up in the oncology clinic and remained relatively symptom free for the next six months. The patient later complained of shortness of breath. A chest X-ray showed presence of pleural effusion, which was drained and this improved his breathing symptoms. The patient was brought into A&E department four weeks later as he had complained of dizziness and headaches. A brain CT was arranged which showed presence of multiple metastases in the brain. He was cared for at home by family members with support from the community palliative care team. He died a month later.

Bronchial carcinoid

The clinical manifestation and radiological appearances of carcinoid tumours may prompt the bronchoscopist to take biopsies as the favoured diagnostic sample as it allows accurate classification of bronchial carcinoids. It is, however, important to recognize the cytological appearance of this tumour, because occasionally it is encountered incidentally and also because the bronchoscopist may choose to use cytological sampling, as these lesions are often vascular and the risk of bleeding may prohibit performing a biopsy. Presence of a cytologist during the procedure also allows for adequacy assessment of biopsies by examining imprints.

Typical carcinoids have uniform sized cells with oval or rounded nuclei and a speckled chromatin pattern (Figure 10.45a).

The cells are generally dispersed and many lose their cytoplasm and appear as bare nuclei (Figure 10.45b), but necrosis and inflammatory changes, which are common features in small cell carcinoma, are not seen. These bare nuclei are also robust and do not show the 'crush' artefact that is seen in small cell carcinoma. Occasional gland-like structures may be seen (Figures 10.45c and d). Bronchial carcinoids react with the typical neuroendocrine markers such as CD56 (Figure 10.45e), but are characterized by having a low proliferation rate; when stained with **Ki67** (a proliferation marker), less than 2% stain positive (Figure 10.45f).

Ki67
This is a useful marker of cell proliferation.

SELF-CHECK 10.8

Why is it important to correctly distinguish a bronchial carcinoid from a small cell anaplastic carcinoma?

Metastatic carcinoma

The lungs are frequently involved in metastatic spread of cancers from elsewhere. Metastatic tumours often have characteristic appearance on CT and chest radiographs

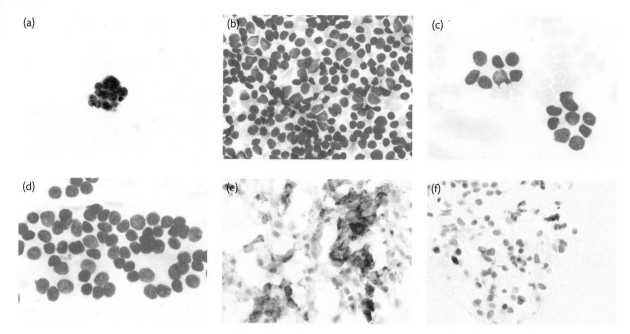

(a) (b) (c)
(d) (e) (f)

FIGURE 10.45

Bronchial carcinoid. **(a)** A uniform population of cells with coarse chromatin. **(b)** Dispersed cells, including many bare nuclei. There is some degree of nuclear moulding, but there is no necrosis or 'crush' artefact as seen in small cell carcinoma. **(c)** Occasional acinar or rosette-like structure may be seen. **(d)** Dispersed regular-sized bare nuclei and acinar structure. **(e)** Reaction with CD56 antibody. **(f)** Reaction with Ki67 antibody.

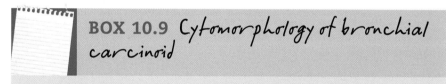

BOX 10.9 *Cytomorphology of bronchial carcinoid*

- Uniform small cells
- Dispersed cells
- Bare nuclei
- Occasional gland-like structures
- No evidence of necrosis or inflammatory change

that may raise the suspicion of a secondary cancer. The cytologist should pay particular attention to the clinical history and review previous cytological and histological samples. Immunocytochemistry is also very helpful in establishing the origin of a tumour as secondary. The majority of lung cancers (and carcinomas of thyroid origin) are reactive with TTF1, which is helpful in differentiating primary lung cancer from secondary malignancy. In some scenarios, it is possible to use the variable expression of cytokeratins to differentiate between tumour origins. For example, most lung cancers react with CK7 and are unreactive with CK20; this is in contrast with colonic cancer, which reacts with CK20 and is unreactive with CK7.

CASE STUDY 10.5

An 82-year-old man with previous history of colon cancer attended his GP as he had developed a chronic cough. The GP arranged for a chest X-ray that showed central shadowing. The patient had a CT scan which showed extensive enlarged sub-carinal lymph nodes. The consultant physician felt that the enlarged lymph nodes could be easily sampled with TBNA technique at bronchoscopy. At bronchoscopy it was noted that the carina had widened and multiple samples were taken from this area. A cytologist attended the procedure and microscopically assessed the quality of the samples during the procedure.

Needle washings were taken for subsequent processing in the laboratory. The cytological preparations showed large malignant cells with features consistent with a non-small cell carcinoma. A cell block was made from the needle washings and sections were stained with a panel of antibodies which included TTF1, CK7, and CK20. The cells showed positive reactivity with TTF1 and CK7 and were unreactive with CK20, thus confirming this lesion to be a new primary lung carcinoma.

The case was discussed at the respiratory MDT and the patient was offered chemotherapy. The cancer was too extensive for surgical resection.

SUMMARY

- The main role of cytology in the lower respiratory tract is in the diagnosis of lung cancer.

- Respiratory cytology is heavily dependent on good sample preparation; the involvement of an appropriately trained cytologist during sample collection at bronchoscopy or FNA procedures ensures sample adequacy.

- Infective agents are occasionally encountered in cytological samples and those with specific features, such a herpes simplex virus, *Pneumocystis jiroveci*, and aspergillus, amongst others, may be reported.

- Lung cancer is divided into two main types: non-small cell and small cell lung cancer. The treatment for these subtypes is very different and it is vital to have an accurate cytological or histological diagnosis. Light microscopy of respiratory samples when combined with appropriate immunocytochemical markers provides the clinician with a highly accurate diagnostic modality that can be used for definitive treatment of the patient.

FURTHER READING

- For definition of statistical methods including database resources, see London Health Observatory website: http://www.lho.org.uk/LHO_Topics/Data/Methodology_and_Sources/AgeStandardisedRates.aspx (accessed 2010).

- Useful information can be accessed via the National Institute for Clinical Excellence (NICE) http://www.nice.org.uk/ (accessed 2010).

- The World Health Organization website include statistics data on lung cancer globally: http://www.who.int/en/ (accessed 2010).

- Bibbo M and Wilbur D (2008) *Comprehensive Cytopathology*, third edition. Saunders, Philadelphia.

- Erozan YS and Ramzy I (2009) *Pulmonary Cytopathology*. Springer, New York.

- Peros-Golubicic T and Sharma OP (2006) *Clinical Atlas of Interstitial Lung Disease*. Springer, London.

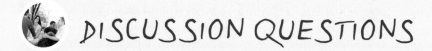

DISCUSSION QUESTIONS

1. Discuss the role of cytology in diagnosis in lower respiratory tract conditions.
2. Discuss the potential problems in interpretation of small cell anaplastic carcinoma.
3. Discuss the role of immunocytochemistry in respiratory cytology.

Answers to the self-check questions, and tips for responding to the discussion questions, are provided in the book's Online Resource Centre.

 Visit www.oxfordtextbooks.co.uk/orc/shambayati/

11

Fine needle aspiration cytology

Behdad Shambayati

In previous chapters you read about exfoliative cytology, the branch of cytology that studies cells which are shed or scraped from epithelial or mesothelial surfaces. Fine needle aspiration cytology (FNA) allows sampling of various solid tissues in the body with the aid of a fine needle.

FNA is a safe diagnostic procedure and significant complications are rare. It is also cost effective, acceptable to patients, and allows for rapid and accurate diagnosis.

Key Points

FNA is a safe, cost effective, diagnostic procedure acceptable to patients and allows for rapid and accurate diagnosis.

FNA cytology for diagnosis of malignancy was developed in the 1970s in Sweden's Karolinska Institute and is now widely used throughout the world for assessment of tumours and tumour-like conditions.

FNA is often used as a first-line approach in assessment of patients with palpable lumps. FNA is applicable to superficial organs such as breast, thyroid, salivary glands, and lymph nodes, and with the aid of ultrasound these superficial organs can be targeted with higher accuracy. More deeply located organs such as lung and liver can also be sampled under radiological guidance, but in recent years the trend has been to obtain a core needle biopsy from these organs for histological diagnosis. Development of **endoscopic ultrasound (EUS)** has allowed physicians to assess gastrointestinal and lung lesions and combined with FNA this has provided a minimally invasive alternative procedure to surgical exploration for assessment of these lesions.

Although there are many morphological similarities between exfoliative and FNA cytology, there are also important differences. The morphological features common to both FNA and exfoliative cytology include assessment of nuclear and cytoplasmic features, which is the cornerstone of cytological diagnosis. In addition to cytological features, interpretation of FNA samples requires knowledge of anatomy, histology, and histopathology.

Endoscopic ultrasound (EUS)

This is a procedure that combines endoscopy with ultrasound. It allows for direct visualization and ultrasound imaging of surrounding tissues and organs.

Key Points

In addition to correct analysis of cytological features, interpretation of FNA samples is dependent on knowledge of relevant anatomy, histology, and histopathology.

FNA is not a replacement for surgical histopathology or core biopsy, and there are certainly times where cytology cannot give a definitive answer and histological assessment is required.

This chapter serves as an introduction to FNA and does not cover the whole of cytology in detail. There are individual textbooks dedicated to the cytology of specific organs. The reader is therefore referred to the reading list at the end of the chapter. The principles and technique of FNA are described and the clinical application of cytology illustrated in various diagnostic scenarios.

Learning objectives

By the end of this chapter you should be able to:

- Describe the principles of fine needle aspiration cytology.
- Describe the technical procedures in obtaining cytological material.
- Give examples of the application of FNA in different diagnostic scenarios.

11.1 Principles and technique of fine needle aspiration

The principles of FNA are the same, whether it is used for obtaining cells from a superficial lesion or from a radiologically guided aspiration: a fine needle is inserted into the solid tissue and with the aid of suction, which is normally provided with a plastic syringe, cells and micro-biopsies are removed. The cells that are aspirated into the hollow barrel of the needle can be discharged onto glass slides and direct cytological preparations made. These may be stained for rapid assessment and diagnosis, or the cells can be placed in transport media for preparation at a later time in the laboratory.

Who should perform the fine needle aspiration procedure?

FNAs should be taken by a person with medical knowledge who has good technical abilities to take and prepare samples. Cytopathologists, radiologists, clinicians, surgeons, and nurses may take FNA specimens. It is important that they have adequate training and sufficient workload to maintain their competence, as success of FNA is dependent on representative cellular samples that are well prepared. Cytopathologists are well placed to aspirate palpable lumps and then to prepare and check the adequacy of the sample. When the sample is collected by a radiologist or a clinician it is important to offer expert support from a biomedical scientist who has experience in non-gynaecological cytology (1).

Equipment

Equipment needed for FNA is inexpensive and already available in most laboratories, or is easily obtainable.

Needles

To reduce bleeding it is important to use as fine a needle as possible. For aspiration of palpable lumps, standard surgical needles 23 gauge (outside diameter 0.6 mm) or 25 gauge (outside diameter 0.5 mm) are suitable. For deeper lesions and EUS guided FNAs, longer specialist needles are available.

Syringes and holders

Standard 10 and 20 ml disposable syringes produce sufficient negative pressure. Many aspirators prefer to use 10 ml syringes as the small size makes them easier to handle.

A syringe holder may be used when aspirating palpable lumps as they allow for one-handed operation; the free hand may be used to immobilize the lump (Figure 11.1).

Glass slides

Frosted end slides are suitable for immediate labelling with a pencil and can also be used as an implement for making direct preparation.

Fixative and transport media and routine staining methods

Direct preparations can be spray fixed for Papanicolaou staining or immersed in 95% alcohol, but if on-site assessment is available, it is recommended to make a maximum of two air-dried direct smears and stain these with Romanowsky stain such as Diff-Quik™ and check for adequacy. The content of the needle and subsequent samples should be rinsed in transport media for processing in the laboratory. There are different transport media available. Some have been specifically designed for use in cytology: these include commercial LBC preservatives such as

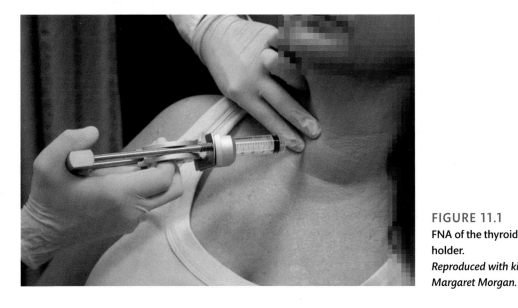

FIGURE 11.1
FNA of the thyroid gland using a syringe holder.
Reproduced with kind permission from Margaret Morgan.

CytoLyt™ or CytoRich™. Others have been adopted for use from the field of tissue culture, including culture medium such as RPMI 1640.

The success of FNA depends on accuracy and specificity, which can be greatly enhanced by high quality preparations and by the application of ancillary techniques. It is therefore important to collect sufficient material for ancillary techniques.

The fine needle aspiration procedure

There is no replacement for learning the FNA technique from an experienced practitioner; the following is intended only to describe the technique for biomedical scientists who attend the procedures.

Palpable lesions

As with all surgical procedures, infection control measures such as washing hands and wearing gloves must be observed. The skin must be cleaned with alcohol swabs. The FNA procedure is illustrated in Figure 11.2. The lump is palpated and immobilized, and the needle inserted into the lump (a). Negative pressure is induced by withdrawing the syringe plunger (b). Whilst the negative pressure is maintained, the needle is moved to and fro within the tissue to dislodge cells (c). When material is seen in the hub of the needle, the plunger is released to its original position (d), and the needle/syringe assembly is withdrawn from the lump (e). The needle is removed from the syringe and the plunger withdrawn to draw air into the syringe (f). The needle is re-attached to the syringe and the small amount of material is gently expelled onto a glass slide (g).

An alternative method has been described which uses a fine needle without an attached syringe. In this method, a needle is held between the index finger and the thumb, and passed through the skin. The needle is rotated and also moved to and fro whilst in the lesion to dislodge cells. To expel the cells a syringe is attached and material is gently expelled onto a glass slide. The cell yield in this method is lower than in the aspiration technique, but this method causes less bleeding and is suitable for aspirating vascular organs such as the thyroid gland (Figure 11.3).

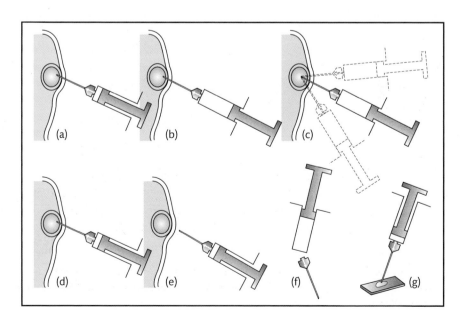

FIGURE 11.2
Diagrammatic illustration of FNA procedure with aspiration.
Reproduced with kind permission from Victoria Burden.

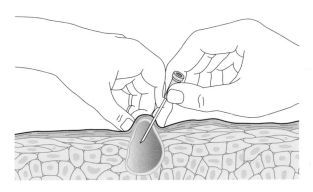

FIGURE 11.3

Diagrammatic illustration of FNA procedure without aspiration.
Reproduced with kind permission from Victoria Burden.

Image guided fine needle aspirations

Image guided aspiration is often carried out by radiologists or clinicians (see below) and close cooperation with the cytology department ensures best results. Offering rapid on-site assessment by appropriately trained biomedical scientists reduces the number of inadequate samples and ensures sufficient material is collected for ancillary techniques.

Ultrasound guided fine needle aspiration

Radiologists often have their own preferences for carrying out the aspiration. To improve manoeuvrability and to aid sampling some operators prefer to use the system illustrated in Figure 11.4a. This arrangement uses a short extension tube between the needle and syringe and allows the radiologist freedom to handle the ultrasound probe, but requires the presence of a second operator to control the syringe and aspiration (Figure 11.4b). It is important to

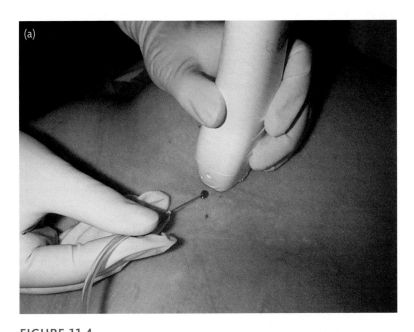

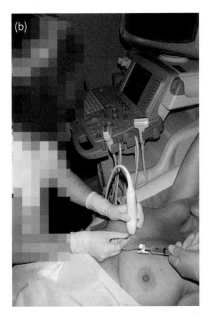

FIGURE 11.4
(a) FNA procedure with an extension tube. This arrangement uses a short extension tube between the needle and syringe and allows the radiologist freedom to handle the ultrasound probe. (b) Ultrasound guided FNA of breast.

avoid contaminating the sample with ultrasound gel, as ultrasound gel can take up cytological stains and obscure cellular detail.

Computed tomography fine needle aspiration

Computed tomography (CT) allows localization and sampling of very small lesions in the lung or abdomen and again cooperation between radiology and cytology departments ensures optimal sample for cytological analysis.

Endoscopic ultrasound-guided fine needle aspiration (EUS-FNA) and endobronchial ultrasound-guided transbronchial needle aspiration (EBUS-TBNA)

Cross reference
Chapter 10, Section 10.2.

EUS-FNA allows physicians to collect diagnostic material from **gastrointestinal** (oesophagus, stomach, and duodenum) and **pancreatobiliary** (pancreas, bile ducts, and gall bladder) systems. This technique has allowed clinicians access to organs that were only previously possible to sample during surgery. EBUS-FNA is used to obtain specimens from mediastinal and other masses and is useful in the diagnosis and staging of lung cancer.

After aspiration, the material can be expelled onto a glass slide using a syringe full of air, as in the technique for palpable masses. Another method which is recommended by some manufacturers of EUS needles is to re-introduce the **stylet** (fine wire) into the needle. This produces a high pressure that helps expel the material drop by drop on a glass slide. This method has the advantage over the use of a syringe in that it allows more control in the expulsion of material from the needle.

EUS-FNA and EBUS-TNBA greatly improve patient management by facilitating diagnosis without the need for diagnostic surgery. These techniques reduce overall healthcare costs by removing the need for hospitalization for surgery.

Sample preparation

Directly prepared smears must be optimally produced. They must be neither too thick nor too thin, and free from crush artefacts. The preparation of direct smears must be learned from an experienced biomedical scientist.

There are various methods that can be used to prepare direct smears. A method commonly used is similar to that used for making blood films. This method is illustrated in Figure 11.5 and is suitable for most aspirates. The spreading slide is placed at an angle of approximately 45° on the specimen slide and brought into contact with the aspirated material. The spreader slide is gently moved along and material is spread via surface tension. If there is any solid matter remaining, this can be gently spread at the end.

Another method is to use a plastic microbiological loop. In this method the loop is used to spread the material in a semicircular fashion along the slides (Figure 11.6). Slides for Diff-Quik stain must be thin enough to air-dry rapidly, and slides for Papanicolaou stain must be quickly fixed by either immersing in 95% alcohol or spray fixing.

The remainder of the material or subsequent aspirates should be discharged into transport media for processing in the laboratory. Once the sample is in the laboratory, consideration should be given to optimal sample preparation for ancillary techniques. Immunocytochemistry on cell blocks (prepared from the remaining FNA material) has greatly enhanced conventional cytology. FNA can also be used to obtain material for flow cytometry which is used in the diagnosis of lymphomas (2). Molecular and/or genetic techniques are increasingly used in

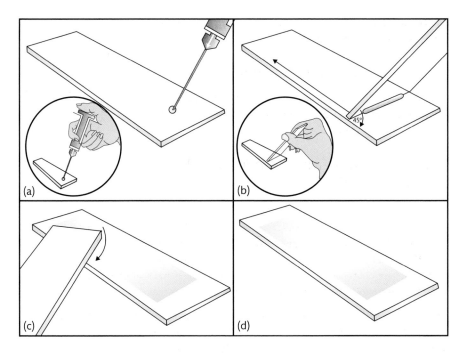

FIGURE 11.5
Direct smearing using a slide.
Reproduced with kind permission from Victoria Burden.

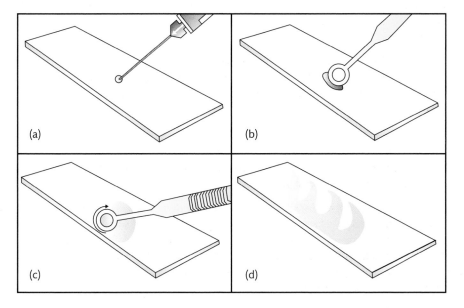

FIGURE 11.6
Direct smearing using a microbiological loop.
Reproduced with kind permission from Victoria Burden.

analysis of material obtained using EUS. One such example is analysis of mutant K-Ras gene in diagnosis of pancreatic cancer (3).

Cross reference
Chapter 13.

11.2 **Reporting terminology**

The FNA report should convey to the clinician in clear and simple language the cytological appearances and, in some cases, the clinical significance of the findings. The aim should be to give a firm diagnostic opinion that a specimen is benign or malignant, although there will be instances when this is not possible. The sample may be hypocellular, contain excessive

BOX 11.1 *Reporting categories in breast cytology*

For full explanation of reporting categories see NHSBSP publication No. 50 (4).

C1 (inadequate)

C2 (benign) this category indicates an adequate sample showing no evidence of significant atypia or malignancy.

C3 (atypia probably benign) the aspirate has all the characteristics of a benign lesion but includes features not commonly seen in benign aspirates, including nuclear pleomorphism, loss of cellular cohesiveness, increased nuclear and cytoplasmic changes, and increased cellularity.

C4 (suspicious of malignancy) the aspirate contains cells with highly atypical features, but a confident diagnosis of malignancy cannot be made as the material may be too scanty, or may have malignant cells intermixed with a majority of benign cells.

C5 (malignant) this category indicates an adequate sample containing cells characteristic of carcinoma or other malignancy.

blood, or the cells fail to exhibit enough diagnostic features. In these circumstances the cellular features should be described and an equivocal, albeit clinically unsatisfactory, report issued. Assessment of specimen adequacy is an important part of FNA interpretation as it expresses the degree of certainty with which a clinician can rely on the cytological report. Cytological classification by numerical reporting schemes have been established for breast (C1–C5) and thyroid samples (Thy 1–Thy 5) (see Boxes 11.1 and 11.2) (4, 5). Similar reporting schemes could be applied for general reporting of all FNAs:

- Unsatisfactory or non-diagnostic
- Negative for malignancy (infections noted)
- Atypia, probably benign (should be used sparingly)
- Suspicious for malignancy (should be used sparingly)
- Malignant cells present—cell type suggested

Key Points

Assessment of specimen adequacy is an important part of FNA interpretation as it expresses the degree of certainty with which a clinician can rely on the cytological report.

11.3 **Assessment of sample adequacy**

The accuracy and cost effectiveness of FNA is heavily dependent on adequately cellular material which is representative of the tissue being aspirated. It has been alluded to throughout this chapter and this book that involvement of cytology departments at the point of sample collection can ensure optimal sample preparation. This can be further enhanced if a cytopathologist or an adequately trained biomedical scientist can offer rapid on-site assessment of specimen adequacy. The attending cytologist prepares direct smears which are air-dried and stained with Diff-Quik™ for immediate microscopic assessment.

The best way to provide this service is to set up a lockable trolley that contains all consumables needed. This portable mini-lab can be wheeled to wards, clinics, and radiology departments where FNA samples are taken. It is important to include a good quality microscope in the trolley, and ideally a microscope with teaching attachment to facilitate training of new staff.

Key Points

In order to provide accurate diagnostic information for optimal clinical management, an FNA sample must be cellular, representative of the underlying lesion under investigation, and correctly smeared and processed. Collection of further material in transport media would allow for application of ancillary techniques. To ensure for optimal collection of sample an adequately trained biomedical scientist should attend FNA procedures and can offer rapid on-site assessment of sample quality.

11.4 Case studies

In this section the application of FNA in different clinical settings will be described, with a thumbnail sketch of the condition. This section is not intended to be descriptive and detailed, and serves only as an introduction to the field of FNA.

Breast

FNA is an effective tool in diagnosing and evaluating breast lumps or masses. Until recently it was the main non-operative diagnostic modality and was part of the **triple approach** in breast cancer diagnosis. Triple approach combines the result of clinical examination and imaging with FNA cytology and/or core biopsy and offers the highest level of diagnostic accuracy. The diagnosis of breast lumps is very accurate using the triple approach when all three modalities are concordant. The recent trend has been to use core biopsy as part of a triple approach as it can differentiate between invasive cancer and *in situ* carcinoma. This can affect the extent of surgery and the ultimate prognosis. Core biopsy is also used for assessing the **oestrogen and progesterone receptor status (ER and PR)** of breast cancer, and also calculating the grade of the tumour. Receptor status and grade of the tumour are some of the prognostic indicators in breast cancer.

Nevertheless, FNA of the breast is a complementary test and still used in specific situations where it is preferable to core biopsy. These include: investigation of axillary lymph nodes and local recurrence of breast cancer, evaluation of cystic areas with atypical features, confirmation of clinically benign lesions, or in diagnosis of breast cancer when carrying out a core biopsy is contraindicated.

FNA is still widely used in confirming the nature of clinically benign masses when surgery is not indicated, particularly in younger women.

SELF-CHECK 11.1

What is the role of FNA in breast disease?

Case Study 11.1 demonstrates use of FNA in breast cancer diagnosis.

CASE STUDY 11.1

A 47-year-old woman attended her GP concerned about a hard lump in her left breast. She mentioned that she had noted the lump a month ago, but as it was not getting bigger she had decided to wait and had postponed attending the GP. The GP examined the patient and confirmed the presence of a lump. She was referred urgently to the local hospital breast clinic.

She was seen one week later in a **one-stop breast clinic**. On examination the lump appeared to be 3 cm in diameter. It was hard in consistency and suspicious of malignancy. She was sent to the radiology department for imaging and core biopsy. The radiologist described the mammogram as a **stellate** (star-like, radiating from centre) white area suggestive of malignancy. The radiologist carried out ultrasound examination which showed a mass measuring roughly $3.5 \times 2 \times 2$ cm, but also noted enlarged lymph nodes in the axilla. The radiologist explained the findings to the patient and told her that she would need to take a core biopsy from the breast lesion and an FNA from the axilla. The patient mentioned to the radiologist that she was on **heparin anticoagulant** therapy. The radiologist told the patient that as she was

taking heparin, the risk of bleeding was increased and she would rather take a fine needle aspirate from the lesion. Fine needle aspirates were taken from both the breast lump and axilla. The samples were prepared directly and stained with MGG. Both samples included numerous highly pleomorphic malignant cells. A preliminary diagnosis of malignancy was given to the surgeon by the cytologist (Figure 11.7).

The patient was anxious to know the results and the surgeon informed her that the preliminary results were suggestive of cancer, but that her case would be discussed at the breast MDT before deciding on her treatment plan. The MDT decision was to offer wide local excision (WLE) and as her axillary lymph node showed evidence of metastatic disease, carry out **axillary clearance**. The microscopical examination of the surgical specimen showed a grade 3 invasive **ductal carcinoma**, which measured 3.3 cm in maximum dimensions, **excision margins** were clear. Four out of nine lymph nodes examined showed involvement by tumour. **Oestrogen and progesterone** receptor (**ER and PR**) status was negative. **HER2** status was positive. The case was discussed again at the breast MDT and the patient was

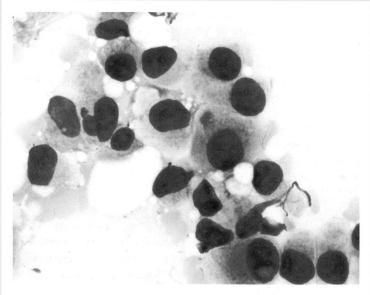

FIGURE 11.7
FNA breast. Poorly cohesive malignant cells with obvious nuclear enlargement and pleomorphism.

offered chemotherapy, followed by a course of radiotherapy to the breast and axilla.

Patients with suspected breast cancer are referred urgently to a specialist breast clinic. Some centres which are adequately resourced offer a one-stop clinic. In a one-stop clinic, the patient can expect a consultation with a specialist, any necessary imaging test, and, if needed, a core biopsy or FNA all in the same day. The radiologist chose to take an FNA sample, as the patient was on anticoagulant therapy and a core biopsy may have caused considerable bleeding. The axillary lymph nodes appeared radiologically involved with tumour indicating that the original tumour was invasive rather than *in situ*. This is one of the main reasons that core biopsy is preferred over FNA.

With the diagnosis established using the 'triple' approach the surgeon could plan the operation. The patient was offered wide local excision. This procedure removes the tumour and a small area of surrounding normal tissue. The resection sample provides important diagnostic and prognostic information:

■ Type of tumour: the resection confirmed that the tumour was invasive and of ductal type; ductal carcinoma is the most common subtype of breast cancer.

■ Grade of the tumour: the patient had a grade 3 tumour. Grading refers to degree of differentiation. Grade 3 is a poorly differentiated tumour.

■ The excision margin: this tumour was completely excised and this reduces the likelihood of local recurrences.

■ Lymph node status: this is an important prognostic factor. There is a relationship between the number of lymph nodes involved and long-term survival.

■ ER and PR status: this predicts response to endocrine therapy, in this case the tumour did not express the receptors and would not respond to the endocrine therapy.

■ HER2 status: HER2 (also called Her2/neu, c-erb-2 or erb-2) refers to **human epidermal growth factor 2**. This a protein found on the surface of some breast cancer cells. This protein is made by a gene called HER2/neu gene. It is thought that 20% of breast cancers express this gene. Breast cancers which express this gene are generally more aggressive, and do not respond to endocrine therapy. Cancers which express this protein respond to drugs such as trastuzumab (Herceptin®). This patient's tumour was HER2 positive and she would be considered for treatment with Herceptin®.

Thyroid

The main indication for FNA of thyroid is assessment of **thyroid nodules**. Thyroid nodules are abnormal growths that appear as a lump. Thyroid nodules are very common, and it is thought up to 10% of adult females and 2% of adult males have palpable lumps. The majority of thyroid nodules are benign and only a small percentage (up to 5%) may be cancerous. Clinical examination, imaging, and measurement of thyroid activity via blood tests provide some helpful clues in assessing the nature of a nodule, but they are not sufficiently accurate to rule out malignancy. FNA has been recognized as the most valuable non-invasive method of assessing patients with thyroid nodules. Nodules that are equivocal or malignant on cytology require surgical excision for histological diagnosis.

Good FNA cytology has reduced the number of diagnostic thyroid operations, thus reducing the risk to the patient and saving valuable resources. Surgical excision of all thyroid nodules is not feasible and happily cytology has now rendered it unnecessary in most cases.

The aspiration technique is similar to other superficial lesions, but requires minor modification as the thyroid gland is very vascular and bleeding can occur. Use of finer needles (25 gauge) and short needling time should minimize bleeding. The non-aspiration technique is a particularly useful technique in thyroid FNA. The number of passes performed has been shown to influence sample adequacy. One study shows 3–5 aspirations provide adequate material for analysis of the lesion (6).

FNA of the thyroid may be performed by a cytopathologist, endocrinologist, surgeon, nuclear medicine physician, oncologist, or radiologist with expertise and interest in thyroid disease.

Cross reference
Chapter 13, Section 13.3.

However, they should be trained and carry out an adequate number of aspirates to maintain their expertise (5).

As previously stated, the usefulness and accuracy of cytological diagnosis depends on obtaining adequate material for diagnosis, and biomedical scientists can play an important role in attending FNA procedures and offering immediate on-site assessment. This will also allow for collection of appropriate material for ancillary testing. In thyroid cytology air-dried MGG stains are generally preferred, but a smaller number of slides should be alcohol fixed and stained with Papanicolaou. As with all FNA samples extra material should be collected in CytoLyt™ or RMPI 1640 for ancillary testing.

A similar reporting system to that used in breast cytology has been developed for reporting thyroid cytology samples. This combines a textual report with a numerical reporting system (see the box below). This combination reporting should help the clinician to make the best decision in the management of the patient.

Key Points

FNA has been recognized as the most valuable non-invasive method of assessing patients with thyroid nodules.

FNA can in most cases classify the nodules into three main groups:

- Colloid nodules
- Thyroiditis
- Neoplastic

Colloid nodules are benign and only rarely require surgical excision. When very large, benign goitres may bother the patient by interfering with breathing or swallowing.

Thyroiditis refers to inflammation of the thyroid gland. There are many types of thyroiditis with different causes and clinical and cytological features. They generally respond to medical treatment.

BOX 11.2 Reporting categories in thyroid FNAs

For full description see British Thyroid Association and Royal College of Physicians Guidelines (5).

Thy1 Non-diagnostic (lack of cells or technical artefact).

Thy2 Non-neoplastic (includes at least six or more groups of cells).

Thy3 Follicular lesion/suspected follicular carcinoma.

Thy4 Suspicious of malignancy (suspicious, but not diagnostic of papillary, medullary, or anaplastic carcinoma or lymphoma).

Thy5 Diagnostic of malignancy (unequivocal features of papillary, medullary, or anaplastic carcinoma, lymphoma or metastatic tumour).

There are different types of neoplastic changes that occur in the thyroid; the type determines treatment and prognosis. These include **follicular adenoma**, **follicular carcinoma**, **papillary carcinoma**, **medullary carcinoma**, **anaplastic carcinoma**, and **thyroid lymphoma**. Please consult the further reading list at the end of the chapter for a comprehensive description on cytopathology of the thyroid.

Thyroid function tests (TFTs)

These are biochemical blood tests which are carried out to measure thyroid activity. Initially the level of TSH is measured and if this elevated or decreased, levels of T4 (thyroxine) and T3 are re-measured.

Intranuclear grooves

This is a term used to describe folds or apparent condensation of chromatin within the nuclei.

SELF-CHECK 11.2

What is the place of FNA in the investigative sequence of thyroid nodules?

Case Study 11.2 demonstrates an example where FNA played an important role in the diagnosis of thyroid cancer.

CASE STUDY 11.2

A 39-year-old woman attended her GP surgery and mentioned that she had noticed swelling in her neck. The GP examined her neck, focusing on the region of the thyroid, and noted that her thyroid was enlarged. The GP suggested **thyroid function tests (TFTs)**. These were carried out and were normal. The GP referred the patient to a surgeon at the local hospital who had a special interest in thyroid cancer.

The patient was seen in a thyroid one-stop clinic where she was examined by the surgeon, who confirmed presence of the lump in the thyroid, but also noted enlarged lymph nodes on the left side. The patient was sent to the radiology department for ultrasound imaging and possible FNA.

The radiologist noted that the left lobe of the thyroid included an area which measured 2 cm in diameter and was partially cystic in nature. He also noted enlarged lymph

nodes on the left side of the neck. He decided to perform an FNA of the lesion and the largest lymph node. The FNA samples were assessed by a biomedical scientist who confirmed the adequacy of the samples. Needle washings were also collected.

The samples were examined in the laboratory later that day. The samples from the thyroid and the lymph node looked morphologically very similar: at low power the preparations were very cellular. They consisted of numerous sheets of enlarged epithelial cells, which had a dense cytoplasm and appeared **papillaroid** (papillary like) (Figure 11.8a). Amongst the sheets occasional **psammoma bodies** were seen (Figure 11.8b). **Intranuclear grooves** could be seen in some nuclei (Figure 11.8c). These features are classic findings for papillary carcinoma of thyroid. A written provisional report was issued and given to the surgeon.

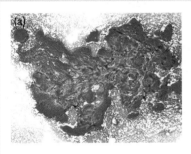 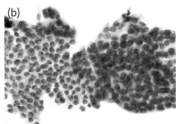 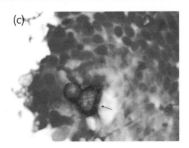

FIGURE 11.8

FNA thyroid. Papillary carcinoma. (a) Low power. Three-dimensional sheets of cells illustrating the papillaroid architecture. (b) Sheet of cells with regular nuclei, many showing longitudinal nuclear grooves (arrows). (c) Psammoma bodies (arrow).

The patient was seen by the surgeon who told her that the FNA samples were satisfactory for diagnosis and her case would be discussed at the specialist head and neck MDT. She was also sent to have a CT scan.

The patient was discussed at the MDT and the decision was made to offer her **total thyroidectomy** and **selective neck dissection** to sample the enlarged lymph nodes.

Histology of the resected sample showed a well-differentiated papillary carcinoma which had not invaded the blood vessels. One of the three sampled lymph nodes showed evidence of papillary carcinoma.

Six weeks after the operation the patient was given radioactive iodine 131 (I-131).

She was given **levothyroxine** and will initially be followed up on a regular basis for the first two years and then annually for life.

Papillary carcinoma is the most common thyroid cancer. This cancer is highly curable by surgery. Papillary carcinomas mostly occur in females (3:1 ratio) with a peak onset between the ages 30 to 50. Metastatic spread to neck nodes occurs in 50% of the cases. The cause of this cancer is unknown, but high dose external radiation to the neck increases the risk of developing papillary carcinoma.

The main treatment is surgery. There are various surgical options available for treating patients with papillary carcinoma, depending on the size and location of the tumour.

This patient was offered total thyroidectomy and lymph node dissection as FNA had confirmed the presence of a tumour in the lymph nodes.

Following the surgical procedure the patient was given I-131. Thyroid tissue absorbs iodine and administration of a high dose of radioactive iodine can be used to destroy any thyroid tissue remaining after the surgery or to treat thyroid cancer that has spread to lymph nodes.

The patient was given **levothyroxine** tablets. Levothyroxine is a synthetic form of **thyroxine**. This serves two purposes: it replaces the natural thyroid hormone that the body can no longer make and suppresses growth of any possible cancer cells that have not been removed or destroyed by surgery or I-131 treatment.

Patients who have been treated for papillary thyroid carcinoma are followed up for life for various reasons; the disease has a long natural history and late recurrences can occur which can be successfully treated, and patients who take levothyroxine need monitoring as there are side effects associated with levothyroxine.

Psammoma bodies

These are microscopic extracellular circular concentric calcium bodies seen in malignant and some benign conditions. Most commonly seen in papillary cancers including thyroid and ovary.

Total thyroidectomy

This is an operation which removes the whole of the thyroid gland.

Selective neck dissection

This is surgical excision commonly employed in treatment of head and neck cancers, which removes lymph nodes on one side of the neck.

Lymph nodes

Enlarged lymph nodes are a common presenting symptom for a variety of benign and malignant conditions. Although surgical excision of an enlarged superficial lymph node is fairly straightforward, it nevertheless requires anaesthesia, theatre time, and other resources. Surgical excision is not necessary in all cases. FNA offers an alternative first-line investigation and in many centres has become an integral part in the initial assessment of a patient presenting with **lymphadenopathy** (swollen/enlarged lymph nodes).

Technical considerations are similar to FNA elsewhere, but it is particularly vital to collect material for ancillary testing. Tuberculosis commonly presents with lymphadenopathy: if there is a strong clinical suspicion of tuberculosis then to minimize the risk of infection by inhalation, direct preparations should not be made; instead the content of the needle should be discharged directly into transport media for subsequent processing in the laboratory.

FNA may not provide a specific diagnosis in every case and a negative result does not exclude malignancy, but it is useful in the following scenarios:

- Assessment of reactive hyperplasia.
- Diagnosis of metastatic malignancy and indication of possible primary site.
- Initial diagnosis of lymphoid malignancy.

Reactive hyperplasia is a benign and reversible enlargement of the lymph node which occurs secondary to a variety of stimuli, including infections. Surgical excision is not indicated unless it continues to increase in size or clinical suspicion for malignancy is high.

Diagnosis of metastatic malignancy is very helpful to the clinician. In patients with known malignancy who later present with enlarged lymph nodes, cytological diagnosis often avoids surgical excision of the lymph node. In patients without a previous diagnosis of malignancy, cytological results supported by immunocytochemistry may reveal the nature and primary site of the tumour. This helps the clinical team, avoids unnecessary investigations, and aids the delivery of appropriate treatment.

FNA has a limited role in diagnosis of lymphomas as cytological diagnosis of lymphoma must be followed by surgical excision of the lymph node. Even so cytology of lymphoma is helpful by excluding other diagnoses such as cancer. FNA is less helpful in classifying lymphoma. Treatment of lymphomas requires accurate classification, which currently can only be achieved by histological and immunological assessment of the node, nevertheless diagnosis of lymphoma by FNA allows for immediate planning for surgical biopsy.

SELF-CHECK 11.3

What is the main role of FNA in lymph node diagnosis?

Case Study 11.3 illustrates the involvement of FNA in diagnosis of a lymphoid malignancy.

CASE STUDY 11.3

A 22-year-old male was referred by his GP to the surgical outpatient department. He was examined by a head and neck surgeon who confirmed enlarged lymph nodes in the supraclavicular region. The patient told the surgeon that during the past few months he had lost 6 kg in weight and noted fever and heavy sweating (night sweats) whilst sleep.

The surgeon performed a fine needle aspiration. Sample adequacy was confirmed on-site by a biomedical scientist. A further FNA was taken for ancillary testing. Direct smears stained by Papanicolaou and MGG showed cellular samples that included numerous lymphoid cells. Amongst the lymphoid cells there were atypical lymphoid cells and occasional large binucleate cells (Figure 11.9a). The background also included variable numbers of eosinophils and plasma cells (Figure 11.9b). These features were suggestive of **Hodgkin's lymphoma**. A cell block was made for immunocytochemistry. Large atypical cells showed positive staining with **CD15** and **CD30** antibodies (Figures 11.9c and d). The immunoreactivity of the atypical cells confirmed the cytological diagnosis of Hodgkin's disease.

The lymph node was excised under anaesthesia and histology demonstrated presence of numerous **Reed-Sternberg cells** (RS cells), confirming the cytological diagnosis of Hodgkin's disease. The node was further described as mixed cellularity subtype.

The patient underwent staging investigation including CT and PET scanning. No further evidence of disease was found in the chest or abdomen. The disease was classified as early stage, and he received a short course of chemotherapy followed by radiotherapy.

Hodgkin's lymphoma, previously known as Hodgkin's disease, is malignancy of the immune system which starts in a lymph node, spreads to nearby lymph nodes, and eventually to other tissues in the body. It is characterized by the presence of Reed-Sternberg and **Hodgkin cells**. Reed-Sternberg cells are large binucleated or sometimes multinucleated cells. Hodgkin cells are large single nucleated cells. These cells are often outnumbered by other 'background' cell types, including lymphocytes, plasma cells, and eosinophils.

Hodgkin's lymphoma is histologically divided into four subtypes. This is based on the morphology and composition of background cells.

The patient's symptoms of fever, night sweats, and weight loss are caused by production **cytokines**. These protein chemical messengers produced by the tumour tissue result in various systemic effects on the body.

Treatment for Hodgkin's lymphoma depends on the stage of the tumour, age of the patient, general health status, and presence of symptoms. Early disease is often treated by radiotherapy alone, but combination of chemotherapy and radiotherapy is sometimes used. More widespread disease is treated with chemotherapy.

Hodgkin's lymphoma has a very high cure rate; in most cases it can be cured.

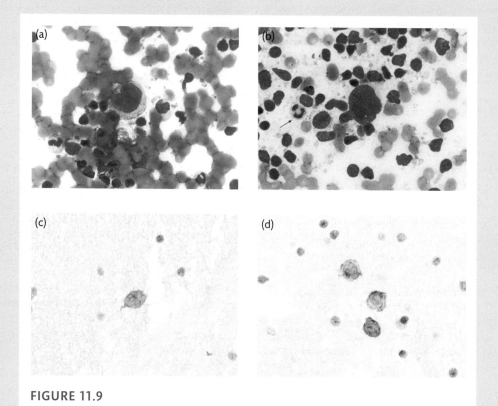

FIGURE 11.9

FNA lymph node. Hodgkin's lymphoma. (a) Large binucleated Reed-Sternberg cell.
(b) Large mononuclear Hodgkin cell and eosinophil (arrow). (c) CD15 positive cell.
(d) CD30 positive cells.

Salivary glands

Salivary glands consist of major and minor salivary glands. The three major salivary glands are the parotid, submandibular, and sublingual. The minor salivary glands occur as numerous microscopic structures spread throughout the oral cavity.

Enlarged salivary glands are easily accessible to FNA and this provides the surgeon with a preoperative diagnosis which the patient's history, clinical examination, and imaging cannot always provide. FNA of salivary glands has been shown to be clinically beneficial and cost effective (7). FNA is preferred to core biopsy, because it has a lower risk of **needle track seeding** of tumours. Core biopsy may also cause formation of a **fistula**.

Nevertheless, some believe that FNA of salivary glands is not efficacious, as most salivary gland lesions ultimately require surgical excision (8). This belief is only partially correct as some patients with non-neoplastic lesions such as **chronic sialadenitis** do not require excision. Also, preoperative diagnosis of benign tumours allows the surgeon to decide, if a patient is not fit for surgery, to postpone or even to avoid surgery altogether. A malignant FNA diagnosis will allow the surgeon to counsel the patient and help in forming a strategy for the operation and other therapy.

Precise classification of salivary gland tumours may not be possible by cytology. There are more than 30 different histological types that exist, some of which can only be accurately typed by examining the architectural details seen on histology. Some salivary gland malignancies are low grade and do not show obvious cytological abnormality.

Cytology, however, is very useful in diagnosing the two most common neoplasms, **pleomorphic adenoma** (see Case Study 11.4) and **Warthin's tumour**. These benign tumours which compromise up to 80% of all salivary gland neoplasms have characteristic cytological features.

Cytology may not be able to subtype malignant salivary gland neoplasms accurately in all cases, but high-grade neoplasms show nuclear abnormality that can be recognized in cytological preparations. Reporting such cases as malignant is usually sufficient for clinical management of the patient.

FNA technique is similar to that of other palpable lesions. Aspirations performed with a gauge 23 to 25 needle should provide sufficient material for direct preparations, for allocation for LBC preparations, and for making cell blocks.

Needle track seeding
This is the growth of tumour cells along the site of biopsy that can occasionally occur after core biopsy and rarely after FNA.

Fistula
An abnormal opening that joins two organs or epithelial surfaces together.

Chronic sialadenitis
This is inflammation of the salivary gland.

Pleomorphic adenoma
This the commonest salivary gland tumour. This benign tumour consists of a variable number of epithelial and myoepithelial cells that sit in the background of myxoid material (mucoid).

Warthin's tumour
This is the second most common benign salivary gland tumour. It consists of transformed epithelial cells (oncocyte-containing numerous mitochondria) within a lymphoid tissue stroma.

SELF-CHECK 11.4

What are the indications for FNA of salivary glands?

Case Study 11.4 is a patient's journey who presented with a salivary gland swelling.

CASE STUDY 11.4

A 29-year-old female was referred to an Ear, Nose, and Throat (ENT) clinic by her GP. The patient had noticed a lump on the side of her right cheek behind the ear just above the jaw bone. The lump had been present for nine months and slowly increasing in size.

The ENT surgeon carried out clinical examination which confirmed a swelling at the angle of her jaw on the right side which could be the parotid gland or alternatively a lymph node. He also carried out a fibre-optic examination of her nasal passageways, the post-nasal space, pharynx, and larynx, which were normal. He referred the patient for ultrasound guided FNA.

The radiologist performed the FNA with a 23 gauge needle. Air-dried and alcohol fixed preparations were made. Air-dried preparations were stained with Diff-Quik™ and showed numerous single and clusters of bland, round to oval cells (Figure 11.10a). The background included abundant **fibrillary stroma** (Figure 11.10b). This was reported as features consistent with pleomorphic adenoma.

Under general anaesthesia the surgeon performed a superficial **parotidectomy**. Post-operatively, the patient developed numbness in her right cheek which improved after three months.

Histological assessment of the surgical sample confirmed that the tumour was a pleomorphic adenoma and was fully excised. The patient was seen post-operatively in clinic and discharged.

Pleomorphic adenoma, the commonest salivary gland tumour, is a benign growth consisting of an admixture of epithelial, myoepithelial cells, and stroma (connective tissue). The FNA features are quite characteristic. It often appears as a mixture of uniform population of epithelial cells amongst stroma. The stroma appears fibrillar (fibre-like) and stains red to purple with Diff-Quik™ stain. On Papanicolaou stain its colour could vary from pale blue to green (Figure 11.10c), but its fibrillary nature can be appreciated.

This tumour if left unresected will continue to grow in size. It can rarely undergo malignant transformation; this is called carcinoma ex-pleomorphic adenoma.

The aim of the treatment is to completely remove the tumour as it has a tendency to recur in time. The surgical procedure carries the risk of facial nerve damage as nerves that control facial movement pass through the parotid gland. This patient underwent a partial parotidectomy; in this case it was only necessary to remove the superficial lobe of the parotid gland.

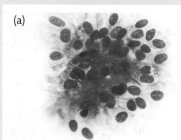 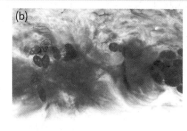 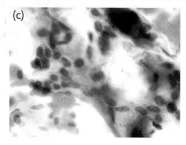

FIGURE 11.10

FNA parotid. Pleomorphic adenoma. (a) Fibrillary stroma intermixed with epithelial and myoepithelial cell. (b) MGG stain clearly demonstrates the frayed edges of the fibrillary stroma. (c) In Papanicolaou stained smears the fibrillary matrix appears pale blue.

Myoepithelial cells

These are smooth muscle-like cells found in the basement membrane of glandular tissue and are involved in secretory activity of the gland.

Gastrointestinal tract

The advent of endoscopy has facilitated better detection and diagnosis of gastrointestinal (GI) lesions. The majority of the lesions that are directly visualized are usually biopsied in preference to cytological sampling.

The advent of endoscopic ultrasound (EUS) has further improved visualization of the gastrointestinal tract and surrounding structures, such as the pancreas and structures within the mediastinum. FNA combined with EUS (EUS-FNA) allows sampling of masses suspicious of malignancy in pancreatic, GI, pelvic, hepatobiliary, or mediastinal lesions. EUS-FNA can also be used for the staging of known GI or pulmonary malignancies. This improvement in technology has provided the cytologist with the opportunity to be involved in diagnosis of GI lesions.

As in FNA of palpable or superficial lesions, the cytologist's involvement during sampling procedures and on-site microscopical assessment will allow for optimal preparation of slides, and reduces the need for unnecessary needle passes. One study compared the diagnostic yield of the samples obtained with and without the presence of a cytotechnologist. The result showed that on-site processing and interpretation of cytological specimens by a cytotechnologist had a significant impact on the diagnostic yield of EUS-FNA. This was independent of the total number of needle passes undertaken for tissue sampling (9).

What are the main indications for FNA in the GI tract?

Case Study 11.5 demonstrates the role of EUS FNA in a patient who had previously been treated for lung cancer and presented with gastric bleeding.

Haematemesis

This is vomiting of blood; the source is generally from the upper GI tract.

Melaena

The passing of black faeces. It is commonly due to bleeding in the upper GI tract, usually in the stomach or duodenum.

CASE STUDY 11.5

A 63-year-old woman, with a history of resected lung cancer two years previously, had experienced haematemesis (vomiting blood) and melaena (dark stools stained with altered blood). Her oncologist referred her to a gastroenterologist for oesophagogastroduodenoscopy (or OGD for short). The OGD showed normal oesophagus and duodenum. A 3 cm polypoidal (a polyp-like) tumour was seen in the stomach. This appeared submucosal. The epithelium overlying the surface was slightly inflamed. No active bleeding could be seen, but the endoscopist decided to carry out an FNA using EUS at a later date when the patient was more stable.

A CT scan was also requested. This also showed presence of a polypoid tumour and no evidence of infiltration through the stomach wall could be seen.

An EUS was carried out and FNA samples were obtained. A biomedical scientist who was present in the endoscopy suite prepared direct smears and stained them with Diff-Quik™. The stained slides were then assessed microscopically for adequacy. The second needle pass was considered satisfactory. A further additional needle pass was taken and this was stored in transport media for subsequent LBC and cell block preparation.

Cytopathological interpretation of direct and LBC preparations showed groups of elongated spindle cells which showed mild nuclear pleomorphism (Figures 11.11a and b).

Immunocytochemistry of the cell block showed that spindle cells stained strongly with **CD117** (Figure 11.11c) and **CD34** (Figure 11.11d). The cytological features and immunocytochemical reaction were those of a gastrointestinal stromal tumour (GIST).

The patient was offered resection. Histological examination confirmed the cytological diagnosis and showed this tumour to have a low rate of mitosis and to be likely to behave as a benign tumour.

Gastrointestinal stromal tumour or GIST is a rare type of **sarcoma** (connective tissue/stromal tumour) that arises in the GI tract, mostly commonly in the wall of the stomach. More specifically it is now known that these tumours arise from a type of cell called **interstitial cells of Cajal**. The interstitial cells of Cajal are thought to be gut 'pace makers' and are involved in **peristalsis** (rhythmic contraction in the GI tract).

Many patients (those with smaller tumours) do not have any symptoms, and GISTs are often discovered when doing tests for other medical conditions. The patients with advanced GIST may present with pain in the abdomen, blood in faeces or vomit, and anaemia due to low red blood cell count.

GISTs can have varied morphology; the majority are composed of spindle cells, some appear epithelioid, and a small number will show mixed cells of both epithelioid and spindle cell morphology.

It is known that GISTs express **c-kit** (also called CD117) a cell membrane-spanning, signalling molecule the expression of which can be demonstrated by immunocytochemistry. Another immunohistochemical marker often found to be positive in GISTs is CD34, a protein also expressed by **haematopoietic stem cells** (stem cells that generate all blood cells).

FNA samples and small needle biopsies may not provide sufficient material to assess the nature of GISTs (whether benign or malignant). Definitive diagnosis is made by examining the resected tumour and determining the proliferation rate of the tumour. A higher proliferation rate increases the likelihood that the tumour may behave as a malignant neoplasm. The proliferation rate may be assessed by counting cells in mitosis or using immunocytochemistry.

Treatment for GIST is usually by surgery. GISTs do not respond to chemotherapy or radiotherapy, but a **biological therapy** such as **imatinib** (commercial name Glivec) is used when the tumour cannot be completely excised. This medication is a tyrosine kinase inhibitor which works by blocking the site of tyrosine kinase and preventing its activity, thereby affecting tumour cell proliferation.

This medication is particularly effective if the tumour expresses CD117.

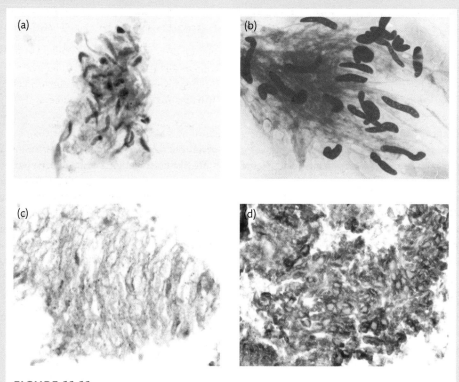

FIGURE 11.11
FNA of gastric mass. Gastrointestinal stromal tumour. (a) and (b) groups of elongated spindle cells which showed mild nuclear pleomorphism. Note the cells with oval bland nuclei in (a). (c) Staining with CD117. (d) Staining with CD34.

Oesophagogastroduoden-oscopy (OGD)

An endoscopic procedure to examine the oesophagus, stomach, and duodenum.

Biological therapy

A form of therapy that uses the body's own immune system to attack cancer cells, or uses medication to interfere with receptors in the body that are involved in cell division.

Cross reference

Chapter 10, Section 10.5.

Respiratory tract

FNA complements other diagnostic tools in malignant respiratory disease. In Chapter 10 the place of FNA in the investigative sequence of respiratory tract lesions was discussed. In this chapter use of FNA is further demonstrated with a case study.

Skin and soft tissues

FNA has found limited application in diagnosis of skin lesions. This is partly due to the relative ease of surgically excising skin lesions and to complex pathology of inflammatory skin lesions that require histopathological assessment.

FNA diagnosis of primary soft tissue tumours has also not gained widespread acceptance, mainly due to the rarity of soft tissue tumours and the vast range of pathology that requires

CASE STUDY 11.6

A 76-year-old male was referred to the respiratory clinic by his GP with episodes of **haemoptysis**. His previous medical history included stage 2 fully excised malignant melanoma treated seven years previously, and type 2 diabetes.

He had a chest CT scan which showed a solid lesion measuring 3.0 × 3.3 cm in the left mid/lower zone, with appearances consistent with neoplasia. There were several moderately enlarged mediastinal lymph nodes.

The chest physician reviewed the CT scans and decided to carry out an endobronchial ultrasound guided transbronchial needle aspiration (EBUS-TBNA) of the mediastinal lymph nodes.

On-site assessment of Diff-Quik™ stained direct smears confirmed adequacy of the first needle pass. A second needle pass was requested and this was stored in transport media for subsequent processing in the laboratory.

Cytological smears showed numerous large dissociated cells amongst the lymphoid background. The nuclei of these cells were eccentrically placed in relation to the cell cytoplasm. Other features included prominent nucleoli (Figure 11.12a). These features were consistent with malignancy and immunocytochemistry was requested on the cell block to further determine the origin.

Tumour cells stained positive with **Melan A** (Figure 11.12b) and **HMB45** (Figure 11.12c). Staining with TTF1 and EMA was negative. The results of the above panel are consistent with metastatic malignant melanoma.

The patient was referred for palliative chemotherapy.

Stage 2 malignant melanoma is a localized tumour, and at the time of diagnosis there was no evidence that it had spread to lymph nodes or other parts of the body; it was therefore considered to be of intermediate risk for local recurrence or distant metastasis. However, malignant melanoma is a highly unpredictable malignancy as it can remain dormant for years and then present as distant metastasis. Malignant melanoma can metastasize to virtually any organ, but the lung is commonly involved. Treatment for advanced melanoma is dependent on the site of the metastasis, the patient's symptoms, and general health.

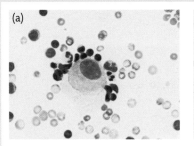
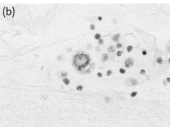
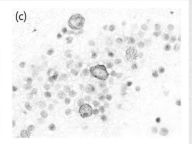

FIGURE 11.12
FNA of mediastinal lymph node. Malignant melanoma. (a) Large cell with eccentrically placed nuclei. Multiple nucleoli also present. (b) Staining with Melan A. (c) Positive staining with HMB45.

tissue architecture for diagnosis. FNA cytology of soft tissue tumours has been described by cytopathologists practising in specialist referral centres, who have gained experience based on correlation of clinical, radiological, cytological, and histological findings. Please consult the further reading list at the end of the chapter for further information.

FNA, however, is very useful in assessing skin and soft tissue lesions suspicious for metastatic disease or tumour recurrences. Case Study 11.7 will demonstrate this scenario.

SELF-CHECK 11.6

What is the most common use of FNA in skin and soft tissue masses?

CASE STUDY 11.7

A 70-year-old woman was referred by her GP to the breast clinic. She presented with a hard lump on her chest wall above her breast. There was no family history of breast cancer. She was an ex-smoker and there was no other significant past medical history.

The surgeon examined her breast, but could not find any masses in the breast, and axillary lymph nodes were normal in size. She was sent for mammogram and ultrasound examination of the breasts, which did not show any convincing mass lesions in either breast. The lesion on the chest wall appeared to be in subcutaneous fat. The radiologist performed an FNA of the lesion and prepared two direct air-dried smears.

The FNA is shown in Figure 11.13. The cytopathology report read as follows: 'A highly cellular aspirate containing numerous clusters of poorly cohesive cells with pleomorphic nuclei which show nuclear moulding. The features are consistent with small cell anaplastic carcinoma. Immunocytochemistry cannot be performed to confirm the cytological findings as only two air-dried slides were submitted to cytology.'

This patient was discussed in breast MDT and referred to respiratory MDT for assessment.

Chest X-ray and CT scans revealed a left upper lobe mass with mediastinal lymphadenopathy. She was seen by the oncologist and offered palliative chemo-radiotherapy. She survived for eight months, but died of brain and liver metastases.

Skin metastases as the first sign of internal malignancy occur infrequently. Small cell anaplastic carcinoma has poor outcome and the course of the disease in this patient may not have been altered significantly by FNA diagnosis of the skin lesion. Nevertheless, she was spared unnecessary further testing and her treatment was started without delay.

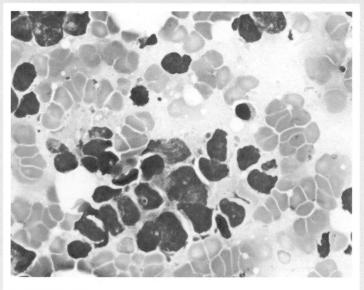

FIGURE 11.13
FNA chest wall. Small cell anaplastic carcinoma.

SUMMARY

- FNA is a safe diagnostic procedure and significant complications are rare.

- FNA is an accurate and cost effective method applicable to lesions that are easily palpable.

- Deep lesions can be sampled with FNA with the aid of ultrasound or CT imaging for guidance.

- Development of endoscopic ultrasound (EUS) has allowed sampling of gastrointestinal, mediastinal, and lung lesions.

- On-site assessment of FNA sample adequacy by a cytopathologist or trained biomedical scientist provides the optimal approach for obtaining adequate material for immediate investigation, for ancillary testing, and for eventual accurate diagnosis.

- FNA is not a replacement for surgical histopathology or core biopsy. Sometimes and in some conditions, cytology cannot give a definitive answer and histological assessment is required.

FURTHER READING

- Kini SR (2008) *Thyroid Cytopathology: a Text and Atlas.* Lippincott Williams & Wilkins, Philadelphia.

- Kocjan G (2001) *Clinical Cytopathology of the Head and Neck: a Text and Atlas*, first edition. Greenwich Medical Media, Cambridge.

- Orell SR, Sterrett GF, and Whitaker D (2005) *Fine Needle Aspiration Cytology*, fourth edition. Churchill Livingstone, New York.

- Sidawy MK and Ali SZ (2007) *Fine Needle Aspiration Cytology: a Volume in Foundations in Diagnostic Pathology*, first edition. Churchill Livingstone, Phildelphia.

- Yang GCH and Tao L-C (2007) *Transabdominal Fine-needle Aspiration Biopsy: a colour atlas and monograph*, second edition (with CD-ROM). World Scientific Publishing, Hackensack, New Jersey.

DISCUSSION QUESTIONS

1. Discuss the provision of an on-site assessment FNA service to a district general hospital.

2. Carry out an appraisal of fine-needle aspiration biopsy of the thyroid.

3. The National Institute of Clinical Excellence (NICE) guidance on cancer services produced the document *Improving Outcomes in Head and Neck Cancers* in 2004. One of the 'key recommendations' was the stipulation that 'diagnostic clinics should be established for patients with neck lumps'. Is this recommendation evidence based?

Answers to the self-check questions, and tips for responding to the discussion questions, are provided in the book's Online Resource Centre.

 Visit www.oxfordtextbooks.co.uk/orc/shambayati/

12

Basic semen analysis

Stephen Blackman and Behdad Shambayati

In this chapter we start by exploring the anatomical basis of semen production and then examine the range of analytical tests that provide the data used for clinical diagnosis. These include investigations following vasectomy procedures and a series of tests specific to infertility samples.

In the pathology department, these related techniques are commonly referred to as semen analysis, which describes a range of tests that utilize skills ranging from microscopy to immunology and include aspects of haematology, biochemistry, microbiology, and of course cytology. As a result, the discipline has no natural home in any specific pathology department. This has, until relatively recently, lead to big variations in both the type of tests that are performed and the reference values that the results are measured against.

In an attempt to address the lack of standardization of procedures the World Health Organization (WHO) first published a *Laboratory Manual for the Examination of Human Semen and Semen-Cervical Mucus Interaction* in 1980; further editions followed in 1987, 1992, and 1999, leading through to the fifth edition published in 2010. In the UK there are two professional bodies: the British Andrology Society (BAS) and the Association of Biomedical Andrologists (ABA). The ABA provides a logbook based training course. An **external quality assurance (EQA) scheme** provided by the **United Kingdom National External Quality Assessment Service (UKNEQAS)** is available to laboratories, and covers sperm count, morphology, motility, and sperm antibody testing.

This chapter is not intended to be a technical manual or reference text because that role is already filled by the WHO publication.

External quality assurance (EQA) scheme
An inter-laboratory comparison designed and operated to assure various quality measures are achieved by participating laboratories.

United Kingdom National External Quality Assessment Service (UKNEQAS)
An external body that oversees various quality assurance schemes.

Learning objectives

After studying this chapter you should be able to:

- Describe the process of spermatozoa production.
- Discuss the composition of semen and the function of its different components.
- Describe the range of laboratory tests that can be applied to a semen sample.
- Discuss how the test results relate to the normal ranges.
- Discuss factors that can lead to sub-optimal results.

12.1 **Spermatozoa production**

Semen is usually thought of as a simple solution of spermatozoa in an aqueous medium; however, its true nature and the lengthy pathway by which it is manufactured is considerably more complex.

The whole process can be thought of as a production line that starts in the testes. Look at Figure 12.1, which summarizes the first stages in spermatozoa production. Notice how a single type B spermatogonium cell produces eight spermatozoa.

This production line activates with the onset of puberty and the first divisions of the **spermatogonia** stem cells in the **seminiferous tubules** of the **testes**. One of the daughter cells becomes a type A spermatogonium to maintain the supply of stem cells, the other becomes a type B spermatogonium and then mitotically divides again to produce two further daughter cells. It is these cells that become primary spermatocytes as they migrate towards the lumen of the seminiferous tubules and attach themselves to a **Sertoli cell**. There they begin a growth, multiplication, and differentiation process known as **spermatogenesis**. They remain in close contact with Sertoli cells, which are thought to provide structural and metabolic support, until spermatogenesis is complete. The cytoplasm of the Sertoli cells extends all the way from the basement membrane to the lumen of the seminiferous tubule. There are **tight junctions**

Spermatogonia
A type of stem cell found in the wall of the seminiferous tubules that undergoes mitosis to produce types A and B spermatogonium cells.

Seminiferous tubules
Found inside lobules in the testes within which spermatozoa are produced. There are about 750 per testicle.

Sertoli cells
Found in the walls of the seminiferous tubules of the testis. They provide an anchoring point and source of nutrition for developing spermatids and also form the blood testes barrier.

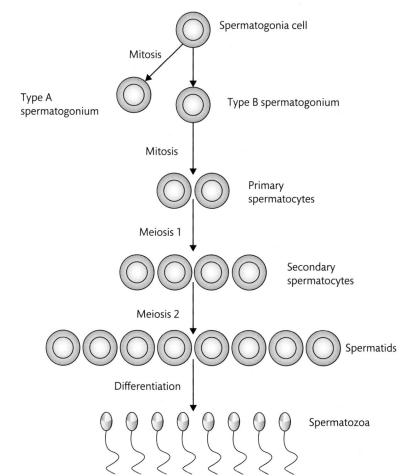

FIGURE 12.1

A schematic overview of the cellular processes leading to spermatogenesis.

between the cells that effectively form a blood/testes barrier, stopping the body's immune defence system detecting and attacking the spermatozoa.

What type of cell division does the type B spermatogonium undergo?

Chromosomes

Structures found in the cell nucleus that contain the genetic material; human cells contain 23 pairs of chromosomes making a total of 46.

Chromatid

Either of the two strands formed when a chromosome duplicates itself during the early stages of cell division, the chromatids are initially joined together by a structure known as a centromere but separate later, one going to each daughter cell.

Interphase

The stage in the development of a cell following mitosis or meiosis, during which the nucleus is not dividing. In cells preparing for a further division the DNA in the nucleus is duplicated during interphase.

Meiosis

A specialized form of cell division in which there are two successive divisions (meiosis I and II) without any chromosome replication in between.

Reductional division

Also called the first meiotic division, where the chromosome number is reduced from diploid (46 chromosomes) to haploid (23 chromosomes).

Equational division

Also called the second meiotic division, where sister chromatids finally split, creating a total of four haploid cells with 23 chromosomes.

Cytoplasmic bridge

A thin strand of cytoplasm linking the spermatids.

Acrosome

A membrane-bound compartment found in the head of a sperm. It contains the enzymes that allow penetration of the outer surface of the egg.

Flagellum

A whip-like extension of certain cells, for example sperm, that functions as an organ of locomotion.

Growth

The new cells start with conventional pairs of **chromosomes** that each have a single **chromatid**. They now enter a period known as **interphase** during which they duplicate this chromosomal material, producing a full set of double chromatid chromosomes. These cells are the primary spermatocytes.

Multiplication

The primary spermatocytes then undergo a two-stage type of cell division known as **meiosis** I and II. The first of these is a **reductional division** where the daughter cells both end up with one double chromatid chromosome from each pair (n). These cells are known as secondary spermatocytes.

The second division follows and is an **equational division** which splits up the double chromatid chromosomes, producing a final generation of daughter cells known as spermatids. These cells are still linked together by small connections known as **cytoplasmic bridges**.

Differentiation

This final stage, also known as **spermiogenesis**, is the process of transforming the spermatids into mature sperm cells with the specialized form and equipment that they require for the tasks they are designed to accomplish. Differentiation involves moulding the spermatid into a more streamlined shape, by shedding of the majority of the cytoplasm and the conversion of the **Golgi apparatus** into a bag of digestive enzymes called the **acrosome**, arranged in front of the nucleus. At the other end of the cell, a long tail-like **flagellum** is formed.

Look at Figure 12.2, which shows the structure of a sperm. Notice how the mitochondria that power the flagellum are packed into the middle area.

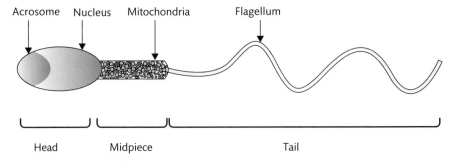

FIGURE 12.2

A cross section of a sperm cell showing its main components.

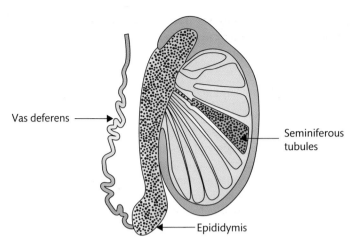

FIGURE 12.3
A cross section of a testicle showing its main functional components.

Labels: Vas deferens, Seminiferous tubules, Epididymis

SELF-CHECK 12.2

What is the Golgi apparatus converted into?

Once differentiation is complete, the newly formed spermatozoa separate from the Sertoli cell that they have been bound to and are released into the lumen of the seminiferous tubule. The whole process takes 74 days.

In Figure 12.3 you can see the complex packing arrangement of seminiferous tubules within the testicle. Note how this is used to provide the greatest possible area for spermatozoa production.

Estimates of the number and length of the seminiferous tubules in a human testicle vary considerably, with numbers ranging from 500 to 1200 tubules and individual lengths of 30–90 cm. Taking mid-range figures of 750 tubules and 60 cm, this gives a total length of about 450 metres per testicle, that is, 900 metres of sperm-producing seminiferous tubule per adult male. Sperm are produced in the testes at a rate of approximately 200 million per day; therefore in one week a fertile man will produce 1.4 billion sperm. These newly formed spermatozoa, suspended in testicular fluid secreted by the Sertoli cells, are pushed out of the seminiferous tubules and into a nursery area called the epididymis by peristaltic contractions of the tubule walls. Whilst in the epididymis they mature and become potentially motile and capable of fertilization.

The mature sperm are moved along by further peristaltic contractions into and through the vas deferens, a long tube that runs from the epididymis up to the bladder and then down towards the **prostate gland**. Near the prostate, the vas deferens widens and forms a gland called the **ampulla**. This is where the sperm are stored ready to be ejaculated. Sperm can be stored inside the ampulla for approximately three weeks after they have matured. Therefore, the entire sperm life cycle inside a man's body comprises ten weeks' total manufacturing time and three weeks' storage lifespan, a total of 13 weeks or about three months.

12.2 **Semen composition**

The fluid known as semen is actually a mixture of secretions produced by several different glands. The components are mixed together at ejaculation to produce seminal fluid, a medium that will stimulate and feed the sperm and also help them survive once they are deposited in the vagina. The majority of an ejaculate is seminal fluid, with sperm only comprising about 5% by volume. In Figure 12.4 you can see the main organs and glands that are involved in semen production. We now examine each of the significant glands involved in this process.

The majority of the seminal fluid is produced by three glands:

Seminal vesicles These are about 5 cm long and are located near the prostate gland, discharging into the ejaculatory duct, just after the ampulla. They produce a sticky fluid that is clear to yellowish in colour and comprises the majority of the seminal fluid by volume, at about 65% of total. The fluid contains a sugar called fructose, which the sperm rely on entirely as their energy source, plus various hormones called prostaglandins (involved in suppressing the immune response against the semen once in the vagina) and proteins that stimulate the sperm to begin swimming. The seminal vesicles also produce a protein called **semenogelin**. This causes coagulation of the semen and gives it its thick, sticky, consistency.

Prostate gland This small gland is about the size and shape of a walnut, surrounding the beginning of the urethra near the bladder. The prostate gland produces prostate fluid, which makes up about 35% of the seminal fluid by volume. Prostate fluid is thin and milky white and gives semen its characteristic appearance. It contains prostate-specific antigens to liquefy the coagulated semen, reversing the action of semenogelin. Sperm must undergo this coagulation and liquefaction process to be able to fertilize the egg. Semen usually liquefies within 5–20 minutes after ejaculation.

Bulbourethral (Cowper's) glands These are small glands about the size of a pea, located near the base of the penis. They produce pre-ejaculatory fluid which is sometimes expelled as a few drops before actual ejaculation, but can also be ejaculated with the rest of semen. It is tiny in volume compared to the other secretions and is completely clear, sticky, yet slick to the touch. Its purpose is to help lubricate and flush out the urethra in preparation for the sperm that will soon be passing through it.

SELF-CHECK 12.3

What is the role of the prostate gland in semen liquefaction?

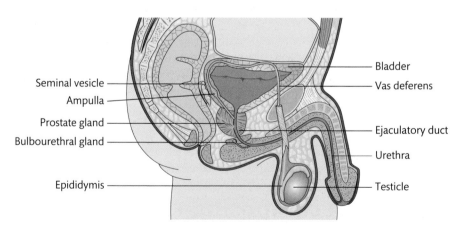

FIGURE 12.4
A diagram of the male reproductive organs.

12.3 Reasons for testing patients

Semen analysis is recommended to patients following a vasectomy and when a fertility problem is suspected. At least 20% of couples experience problems conceiving, and in about 40% of cases this is due to the male partner. A simple semen test can prevent a lot of unnecessary investigation of the female partner. Approximately 10–20% of all men have sperm counts below the WHO reference value and 2–3% of all males produce no sperm at all.

The following are some of the risk factors which may lead to poor semen quality:

- A history of reproductive tract infections or sexually transmitted diseases.
- Exposure to environmental pollutants, e.g. agricultural chemicals.
- Male athletes exposed to conditions which might lower sperm counts.
- Alcohol abuse.
- Use of tobacco/recreational drugs.
- Malignancy.
- Medicinal drugs, especially steroids.
- Previous testicular surgery or recent trauma.

12.4 Specimen requirements

We have examined how the semen is produced and now we can start to prepare for examining it in the laboratory.

Sample collection

Pathology samples vary widely in type and collection method and all have different time periods for which the sample can be regarded as viable. In the case of semen, a maximum period of one hour between ejaculation and analysis is permissible, but the sample must be kept warm during that period. A wide temperature range between 20°C and 40°C is acceptable and this can usually be achieved on even a cold winter's day by transporting the sample container in an internal coat pocket. Specimen containers are also important since some materials are harmful to sperm and reduce motility (in practice, the polystyrene white capped containers known as 'sputum pots' are ideal and rarely cause problems).

In order to produce a semen sample that is suitable for analysis, the patient must follow a set of instructions preferably given in written form:

- Collect a request form, sample container, and specimen bag from the requesting clinician.
- Abstain from sex for a period of at least two days, but not more than seven days, before producing a sample.
- Produce a sample by masturbation only.
- Collect the whole sample in the container provided.
- Write the date and time of sample production on the container.
- See that the laboratory gets the sample before it is one hour old.
- Keep it warm under your clothing while in transit.

Contamination of the sample is not uncommon and it can sometimes be harmful to sperm quality; examples include:

- Blood
- Carpet fibres
- Hair
- Soap
- Lubricants (some of these contain spermicides)
- Talc

SELF-CHECK 12.4

What is the temperature range that is acceptable during transport of the specimen to the laboratory? How could this be best achieved?

Safe handling of specimens

Human semen may contain significantly harmful infectious agents, most notably viruses like HIV and hepatitis B, so all samples arriving in the laboratory must be treated as potentially infectious. A laboratory coat and gloves must be worn whenever samples are handled; and procedures that may produce an aerosol, for example vortexing of open containers, should be avoided. Other procedures, such as pipetting samples, must be performed in a class 1 microbiological safety cabinet.

12.5 **Laboratory equipment**

The list of desirable equipment for a laboratory performing semen analysis (see below) is a long one; however, there are some absolute essentials:

- Incubator (specimen must be kept warm).
- Microscope with x100 oil objective (essential for morphology assessment).
- Bench top warming plate (wet preparations containing live sperm must be kept warm).
- **Haemocytometer**, preferably an **improved Neubauer haemocytometer** (Figure 12.5) with lots of spare coverslips (essential for counting sperm).
- Ph papers or test strips to cover the range Ph 6–10.
- Laboratory counter, a simple tally counter could be used.
- Calibrated plastic **pastettes** up to 5 ml, or some other way to measure semen volume.
- Selection of conventional pipettes to dispense volumes from 10 µl to 1000 µl.
- Microcentrifuge tubes.
- Glass slides and coverslips.

Other optional (but very useful) items of equipment include:

- Phase contrast microscope with x10, x20, x40 phase objectives, by far the best way to observe unstained sperm when measuring motility or performing a count and far better than a conventional microscope with the condenser racked down.
- Heated stage for microscope is best at keeping sperm swimming when you are measuring motility.
- Vortex mixer.

Haemocytometer
A device which was originally designed for counting blood cells which can also be used for counting other types of cells. Improved Neubauer haemocytometer–a design of haemocytometer.

Pastette
A disposable plastic pipette.

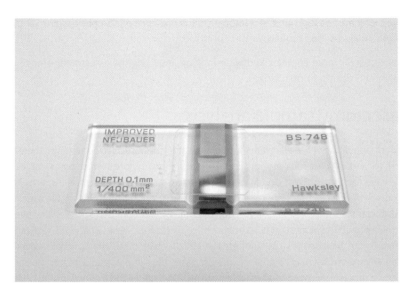

FIGURE 12.5
An improved Neubauer haemocytometer.

- Multi-parameter electronic counter is best to use when assessing motility and morphology and it really helps to get the job done more quickly.
- Positive displacement pipette with a 10–100 μl range; semen is a viscous liquid that is hard to dispense accurately with a conventional pipette.

The following reagents are also required:

- Semen diluent solution consisting of 5 g of sodium bicarbonate ($NaHCO_3$) and 1 ml of concentrated formaldehyde solution, made up to 100 ml with distilled water.
- Rapid Giemsa type stain solutions.
- Anti-sperm antibodies (MAR) test kit.
- Hypochlorite solution (5.25 g/l) for disinfecting equipment; this is equivalent to a 1/10 dilution of household bleach.

SELF-CHECK 12.5

What is the best type of haemocytometer to use for semen analysis?

SELF-CHECK 12.6

What type of pipette is best for dispensing semen?

SELF-CHECK 12.7

Which disinfectant agent is recommended for sterilizing equipment?

12.6 **Types of sample**

Essentially, diagnostic semen samples are of two types: post-vasectomy or infertility. Each has different procedures.

The vasectomy operation involves cutting the vas deferens tubes and sealing them either with stitches or with heat (cauterization), thereby stopping any sperm ending up in the ejaculate.

The objective of the post-vasectomy test is to confirm the presence or absence of sperm and, if any are found, to assess their motility.

The infertility test is far more complex and wide-ranging, involving the assessment of many different parameters against normal ranges.

Post-vasectomy sample analysis

Sperm do not disappear from the ejaculate immediately after the vasectomy operation, therefore it is best to wait at least four months and 24 ejaculations before testing any samples. Occasionally, vasectomy operations do fail, leaving the man with millions of motile sperm in his semen. However, most samples are clear of sperm or have only scanty non-motile sperm remaining. Unfortunately, the patient cannot be given the all clear until two samples are seen that are completely free of sperm, that is, **azoospermic**, two to four weeks apart. The test can be complicated by the presence of large numbers of epithelial cells or inflammatory cells, which might obscure any sperm that may be present. In these cases, the sample should be reported as inadequate and a repeat requested. It is best to examine both a dilute and a concentrated (that is, a centrifuged deposit) sample, since the number of sperm may be very low.

The Method box below shows how the test is conducted.

The vasectomy operation is potentially reversible, since the procedure has simply interrupted the flow of sperm from the testes through the vas deferens and thus prevented any sperm from being ejaculated. Theoretically, if the tubes are surgically reconnected, the flow of sperm can resume.

One of the reasons why vasectomy reversals have limited success, even if the vas deferens are successfully reconnected, is that the operation exposes some sperm to the immune system, leading to the production of anti-sperm antibodies. As is discussed later, these antibodies are potentially a cause of infertility.

Azoospermic
A condition where no spermatozoa can be found in the semen.

SELF-CHECK 12.8

What is the minimum post-operative time before testing a post-vasectomy sample? At what stage can the patient be given the 'all clear' following a vasectomy operation?

 METHOD *Post-vasectomy method*

- Make a wet preparation with one drop of semen on a clean slide and a 22 mm coverslip; examine it using a phase contrast microscope with the x40 objective.

- If there are scanty non-motile sperm present, note the number of sperm in each high power field (x40 objective).

- If there are any motile sperm present, note the number of sperm in each high power field (x40 objective) and record the percentage that are motile.

- If there are no visible sperm, spin the sample in a centrifuge at 3000 g for 15 minutes.

- Make another wet preparation with one drop of the centrifuged deposit on a clean slide and a 22 mm coverslip; examine it using a phase contrast microscope with the x40 objective.

- If there are scanty non-motile sperm present, note the number of sperm in each centrifuge deposit high power field (x40 objective).

- If there are any motile sperm present, note the number of sperm in each centrifuge deposit high power field (x40 objective) and record the percentage that are motile.

What can complicate a successful vasectomy reversal operation?

Infertility sample analysis

The infertility sample assessment is far more complex and includes a whole range of tests designed to give an in-depth profile of semen quality. The World Health Organization, in its latest (2010) manual, recommends that the following tests should be routinely performed on all infertility samples:

1. Liquefaction
2. Appearance
3. Volume (in ml)
4. Viscosity
5. pH
6. Motility
7. Morphology (percentage of normal sperm)
8. Sperm count and **leucocytes** count (millions per ml)
9. Anti-sperm antibodies
10. Vitality (percentage of live sperm)

The list can appear rather intimidating and initially looks like several hours of work per sample, but fortunately most tests are fairly simple and quick to perform. It is best to begin by separating them into two categories, namely macroscopic, for example measuring volume, and microscopic, that is, all the complex analyses that require a microscope. It is best to perform testing of these samples at 37°C to optimize conditions for the motile sperm. Start by placing the newly arrived sample in a 37°C incubator and leave for 15 minutes to warm up.

The ejaculate usually contains cells other than sperm, usually referred to as **round cells**. These cells are usually either epithelial cells or have originated elsewhere in the genitourinary tract, for example the prostate. However, sometimes leucocytes, mostly **polymorphs** (pus cells), are found which are easy to identify on the morphology slide. If numerous polymorphs are observed, they can be counted using the haemocytometer, at the same time as the sperm concentration is assessed.

Round cells
This describes other types of cells found in semen, such as immature sperm cells, inflammatory cells, or epithelial in origin.

12.7 Standard laboratory test for evaluation of semen

1. Liquefaction

A normal semen sample will liquefy within 15 minutes at room temperature, but may take up to one hour. Usually the process is complete by the time the sample is removed from the incubator for testing.

2. Appearance

Semen usually has a homogenous cloudy, whitish grey, slightly opalescent appearance. The sample may appear clearer if sperm density is very low or reddish brown if blood is present.

3. Volume

Ejaculate volume is easy to measure using a graduated pastette (up to 5 ml ones are available) or similar graduated containers. It is best to avoid plastic syringes because some affect sperm motility. Sample volumes may range from 0.1 to 10 ml.

4. Viscosity

The viscosity or consistency of the sample is another easy test. A normal sample will form discrete drops when gently expelled from the volume measuring pastette, while an abnormally viscous sample will form a thread more than 2 cm long.

5. pH

The pH should be measured within one hour of ejaculation using pH paper that covers the range 6.1–10. Add a drop of semen to the paper with the pastette, leave for 30 seconds, then compare colour change to calibration strip.

SELF-CHECK 12.10

What effect would the presence of blood have on the appearance of the semen sample?

Initial microscopic evaluation of the sample

We start this series of tests with the production of a standardized wet preparation slide that is used to calculate the dilution for the counting chamber, assess sperm motility, and finally produce a stained slide for morphology assessment. It is best to keep a supply of warm slides in the semen sample incubator and to observe the wet preparation slide on a phase contrast microscope with a heated stage.

You can now use the slide to begin testing.

6. Motility

Using the same slide, you proceed by performing a motility assessment. This is based on a simple grading system which places each individual observed sperm into one of four categories:

(a) Rapid progressive motility: sperm swimming progressively at a speed greater or equal to 25 μm per second (which is about half the length of a sperm's tail per second).

(b) Sluggish progressive motility: sperm swimming progressively at a speed slower than 25 μm per second.

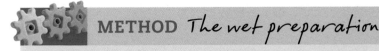

METHOD *The wet preparation*

- Vortex the sample for 30 seconds to ensure it is thoroughly mixed.
- Using a positive displacement pipette, place 10 μl of the sample on a warm clean glass slide and cover it with a 22 × 22 mm coverslip. You should try to avoid getting any air bubbles under the coverslip.
- Finally, place the slide on the heated microscope stage and leave it to settle for one minute.

METHOD Calculating the counting chamber dilution

■ Examine the slide using the x40 phase objective and estimate the average number of sperm per high power field.

■ Using Table 12.1, you can now calculate the dilution factor for the counting chamber; make a note of this ready for use when preparing the haemocytometer.

■ If there are no visible sperm in the wet preparation, then the whole ejaculate should be spun in a centrifuge at 3000 g for 15 minutes and a second wet preparation made from the deposit. Any sperm found in the deposit can be reported as for a post-vasectomy sample, that is, numbers per high power field and percentage motile.

TABLE 12.1 Counting chamber, dilution factors.

Spermatozoa per high power field	Dilution factor (semen: diluent)	Dilution volume (semen: diluent)
>200	1:50	10 µl semen 490 µl diluent
40–200	1:20	25 µl semen 475 µl diluent
15–40	1:10	50 µl semen 450 µl diluent
<15	1:5	100 µl semen 400 µl diluent

METHOD Performing the motility test

■ Systematically observe and grade all of the sperm in several high power fields, that is, x40 phase objective, until at least 200 sperm have been counted (you will find it easier to do this using a multi-parameter electronic counter and it gets the job done more quickly).

■ Express the results as a percentage score for each of the four motility categories. For example:

– Rapid 25%

– Sluggish 35%

– Non-progressive 20%

– Immotile 20%

(c) Non-progressive motility: sperm that are moving but not going anywhere, either swimming in circles or just twitching.

(d) Immotility: sperm that show no signs of movement at all.

You must keep the wet preparation slide warm at this stage; otherwise sperm motility will rapidly decline.

Preparing the morphology slide

There are many staining methods that could be used at this stage and everybody has their favourite. However, it is worth remembering that you are going to observe the slide using an x100 oil immersion objective.

A dry rapid Giemsa stain type preparation is ideal for this since it can be screened without coverslipping as soon as it is dry. The staining times in the following method should be adjusted for the type of kit used and personal preference.

7. Morphology (percentage of normal sperm)

The surprising thing that you will find about this part of the test is that the majority of the sperm seen on the morphology slide are always abnormal. Each new version of the WHO manual has reduced the normal range for morphology and the latest 2010 version has further reduced the lower reference limit to just 4% normal sperm. It is in fact far easier to define the properties of a normal sperm than to describe the numerous defects that can arise.

In summary, the head should be oval in shape and about 4.1 µm long by 2.8 µm wide, with a well-defined acrosomal region occupying about 40–70% of the head area. The mid-piece should be the same length as the head and about 0.6 µm wide, while the tail should be 45 µm long: it may be looped back on itself but must not have sharp angles. Clearly, it is not practical to measure 200 individual sperm in every slide, so a degree of observer experience and expertise is required, much like the rest of cytology. Figure 12.6 shows an example of normal spermatozoa and Figure 12.7 shows some examples of abnormal spermatozoa.

SELF-CHECK 12.11

What are the characteristics of the head of a normal sperm?

METHOD *Staining the morphology slide*

- **Remove the coverslip from the wet preparation slide and discard it into a sharps bin.**
- **Leave the slide to air dry for five minutes.**
- **Fix the slide for five minutes in methanol.**
- **Dip the slide slowly five times into eosin solution.**
- **Dip the slide slowly five times into methylene blue solution.**
- **Rinse the slide thoroughly in buffer solution and then leave it to air dry for five minutes.**

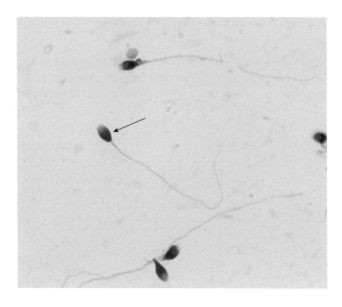

FIGURE 12.6
An example of normal sperm (arrow).

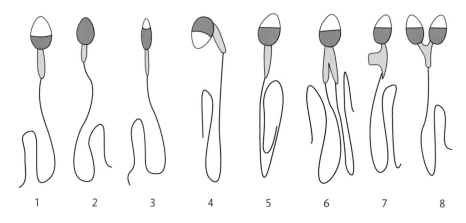

1 A normal sperm

2 No acrosome present

3 Thin head region

4 Offset head/midpiece

5 Coiled tail

6 Two tails

7 Midpiece cytoplasrnic droplet

8 Two heads

FIGURE 12.7
A few examples of abnormal sperm morphology.

8. Sperm count (measuring the sperm concentration)

The improved Neubauer haemocytometer (counting chamber) is really a haematology tool, but you can use it to measure the concentration of any types of cells in a fluid, even sperm. The counter set includes a thick glass slide engraved with two recessed counting grids (wells) and a separate coverslip that goes on top of the counting grids to form two counting chambers, which are then filled with the dilute solution of cells.

Look at Figure 12.8 and note that each counting grid contains 25 large squares bordered by three boundary lines; each of these squares is further subdivided into 16 smaller squares, but these are not used for this test. All of the spermatozoa in the central square are counted, along with those whose heads lie *between* the central and inner boundary lines. Spermatozoa whose heads *overlap* the centre line are counted only if the line is on the *top or left-hand side* of the square.

To perform the sperm count, start by preparing a solution of diluted semen using the data from the original wet preparation slide and Table 12.1. For example, if 60 sperm per high power field were observed then the appropriate dilution to use is 1 in 20, that is, 25 µl of semen and 475 µl of diluent solution.

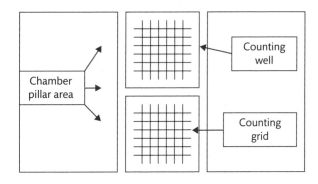

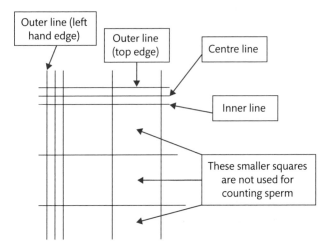

FIGURE 12.8
Diagram showing components of a haemocytometer.

■ Place the freshly stained morphology slide on the microscope stage.

■ Using the x10 objective look for an area of the slide that contains plenty of visible sperm.

■ Add a drop of immersion oil to the slide and switch to the x100 oil objective.

■ Count at least 200 sperm and score as either normal or abnormal, preferably using a multi-parameter electronic counter.

■ Record the percentage of normal sperm.

■ If numerous polymorph cells are seen, make a note of the fact and use the haemocy-tometer preparation described in the next section to perform a count.

METHOD *Preparing the semen dilution*

- Mix the semen sample thoroughly by vortexing.
- Prepare the chosen dilution using a positive displacement pipette for the semen and a conventional pipette for the diluent.
- Mix thoroughly by vortexing.

METHOD *Assembling and filling of an improved Neubauer haemocytometer (WHO manual, fifth edition, 2010)*

- Moisten the counting chamber by breathing on it.
- Press the coverslip firmly onto the chamber pillars (see Figure 12.6) until iridescence (Newton's rings) is observed between the two glass surfaces.
- Vortex the diluted semen.

- Fill each counting chamber with 10 µl of diluted semen, using a pipette.
- Leave to settle for five minutes or until the spermatozoa have settled in one plane.

METHOD *Working out how many squares to count*

- Place the haemocytometer on the microscope stage.
- Using the x40 phase objective, examine the top counting grid and estimate the average number of sperm in each large square (usually a quick count of the top left hand square will do).
- Use this number as a basis to determine the number of squares to count overall for the test; as per Table 12.2.

TABLE 12.2 Counting chamber, number of large squares to count.

Number of sperm in each large square	Number of large squares to count
Less than 10 (<10)	All 25 squares
10–40	10 squares (2 rows of 5)
More than 40 (>40)	5 squares (4 corners and centre)

Note that if numerous polymorphs are seen on the morphology slide, you should count them as well, using the same process.

This section is a little complex, so here are some self-check questions to make sure that you can follow the process.

METHOD *Counting the sperm*

- Count the number of sperm in the appropriate number of squares for the top counting chamber.
- Count the number of sperm in the same squares in the bottom counting chamber.
- Add together the counts from both chambers and then divide by two to calculate the average (if the two counts vary by more than 10% prepare another chamber and repeat the test).
- Note the result.
- Look up the dilution used and the number of squares counted in Table 12.3 to determine the dilution correction factor.
- Divide the average number of sperm counted by the dilution correction factor to produce the final result in millions per ml.

TABLE 12.3 Determination of the dilution correction factor.

Dilution (semen/diluent)	Number of large squares counted		
	25	10	5
1:5	20	8	4
1:10	10	4	2
1:20	5	2	1
1:50	2	0.8	0.4

SELF-CHECK 12.12

You are presented with a wet preparation slide of fresh semen. When you observe it on the phase contrast microscope you see 50 sperm in each high power field.

What dilution should you prepare?

SELF-CHECK 12.13

When you observe the counting chamber, you find 20 sperm in each large square. How many squares should you count?

SELF-CHECK 12.14

How many sperm do you expect to count?

SELF-CHECK 12.15

Now when you combine the answers from questions 1 and 2, which dilution correction factor should you use?

SELF-CHECK 12.16
What is the sperm count to be reported in millions per ml?

9. Anti-sperm antibodies

As you have seen in the section on sperm production, the Sertoli cells in the testes are formed into a tightly linked layer of cells known as the blood-testes barrier. This stops the immune system cells from detecting the highly antigenic sperm located within the male reproductive tract.

Under some circumstances, for example through injury to the testicles, the barrier can be breached, thereby exposing the sperm to the immune cells and triggering an antibody response. These anti-sperm antibodies then attach themselves to different parts of the sperm and interfere with male fertility in a number of ways. They may bind to the tail and immobilize sperm, cause them to clump together, limit their ability to pass through the cervical mucus, or prevent them from binding to and penetrating the egg.

IgG antibodies are the most common. They are usually produced after a vasectomy and may also be seen after testicular injury or an infection.

10. The MAR (mixed antiglobulin reaction) test

The MAR test is performed by mixing fresh semen with latex particles that have been coated with human IgG antibodies. To this mixture is added a second solution containing **antihuman IgG antibodies**. Therefore, in a positive test we will have both latex particles and sperm coated with IgG. When the antihuman IgG antibodies are added to the mixture, they bind to both latex particles and sperm, effectively sticking them together. The strength of the reaction depends on the amount of anti-sperm antibodies present. Results will range from individual sperm with single particles bound to their tails, to total agglutination where all the motile sperm and latex particles are stuck together.

METHOD *Performing the MAR test*

Note: procedures may vary for different manufacturers' test kits.

- Allow all reagents and specimen to come to room temperature.
- Use a positive displacement pipette to place 10 µl of fresh semen on a clean glass slide.
- Use a standard pipette to add 10 µl of latex particles to the slide.
- Using the same pipette (with a fresh tip), add 10 µl of antiserum to the slide.
- Then mix the semen and latex reagent together.
- Finally, mix in the antiserum reagent.

- Cover with a 22 × 40 mm coverslip.
- After 2–3 minutes, examine the slide on the phase contrast microscope and check for any motile sperm with latex particles attached.
- If any are seen, count 200 spermatozoa and calculate the percentage of reactive sperm. Note the result.
- If no attachment is noted, check the slide again after ten minutes. By this time the latex particles themselves should have formed aggregates, thereby acting as a control for the viability of the reagents.
- If no aggregates have formed, then the test kit has failed and should be discarded.

The significance of the test result depends on the percentage of reactive sperm:

0–9% reactive: MAR test negative.

10–39% reactive: MAR test borderline positive, suspicious of immunological infertility.

>40% reactive: MAR test positive, immunological infertility is highly probable.

SELF-CHECK 12.17

Which type of cells form the blood-testes barrier?

SELF-CHECK 12.18

Which type of anti-sperm antibodies are most common?

SELF-CHECK 12.19

Why should a negative MAR test slide be checked after ten minutes?

---

11. Vitality (percentage of live sperm)

There are several ways of testing sperm to see if they are alive or dead. The simplest methods are based on the principle that dead cells with a damaged plasma membrane take up certain stains, while living cells do not. Vitality testing should be performed if the proportion of immotile sperm, as measured during the motility test, exceeds 50%.

Eosin-nigrosin method

As you will see, dead sperm are stained red by the eosin, while the nigrosin provides a dark background, making the live, that is, unstained, sperm visible.

The solutions you will need to use are as follows:

- Eosin Y, 1% solution, that is 1 g in 100 ml of distilled water.
- Nigrosin, 10% solution, that is 10 g in 100 ml of distilled water.

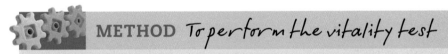 METHOD To perform the vitality test

- Mix one drop of fresh semen with two drops of 1% eosin Y.
- Wait 30 seconds.
- Mix in three drops of 10% nigrosin solution.
- Immediately place one drop of the mixture on a glass slide.
- Prepare a blood film type smear.
- Allow to air-dry.
- Examine under oil immersion using an x100 objective.
- Count 200 sperm, calculate the percentage of live, that is, unstained, sperm.
- Note the result.

Which type of sperm is stained by the eosin Y solution?

Why does the stain solution penetrate these sperm?

12.8 Interpreting the test results

Fortunately, the WHO manual includes a set of reference ranges for infertility tests, establishing acceptable levels. You can see these in Table 12.4. However, since only a single sperm is required to fertilize the egg, men with semen scores well below the values given can potentially still be fertile.

Semen is a mixture of secretions delivered by an intricate process, so if the results are out of the reference ranges, it is worth examining some of the tests in a little more detail in order to consider what the problem(s) might be.

Insufficient semen volume

There are two issues to consider here: namely is the ejaculate volume genuinely low or did some get lost during production. Patients should always be asked to confirm that the entire specimen was collected in the sample container.

Successful ejaculation of semen depends on the closure of the urethral sphincter which controls urine release from the bladder. If this fails to occur, some or all of the semen passes into the bladder, a process known as **retrograde ejaculation**. Fortunately, this is easy to test for. A cytological assessment of a post-coital urine sample will reveal the presence of millions of sperm in the urine.

TABLE 12.4 **Infertility test result lower reference limits (WHO manual 2010).**

Test parameter	Reference values
Semen volume	1.5 ml or more
Total number of sperm per ejaculate	39 million
Sperm concentration	15 million sperm per ml
Total motility including non-progressive sperm	40%
Progressive motility	32%
Vitality (live spermatozoa)	58%
Morphology (% normal)	4%
pH	7.2 or higher
White blood cells	Fewer than 1 million per ml
Sperm antibodies	Less than 50% of sperm reactive (individual manufacturers' test kits may vary)

Low pH

Semen pH should be a little alkaline, that is 7.2 or above. This shows the balance between prostate secretions which are acidic and alkaline seminal vesicle fluid. A low pH with a large semen volume may simply indicate a rather generous contribution from the prostate gland. If, however, there are no sperm present in the ejaculate (azoospermia), then it may indicate the absence of any seminal vesicle component, due to obstruction of the ejaculatory ducts.

Low sperm concentration

A large semen volume will have a diluting effect that will reduce the sperm count, so this parameter should always be considered alongside the total number of sperm in the ejaculate. A man's sperm count can also vary considerably over time, so a count must be consistently low across several samples before it becomes a definite cause for concern.

A consistently low count, with other factors recording as normal, indicates poor functioning of the testes. This can be due to a variety of causes, some of which are *environmental*, such as some of the following risk factors:

- A history of reproductive tract infections or sexually transmitted diseases.
- Exposure to environmental pollutants, for example agricultural chemicals.
- Male athletes exposed to conditions which might lower sperm counts.
- Alcohol abuse.
- Use of tobacco/recreational drugs.
- Medicinal drugs, especially steroids.
- Previous testicular surgery or recent trauma.

It is also worth remembering that the testicles are located outside the body for a good reason, that is, temperature control. Overheating is to be avoided. A cold shower rather than a hot bath may be an aid to conception.

Low motility or morphology

The same factors can also apply here; sperm that have been chemically poisoned or poached in a hot bath cannot really be expected to look good or swim well.

However, other factors involved in sample collection and transport, such as lubricants and inappropriate specimen containers, should be discounted first. Patients should always be provided with a suitable container, along with instructions on how to produce and collect the sample.

When motility is being assessed, a warm microscope stage is important. Sperm are designed to be swimming at body temperature, therefore a cold microscope on a winter morning can easily be 20 degrees below that.

Low vitality

Abuse does, of course, kill sperm; however, the vitality test can also indicate if the immotile sperm are alive but unable to swim because they have faulty flagella.

These immotile, but still living, sperm can still be used in some assisted conception techniques, for example intracytoplasmic sperm injection, where a single sperm is injected directly into an egg.

Anti-sperm antibodies

Once the blood-testes barrier has been breached and anti-sperm antibodies produced, they are very difficult to remove. The use of corticosteroid treatments only temporarily lowers the production of antibodies and has potent side effects. After a vasectomy, at least 70% of men have anti-sperm antibodies. Despite this, pregnancies can often still be achieved after a vasectomy reversal. Clearly the situation is more complex than indicated by the MAR test result alone.

As long as visibly large numbers of motile sperm are still present in the sample, the effects are not always catastrophic.

SELF-CHECK 12.22

What can occur if the urethral sphincter fails to close during ejaculation?

CASE STUDY 12.1

On return from a Mediterranean holiday a 20-year-old male patient with a short history of discoloration of semen was submitted for testing by his GP. On analysis it was found that the semen was reddish brown in colour and although all elements of his sperm count were in the normal range there were numerous red blood cells and leucocytes (five million per ml) present.

Full test results:

Liquefaction: Complete
Appearance: Reddish brown
Volume (in ml): 3 ml
Viscosity: Normal
pH: 7.9
Motility:
 Rapid 35%
 Sluggish 15%
 Non-progressive 10%
 Immotile 40%
Morphology (percentage of normal sperm): 20%
Sperm and leucocyte count (millions per ml):
 Sperm 120 million per ml
 Leucocytes 5 million per ml
Anti-sperm antibodies: 0%
Vitality (percentage of live sperm): 60%

With indicators of inflammation and infection present the GP referred the patient to an STD clinic for further testing. Microbiological assessment confirmed infection with the sexually transmitted disease Chlamydia, which was successfully treated with antibiotics. Subsequent semen samples were clear of red blood cells and leucocytes and also MAR negative, indicating that the patient had been fortunate and that the infection had not caused any long-term damage to his fertility.

CASE STUDY 12.2

A 45-year-old male and his younger partner attended their GP after trying unsuccessfully to conceive for 12 months despite both having conceived in previous relationships; the GP requested an infertility sample and referred the couple to the fertility clinic. The initial infertility semen analysis showed semen volume low, at only 0.25 ml, but all other parameters in normal range, and this was confirmed by a second test performed in the clinic. All other tests were normal and the couple both seemed to be in good health.

Full test results:

Liquefaction: Complete
Appearance: Cloudy white
Volume (in ml): 0.25 ml
Viscosity: Normal
pH: 7.9
Motility:
 Rapid 30%
 Sluggish 25%
 Non-progressive 13%
 Immotile 28%
Morphology (percentage of normal sperm): 17%
Sperm and leucocyte count (millions per ml):
 Sperm 88 million per ml
 Leucocytes 0.1 million per ml
Anti-sperm antibodies: 0%
Vitality (percentage of live sperm): 73%

On a second visit to the clinic the male patient recalled that he used to produce a bigger ejaculate when he was younger, leading to a clinical suspicion of retrograde ejaculation. This diagnosis was subsequently confirmed through the presence of sperm in a post-coital urine sample. After discussion at MDT, which focused on making the best use of the male partner's small ejaculate of good quality sperm, the couple were recommended for IVF treatment. This was successful and they became the proud parents of twin girls nine months later.

CASE STUDY 12.3

A 25-year-old male and his partner attended their GP after trying unsuccessfully to conceive for 18 months despite the female partner having a child from a previous relationship. On examination the male partner had a persistent cough and reported frequent chest infections but otherwise both appeared to be in good overall health. The GP requested an infertility sample and referred the couple to the fertility clinic. The initial infertility semen analysis showed a normal count, but no motile sperm were present, although the vitality test indicated that 66% of the sperm were alive. This was confirmed by a second test performed in the clinic.

Full test results:

Liquefaction: Complete
Appearance: Cloudy white
Volume (in ml): 2.25 ml
Viscosity: Normal
pH: 8.1
Motility:
  Rapid 0%
  Sluggish 0%
  Non-progressive 0%
  Immotile 100%
Morphology (percentage of normal sperm): 17%
Sperm and leucocyte count (millions per ml):
  Sperm 88 million per ml
  Leucocytes 0.1 million per ml
Anti-sperm antibodies: Test not performed due to absence of any motile sperm.
Vitality (percentage of live sperm): 66%

Further investigation of the male partner found that his persistent cough and suscepti-bility to chest infections were due to a genetic condition known as immotile cilia syn-drome (ICS) which also accounted for his live but immotile sperm. After discussion at MDT, which focused on finding a solution to the lack of sperm motility, the couple were recommended for an IVF treatment known as intracytoplasmic sperm injection, where a single sperm is injected directly into an egg cell. The treatment proved effective, leading to a normal pregnancy and a healthy baby.

SUMMARY

- Semen is a complex fluid consisting of spermatozoa and numerous organic and inorganic constituents that provide nutrition and protection for the spermatozoa during their jour-ney through the female genital tract.

- Semen analysis is recommended for patients following a vasectomy, to ensure that no further sperm remain in the male reproductive system, and when a fertility problem is suspected.

- Semen analysis is a simple test that can prevent a lot of unnecessary investigation of the female partner.

- Semen analysis comprises a set of qualitative and quantitative measurements of sperma-tozoa and seminal fluid that helps the clinician to estimate semen quality.

- Standard tests include measurement of volume, pH, sperm, total sperm count, motility, morphology, vitality, and MAR test.

- Normal values of semen parameters issued by the World Health Organization (WHO) in 2010 are generally used as reference.

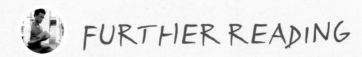

FURTHER READING

- **WHO (2010)** WHO Laboratory Manual for the Examination of Human Semen and Sperm-cervical Mucus Interaction, fifth edition. WHO, Geneva.

- On the website of the ABA at http://www.aba.uk.net/ is a very useful reference document entitled 'ABA Guidelines for Good Practice'.

- For the more technically minded there is an article on the theory behind phase contrast microscopy at: http://en.wikipedia.org/wiki/Phase_contrast_microscopy, or if that is a bit too specialized try: http://www.olympusmicro.com/ for a really wide ranging microscopy resource.

DISCUSSION QUESTIONS

1. Have a look around the laboratory area where you are currently working or studying and compile a list of the laboratory equipment that is available to you which you would use in semen analysis. Discuss what else you would need to do to set up a successful semen analysis service.

2. Discuss the essential features of a patient information leaflet that will cover all of the details required for the successful production and safe transport of semen samples to the laboratory. Produce a simple, compact sample leaflet that will fit on one side of A4.

3. Discuss the lifestyle questions that could be asked of a young male patient presenting with a sub-optimal infertility screen profile and the advice that might be given to help him improve his fertility.

Answers to the self-check questions, and tips for responding to the discussion questions, are provided in the book's Online Resource Centre.

 Visit www.oxfordtextbooks.co.uk/orc/shambayati/

Advances in cytopathology

Margaret Morgan

In the previous twelve chapters you have been introduced to gynaecological and non-gynaecological cytology, including semenology and FNA. This has aimed to give you an understanding of the basic principles and diversity of cytology and the complexities which exist around delivering a successful screening programme, preparatory techniques, common applications; where relevant, investigations and treatments have been discussed.

In this chapter we consider automated screening and what role it could undertake in a cervical screening programme. We will discuss the overall impact of HPV testing, the cervical cancer vaccine, and the effect they will have on the future of cervical cytology.

Immunocytochemistry (ICC) has been in use for many years so the intention in this chapter is not to revisit routine ICC but to explore how new markers can be used to give better diagnostic and predictive information.

Molecular techniques are emerging as useful tests to be used alongside conventional cytomorphology. Many cancers have an underlying genetic cause; analysis of DNA and RNA at the molecular level is yielding more information and contributing to the diagnosis of disease and subclassification of tumours.

Most of what will be discussed in this chapter is not intended to fully replace conventional cytological techniques but to build on existing expertise and extend the use of techniques to help clinicians, surgeons, and **oncologists** provide better treatment and improve survival rates of patients with cancer.

> **Oncologist**
> A medically qualified doctor who specializes in the study and treatment of cancer.

Learning objectives

At the end of this chapter you should be able to:

- Describe the principles of automated screening.
- Explain the rationale and benefit of vaccinating young women against high-risk HPV.
- Discuss how recent advances in immunocytochemical techniques have contributed to management of cancer patients.
- Describe the principles behind flow cytometry.

- Describe the polymerase chain reaction (PCR) and why it has revolutionized molecular diagnostics.
- Discuss *in situ* hybridization techniques which can be used to complement diagnosis.
- Describe the principle behind tissue microarray technology.
- Describe some key applications of molecular techniques in current cytology practice.

13.1 Automated cervical screening

Liquid based cytology

Compared to other disciplines cervical cytology has been slow to benefit from automation, relying on traditional methods such as direct spread and centrifugation techniques to produce samples. It is one of the last high volume tests to benefit from automation.

The main obstacles to early automation were cell overlap, debris, and blood. The introduction of liquid based cytology has minimized these problems. Before continuing it may be worth revisiting Chapter 2 on preparatory techniques. In principle, LBC slides are prepared in such a way that a **near monolayer** of cells is obtained (Figure 13.1). This makes LBC preparations more suited to imaging than the conventional smear. That said, some systems can image conventionally prepared cervical smears.

Near monolayer
A term used in LBC that refers to a cellular preparation which comprises only one to two layers of cells.

Cross references
Chapter 2.
Chapter 3, Section 3.1.

SELF-CHECK 13.1

What impact has the introduction of liquid based cytology had on automated screening?

Automated screening

For the last few decades much effort and capital resource has gone into developing image analysis systems that could capture and analyse all that a cytologist's brain assesses. Most automated systems were introduced as quality control systems prior to being used as primary

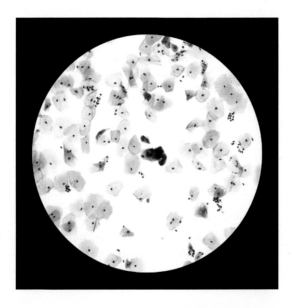

FIGURE 13.1
ThinPrep®, near monolayer of cells making assessment of nuclei easier.
Reproduced with kind permission from Hologic Corporation.

screening devices. Current automated screening systems tend to adopt a more interactive approach, combining human interpretive skills with machine capability. Robust evidence is needed to ascertain if automated systems can at least match in terms of accuracy, or ideally exceed, manual screening.

The two systems with **FDA** approval and the lion's share of the global market are the ThinPrep® Imaging System (TPIS) (Figure 13.2a) and the Becton Dickinson FocalPoint GS Imaging System (BDFP) (Figure 13.2b).

Different systems are in existence but the principles are the same. The cells considered of interest are located, coordinates logged electronically, scanned, and the image presented to a cytologist for further interpretation. For the most part the cells reviewed will be classified as normal by the cytologist. The use of location guidance software makes it easy for cells to be relocated on the slide at any time. Cells are categorized on morphological parameters such as nuclear size, nuclear cytoplasmic ratio, chromatin pattern, and distribution, as well as the DNA content of each cell. Abnormal DNA content or detection of **non-diploid** cells is indicative of an increased risk of an abnormality. The location of cells within the microscopic **fields of view** (FOV) will be stored for future reference.

Cells identified by the imager will be presented to the operator and transferred from the imaging station to the review station, which enables the cytologist to make the final interpretive

United States Food and Drug Administration (FDA)
This organization regulates and supervises the food and drug industry on behalf of the public.

Cross reference
Chapter 4, Section 4.4.

Fields of view (FOV)
Areas selected by the imager for manual review and deemed of interest, not necessarily abnormal.

FIGURE 13.2
(a) ThinPrep® imager to the right linked to an automated microscope and a computer screen.
Reproduced with kind permission from Hologic Corporation.
(b) BD FocalPoint GS imaging system slide profiler scans conventional and LBC slides ranking slides based on the probability of containing an abnormality.
Reproduced with kind permission from Becton Dickinson.

 BOX 13.1 *Definition of ploidy*

Ploidy is the number of complete sets of chromosomes in a biological cell. In humans, the cells that compose the body are diploid, containing 23 pairs of homologous (similar) chromosomes. Ploidy analysis is a test that measures the amount of DNA in tumour cells. **Diploid** cancers have a normal number of chromosomes and tend to have a more favourable outcome than **aneuploid** (abnormal number of chromosomes) cancers.

Cross reference
Chapter 2.

decision and accept the systems selection of cells, or choose to rescreen the entire slide manually before issuing a final negative or unsatisfactory report (Figure 13.3a). If the cells are considered abnormal they are referred to a cytopathologist or a consultant biomedical scientist to report and advise on patient management (Figure 13.3b).

A unique feature of the BDFP system is the existence of two categories: cases requiring 'manual review' (MR) and cases requiring 'no further review' (NFR).

The latter is currently set at a maximum of 25% by the FDA and permits 25% of slides to be signed out of the laboratory as negative without any further intervention or review by the cytologist. The graph in Figure 13.4 shows the distribution of cells for review and non-review. Most publications agree that the **high negative predictive** value of the NFR category is sufficient to eliminate the need for manual quality control (1).

High negative predictive value
A marker of accuracy that is the number of tests considered negative by the imager which have been correctly classified.

The use of automated screening systems varies from country to country and indeed from laboratory to laboratory, some continuing to use them as a quality control device, whilst others are using them for primary screening. A decision to use automated screening is most likely to

(a)

(b)

FIGURE 13.3
(a) BD FocalPoint GS workstation for cytologists comprising an automated microscope linked to a computer terminal.
Reproduced with kind permission from Becton Dickinson.
(b) A single microscopic FOV showing high-grade (severe) squamous dyskaryosis located by ThinPrep® imager and presented to cytologist for review and assessment.
Reproduced with kind permission from Hologic Corporation.

FIGURE 13.4
Graph showing distribution of cells, with A representing the no further review and those on far right (B and C) requiring a manual rescreen.
Reproduced with kind permission from Becton Dickinson.

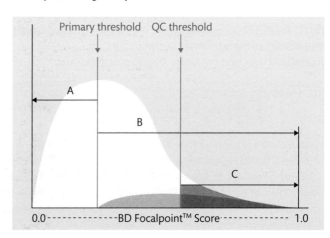

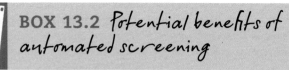

BOX 13.2 *Potential benefits of automated screening*

- Improved sensitivity and specificity.
- Reduced false negatives.
- Reduced false positives and risk of over-treatment, especially in women of child-bearing age.
- Objective and standardized measurements, eliminating human bias and subjectivity.
- Increased productivity, no fatigue or attention deficit.
- Greater job satisfaction for the cytologist by concentrating their skills set on slides requiring more expert interpretation.
- Cost effective.

be determined by the quality of manual screening, the ability to recruit and retain cytologists, productivity gains, and the overall cost effectiveness of such systems (2–7).

SELF-CHECK 13.2

What are the potential benefits of automated screening?

Quality control and automated screening

Regardless of what systems are in use, quality control must be rigorous and meet all the requirements of internal and external agencies, such as **CPA** and **QARCs** and the women being screened.

In keeping with CPA and QARC and to reflect good working practices, accurate and robust processes must exist around receipt, processing, and interpretation of cervical samples. This will necessitate accurate and up-to-date standard operating protocols. Supervisory staff and managers will have a key role in maintaining quality management within the laboratory.

One example of quality control (QC) is the sample rejection capabilities of automated screening devices. They are designed to reject samples which are not 'fit for imaging' due to technical reasons or poor cellularity. To maintain quality and reduce the cost burden it is important to monitor unsatisfactory rates and take corrective action. This may be with a sample taker, an in-house technical problem, or a design failure.

Cross reference
Chapter 3.

Clinical Pathology Accreditation Ltd (CPA)
A regulatory body which oversees quality in UK, conducting peer review visits every three years to registered laboratories.

Quality Assurance Reference Centres (QARCs)
Regional centres which oversee the quality of cervical screening and colposcopy services on behalf of the NHSCSP.

Key Points

All laboratories screened on behalf of NHSCSP are required to participate in regionally run technical external quality assurance (TEQA) schemes. Stains for automated systems will need to meet the quality standards, which may need to be revised if automation replaces manual screening.

Training staff and assessing competencies play a key role in delivering a high quality service. The FDA has made it mandatory for all staff using new devices to undergo a manufacturer's bespoke training programme to cover technical and morphological elements. Such programmes will cover demonstrations, day-to-day operations, basic troubleshooting and a familiarization with morphological images presented by the imaging systems. This approach ensures compliance and standardization among users.

Productivity and cost effectiveness of automated screening

Main drivers for automated screening are improved sensitivity and specificity, reducing false negative and false positive tests, increased productivity, and an overall cost benefit. Capital equipment and consumable costs are relatively easy to calculate on a cost per test basis. However, other or hidden costs (transport, IT, quality assurance), which are not so well defined, still need to be factored in. The way in which these systems will be used will be central to determining cost. There is still some conflicting evidence around productivity and cost benefit. Improving the overall quality of the screening service must be central in the decision to convert from manual to automated screening; cost benefit will also be a key determining factor. The other key players in deciding future strategy for cervical screening are HPV testing and the cervical cancer vaccine: these will be considered in the next section.

13.2 Telepathology

Web-based telepathology is not new, but underperforming IT systems have made it difficult to integrate into the workplace. Improving IT links and connectivity are bringing telepathology to the forefront as services are reconfigured and the need to interface with remote users is becoming more urgent.

Applications include primary diagnosis or second opinions and as an educational tool in distance learning and quality control. Recent advances have permitted a move away from static images transmitted over phone lines and a slow Internet to robotic microscopy stations allowing remote control of screening and the transmission of whole slide images, making it similar to looking at glass slides. This technology has been tested to some degree in automated cervical screening by transmitting images of the highest risk fields of view from a hub to a spoke laboratory. Use of such systems could allow centralization of cytology laboratories and associated resources, and optimize workflow. As with any IT strategy robust back-up systems must be in place to ensure patient confidentiality and well-being is not compromised.

13.3 The role of HPV testing and the cervical cancer vaccine

High-risk human papilloma virus (hrHPV) has been identified as the causative agent in virtually all cervical cancers. Its role in the development of cancer (**carcinogenesis**) has been covered in an earlier chapter. In this section we will consider the benefits and impact of testing for and immunizing against high-risk HPV (hrHPV).

Cross reference

Chapter 5.

High-risk human papilloma virus testing

As persistent hrHPV types (types 16 and 18) are implicated in the development of cervical cancer and precancerous abnormalities it makes sense to identify women most at risk. Early detection and treatment of disease is more effective and significantly reduces the risk of cervical cancer.

Cross reference
Chapter 5, Section 5.1.

Many techniques can be used to detect DNA and RNA in HPV infected cells. The most popular methods are DNA amplification using HC2, PCR, and **invader technology**, which will be discussed in Section 13.6.

The high negative predictive value of HPV testing has led to it being considered as a primary screening replacement for cytology. Studies have shown that despite a high sensitivity there is a low specificity, as HPV is widespread in under 35-year-olds. As a screening test it would be oversensitive and detect too many cases of transient virus which would naturally clear the body given time. Several large trials have taken place including ARTISTIC and TOMBOLA to test the clinical and psychological effect of HPV testing in women. One of the main findings from ARTISTIC was that it would not be cost effective to screen with cytology and HPV combined. HPV testing was of benefit in a **triage** setting or as a primary screening tool triaged by cytology (8).

Triage refers to separating or streamlining women who are hrHPV positive.

Key Points

ARTISTIC (a randomized trial of human papillomavirus (HPV) testing in primary cervical screening) began in 2001 and set out to investigate the role of HPV testing within the national cervical screening programme, including benefits, psychological effects on women, and overall cost effectiveness. The HTA report trial concluded that HPV testing did not add to the effectiveness of cytology (LBC) over two screening rounds and that addition of HPV testing to LBC would not be cost effective. HPV testing used as triage or as an initial test triaged by cytology would be cheaper than cytology. HPV testing was shown to have a high negative value and suitable for high throughput testing with the use of automated platforms. The report acknowledged that replacing cytology with HPV primary screening would require major contraction and reconfiguration of cytology services.

TOMBOLA (Trial of Management of Borderline and Other Low grade Abnormal smears) was a randomized-controlled study to determine the most appropriate way to deal with positive HPV results and associated psychological issues. Final results showed 22% of women with CIN2 or worse were HPV negative. Conversely, 40% of women with HPV positivity showed no evidence of CIN. The authors concluded that in younger women a single HPV test would not be useful in determining who should be sent to colposcopy. In women over 40 a negative HPV test could be used as a test of cure and rule out further investigation.

HPV testing is being carried out as an additional test in many instances to guide the management of women with low-grade abnormalities and as a **test of cure** following treatment. The role of HPV testing will continue to develop as techniques improve and the debate continues as to whether hrHPV could replace cytology as a primary screening test.

Test of cure in this scenario refers to HPV testing and whether the HPV virus has cleared the woman's system.

The overall benefit to the screening programme is that only women with persistent hrHPV will require colposcopy. A negative hrHPV status will allow women to be returned to **normal recall** and possibly for longer screening intervals, avoiding the need for repeat tests and treatments.

SELF-CHECK 13.3

Why is HPV testing a less than ideal primary screening tool in the under 35-year-olds?

Cervical cancer vaccine

The medical, scientific, and research communities stand united in their belief that immunizing against hrHPV will drastically reduce incidence of cervical cancer.

CERVARIX® (GSK Figure 13.5a) and **GARDASIL®** (Merck Figure 13.5b) are currently the two leading vaccines which offer protection against cervical cancer by immunizing against hrHPV. CERVARIX® is an FDA approved bivalent vaccine (protects against two hrHPV types 16 and 18). In 2008, the UK Government took the decision to vaccinate all girls from age 12 with CERVARIX®.

Gardasil® is a quadravalent vaccine (it protects against four types 6, 11, 16, and 18). Like any vaccine they only offer protection to individuals who have not yet been infected with the virus, hence the decision to vaccinate young girls. The debate on vaccinating men remains open.

There are some reports of adverse reactions in young women, uncertainty about the duration of immunity, and concerns that the prevalence of different HPV types varies across continents, for example sub-Saharan Africa (9–10). In time, the true value of the vaccines will become known. It is likely that further vaccine development and HPV testing (self-sampling and testing) are likely to offer the best hope for poorly resourced settings. For these reasons alone it is imperative for scientists to continue to map HPV subtypes across the world so that preventative vaccines can be produced which will work in any geographical area. Drug companies will come under increasing pressure to fund and develop bespoke (customized) vaccines.

Key Points

Persistent hrHPV infection is the causative agent in most cervical cancers. However, current HPV vaccines only offer protection against two high-risk types (16 and 18) and *do not protect against all cervical cancers*. It is important for women to continue with regular cervical screening. The NHS cervical screening programme will continue to offer a cervical screen to women from the age of 25.

(a)

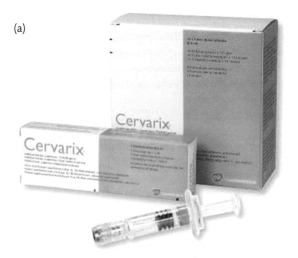

(b)

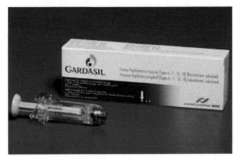

FIGURE 13.5
(a) Cervarix bivalent vaccine is FDA approved and used to immunize girls in the UK.
(b) Gardasil quadravalent vaccine manufactured by MERCK.
Reproduced with kind permission from MERCK.

SELF-CHECK 13.4

What impact will the vaccine have on the cervical screening programme?

13.4 **Future trends in immunocytochemistry**

Immunocytochemistry (ICC) has been in routine use for many years as an **adjuvant** to conventional morphology. Panels of antibodies are regularly used to help the pathologist establish a final and accurate diagnosis.

Diagnosing cancer on cytological and histological morphology alone is no longer sufficient to provide the prognostic and predictive information required by clinicians to manage patients. Differentiating benign from malignant is the first line of cytological investigation. Characterizing difficult tumour cell types by performing ICC is regarded as standard and best practice. More recently, ICC has been found to be useful in determining if some forms of **targeted therapies** will work on certain types of tumours. Cell block preparations offer similar advantages to histological samples, provided adequate material is available. Some manufacturers have developed automated rapid cell block systems to maximize cellularity and enable cytology laboratories to process cell blocks independently from histology.

It is beyond the scope of this chapter to cover the plethora of biomarkers commercially available. Relevant immunocytochemical markers have been referred to in previous chapters. In this section we will concentrate on the use of ICC to determine which patients are suitable for particular targeted therapies, **HER2** status being a good example. We will also consider briefly the important role certain proteins (minichromosome maintenance proteins (MCMs)) and enzymes (Topoisomerase II alpha (**TOP2A**)) play in the regulation of the cell cycle and replication of DNA. Over-expression of MCM and TOP2A occurs during **aberrant S-phase induction** (Figure 13.6) and can be detected using immunocytochemical tests.

Adjuvant
Supplementary test or tests.

Targeted therapies
These are aimed at selectively killing cancer cells whilst preserving surrounding healthy tissues.

Cross reference
Chapters 2 and 8–11.

> *Key Points*
>
> The cell cycle as demonstrated in Figure 13.6 comprises four phases: S phase, M phase, and G1 and G2 phases. During S phase, DNA is replicated and chromosomes are duplicated in the nucleus. An aberrant S-phase induction occurs when there is a premature or delayed entry into S phase of the cell cycle from G1/S checkpoint where DNA is checked for errors prior to replication.

HER2 status in women diagnosed with breast cancer

HER2 receptors are proteins located on cell membranes and are responsible for regulating cell growth by switching genes on and off. Tumours which are HER2 positive are over-producing HER2 protein. HER2 negativity indicates HER2 proteins are not responsible for the cancer. Approximately 20% of breast cancers express HER2 and are associated with a poor prognosis.

HER2 status can be demonstrated by several methods including **FISH** and **CISH**, though ICC, which remains the most popular as it is readily available as a kit and can be easily performed in a routine laboratory. Material must be formalin fixed so cell blocks are the most

Chromagenic *in situ* hybridization (CISH)
This uses a coloured substrate rather than a fluorchrome.

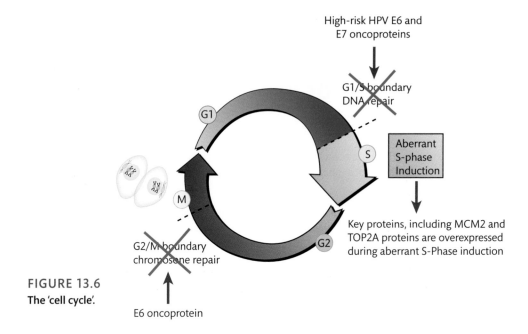

FIGURE 13.6
The 'cell cycle'.

Cross reference
Section 13.9.

Herceptin (trastuzumab)
This is a drug used to treat HER2 positive cancers.

HER2
This is a protein made by HER2/neu gene. Receptors on the protein can bind to cancer cells causing them to divide and grow.

Cross reference
Chapter 11.

Equivocal
This term is used in cytopathology reporting to indicate uncertainty.

Biomarker
Provides information at the genetic level by determining changes in DNA sequence which are associated with a specific disease or predisposition to a disease.

appropriate after histological core samples. Breast samples are scored on a scale of 0–3, with 0 being negative, 1 weakly negative, and 3 strongly positive. Cases with an intermediate score of 2 can be sent for FISH analysis.

Herceptin (trastuzumab) is a drug which can be used to treat **HER2** positive breast cancers, hence the importance of establishing HER2 status to select patients for this kind of treatment. The trastuzumab antibody binds selectively to the HER2 receptor protein thereby causing the cancer cells to cease reproducing. HER2 positive breast tumours respond better to treatment and have a better survival rate than HER2 negative tumours. Intravenous administration of herceptin is regarded as a targeted therapy as it targets HER2 protein production. Herceptin can help treatment of the disease and patient management by suppressing tumour cells and reducing the size of the tumour prior to surgery. Also, by binding to HER2 receptor protein herceptin inactivates tumour cells in metastatic deposits and helps prevent tumour recurrence, thereby prolonging life (11).

SELF-CHECK 13.5

Why is determining HER2 status important in patients with breast cancer?

p16

The **p16** protein also known as CDKN2A or INK4a is a tumour suppressor gene located on chromosome 9p21. It plays an important role in regulating the normal cell cycle and aims to specifically differentiate persistent hrHPV from transient hrHPV. Over expression of p16 is seen in cells where hrHPV oncogenes have initiated cell transformation. It is considered significant in all cancer and intraepithelial lesions of the cervix. Used in combination with the Pap stain, p16 can prove useful in interpreting **equivocal** cases as it provides **biomarker** information without destroying morphological features (Figure 13.7a). p16 is demonstrated as a brown cytoplasmic staining (Figure 13.7b). Sensitivity and specificity of p16 staining increases

(a)   (b)

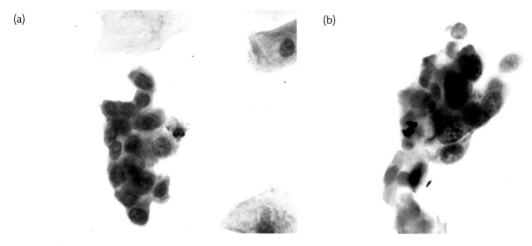

FIGURE 13.7
(a) Pap sample showing abnormal changes of uncertain dyskaryotic grade in squamous cells.
(b) High-grade squamous cell dyskaryosis showing strong positive cytoplasmic staining with p16.

with grade of neoplasia, with some studies showing a 100% sensitivity for CIN 3 (12, 13). Immunostaining for p16 works well on LBC samples and cell block preparations, comes as a standardized kit, and can be performed in any cytology laboratory. p16 can stain *Trichomonas vaginalis*, endometrial cells, and metaplastic cells; therefore it is important to compare and interpret morphology alongside ICC findings.

Ki67

Immunostaining for Ki67 is a cell proliferation marker which can be used alongside other biomarkers such as p16 to give a dual staining result (as seen in Figure 13.8). Both markers are expressed more in CIN2 plus lesion. Ki67 positivity in tumour cells can act as a prognostic

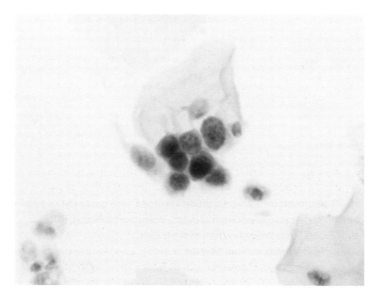

FIGURE 13.8
p16 brown cytoplasmic staining and Ki67 red nuclear stain.
Reproduced with kind permission from Mr Nick Dudding, East Pennine Cytology Training Centre.

marker and help predict the likelihood of survival and recurrence in certain tumour types. Like p16 it is available as a diagnostic kit, making it suitable for use in a routine laboratory setting. Ki67 is demonstrated as a red nuclear stain, contrasting well with the brown cytoplasmic staining of p16 (14).

Topoisomerase II alpha

The TOP2A gene encodes **topoisomerase**, an enzyme which alters DNA during transcription and replication. TOP2A gene amplification is thought to be predictive of a response to a type of inhibitor which includes the **anthracyclines**.

Activity can be demonstrated using ICC, FISH, and PCR.

Key Points

Minichromosome maintenance proteins (MCMs) have a key role to play in the initiation and elongation of the DNA process which takes place during cell replication.

ProExC is a commercially available kit which can be used with standard ICC techniques to detect the presence of key proteins minichromosome maintenance protein 2 (MCM) and Topoisomerase ii alpha (TOP2A), which are both over-expressed during **aberrant S-phase induction** (Figure 13.6). ProExC can provide the additional information required to diagnose a high-grade cervical lesion when the morphology findings on Pap are equivocal (Figure 13.9a); ProExC staining is confirmed in Figure 13.9b and the biopsy supporting the cytological diagnosis of CIN3 with H&E and ProExC staining is shown in Figure 13.9c (15).

Human papilloma virus L1 capsid protein

Most low-grade cervical lesions CIN1 express HPV L1 capsid protein, but in high-grade lesions CIN2–3 the HPV L1 capsid protein is missing. Low to moderate dysplastic squamous lesions

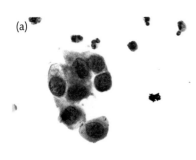
(a)

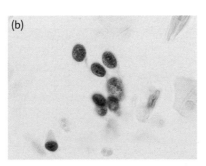
(b)

(c)

FIGURE 13.9
(a) A Pap sample showing atypical squamous cells, grade uncertain, high grade dyskaryosis.
(b) BD ProEx C ICC stain on same sample. (c) Cervical biopsy stained with ProExC (CIN 3),
insert shows haematoxylin and eosin stained section of same case.
(b) and (c) reproduced with kind permission from Mr Nick Dudding, East Pennine Cytology Training Centre.

with hrHPV but without immunochemically detectable HPV L1 capsid protein are significantly more likely to progress than L1 positive cases.

Cytoactiv is an example of a prognostic marker which stains the L1 capsid protein associated with HPV infection. The protein is synthesized in the cytoplasm and makes its way to the nucleus. Using an immunocytochemical kit the active protein can be demonstrated as red nuclear staining using a standard light microsope. One stained cell is sufficient to be considered positive and is an indicator that this lesion is more likely to regress than progress.

The morphology of the cell is also preserved, which is an additional benefit to the interpreter. It is also possible to perform dual staining with p16.

CD117

CD117 or c-kit mutations are a feature of gastrointestinal stromal tumours (**GIST**): overexpression can be demonstrated on formalin fixed cytological samples (air-dried post-fixed formalin slides or cell block preparations).

Cross reference
Chapter 11.

13.5 **Flow cytometry**

Flow cytometry analysis (FCA) is usually carried out in a specialist haematology laboratory. It measures multiple physical and chemical characteristics of cells, as they flow in a fluid stream through a beam of light. The properties measured include cell size, granularity or internal complexity, and relative fluorescence intensity. Modern flow cytometers can analyse thousands of cells per second in 'real time' and have multiple lasers and fluorescence detectors (Figure 13.10). Cytological samples lend themselves easily to FCA. Cells from solid tissue can be separated before analysis.

FCA in itself is fast, taking less than five minutes to obtain a reading; the time consuming part is the preparation stage, which requires multiple washings and addition of the probes. FCA is routinely used in the diagnosis of haematological disorders, particularly **B and T cell**

B cells
Lymphocytes which develop from bone marrow and are responsible for producing antibodies when they reach maturity.

T cells
Lymphocytes originating from the thymus gland. They play a key role in cell mediated immunity.

FIGURE 13.10
Flow cytometer.
Reproduced with kind permission from Specialis Haematology Laboratory, GSTS Pathology.

clonality studies, and to monitor responses to treatment in HIV patients. Using FCA it is possible to look at the ratio of **CD4 helper cells** to establish if a patient has developed full blown AIDS.

FCA is emerging as an adjuvant tool to assist in cytological diagnosis. This kind of information helps select the best form of patient treatment and cure, or at least prolongs survival. If a Pap or **MGG** preparation from an FNA sample or serous effusion, or indeed any sample, comprises a population of lymphoid cells it is prudent to send washings for FCA.

FCA is most gainfully employed when the cytology is **equivocal**. The most important advantage of FCA compared to immunocytochemistry is that it can analyse single cells and requires a very small sample for analysis. Many of the antibodies available for diagnosing lymphomas work with ICC and FCA.

The findings are interpreted in a similar way to cervical cytology, with the more complex and abnormal samples being referred up to a consultant haematologist for a final opinion. Samples (usually serous effusions and FNA washings) sent from cytology will be reported back to the consultant cytopathologist for interpretation alongside the cytomorphological findings and a final report issued, taking into consideration the FCA, morphological, and immunocytchemical results.

13.6 **Molecular techniques**

Most of the methods which will be described below have been used across pathology for many years, particularly in other disciplines. It is only recently that their potential in cytology is being explored, with some techniques being used more extensively than others (16). All of the techniques discussed below can be applied to cell block preparations.

It is accepted that DNA and RNA are better preserved in LBC samples and can be stored for future molecular testing. This could be beneficial as HPV genotyping comes into use.

FNA cytology will have an increasing role as it is a fast, easy, and non-invasive way to obtain diagnostic and additional material for molecular testing. FNA sampling is an effective tissue sparing method for **tissue banking** and monitoring tumours which have undergone therapy.

> ### Key Points
> Formalin fixation is known to chop, cross bind, and degrade DNA. DNA and RNA are easily extracted from cytology samples as is evidenced by a higher DNA extraction success rate than in formalin fixed paraffin embedded (FFPE) blocks. The inadequate rate for automated DNA extraction in cytology samples is 0.3% compared to 14% in FFPE (17).

The main objective of molecular analysis is either to assess alterations in DNA sequence or to quantify the amount of DNA and RNA in a sample. Thanks to the great achievement of the **Human Genome Project**, DNA sequencing has become a powerful tool which has changed the course of medical science and is the key to understanding the genetic code (sequence of bases within DNA) to allow detection, diagnosis, and treatment of multiple conditions including cancer. **DNA sequencing** employs a reaction similar to that of polymerase chain reaction (PCR), which is described in detail in the next section.

Before discussing molecular techniques below we will revisit some basic genetic terminology.

BOX 13.3 *Structure of DNA*

DNA is a macromolecule which forms a double helix and is composed of four nitrogen rich bases, adenine, guanine, cytosine, and thymine, held together by deoxyribose sugars and phosphates. DNA stores vast quantities of genetic information due to its supercoiling effect. To pass on genetic information DNA needs to replicate. This process takes place during interphase, which is the period of cell growth between divisions. DNA is unable to leave the cell so the sequence is copied to **messenger RNA**; this process is called **transcription**. The messenger RNA then leaves the nucleus and enters the cytoplasm where triplets of DNA bases instruct the cell on which amino acids to synthesize. This process is known as **translation** and is carried out with the help of ribosomes, which read the genetic code from messenger RNA and transport amino acids to the growing protein.

BOX 13.4 *Definition of a gene*

A gene is a basic functional unit of heredity and occurs as a section of **DNA** within a chromosome. The coding part of a gene, which determines what the gene does, is called an **exon**; and the non-coding sequence, an **intron**, controls when a gene is active or expressed. Chromosomes are thread-like structures made up of DNA and housed within the nucleus of a cell (Figure 13.11). Humans have 46 chromosomes made up of 23 matching pairs (**diploid**). Cancers may sometimes be diploid and for the most part these have a better prognosis than their **aneuploid** counterparts.

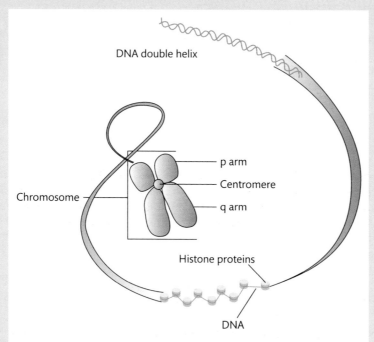

DNA double helix

p arm

Centromere

Chromosome

q arm

Histone proteins

DNA

FIGURE 13.11
Diagrammatic illustration of a chromosome showing centromere, p and q arms.

BOX 13.5 *Definition of a mutation*

Mutations are changes in DNA sequence which may result in disease and can occur for many reasons: some are spontaneous; others are induced by chemical agents or exposure to radiation. **Somatic mutation** refers to a mutation occurring in a non-germline cell and this is the type we will be primarily concerned with. Although somatic mutations can give rise to cancer, one advantage is that they are seldom transferred from one generation to the next. **Length mutations** include deletions, insertions, duplications, and trinucleotide repeats (short sequences repeated tandemly). **Point mutations** occur when a single nucleotide or base pair is replaced by a different nucleotide/s. A **transition mutation** occurs when a purine (adenine or guanine) is replaced by another purine. A **tranversion** is the name given when a purine is substituted by a pyrimidine (CpG). Certain regions of the **genome** are more prone to mutations than others. CpG sites (islands) are common hotspots for mutations to occur.

Key Points

DNA methylation imbalance is known to play an important role in cancer and tumourgenesis, as it causes inactivation of tumour suppressor genes. Detection of changes with PCR is a useful diagnostic tool.

Amplification techniques

Amplification
An increase in the number of copies of a gene in a cell.

In vitro
An artificial environment outside a living organism.

These can be further subclassified into **signal** (HC2) and **target amplification** (PCR). Signal amplification, as the name suggests, amplifies the signal in the presence of target DNA or RNA without increasing the amount of nucleic acids. The latter increases the number of copies of target DNA or RNA by *in vitro* synthesis of nucleic acids.

SELF-CHECK 13.6

Explain the difference between signal and target amplification.

Hybrid capture 2

The HC2 test is an FDA approved signal amplification method that detects 13 HPV types, including hrHPV types 16 and 18. It is an *in vitro* nucleic acid hybridization assay that uses antibody capture and a qualitative chemiluminescent signal detection of HPV types (Figure 13.12).

Polymerase chain reaction

Polymerase chain reaction (PCR) enables the amplification of a specific DNA strand *in vitro*, allowing millions of copies of a pre-selected DNA sequence to be generated.

Polymerase chain reaction has revolutionized molecular diagnostics; it is a highly sensitive technique, easy to set up, and with real time PCR results can be turned around in hours.

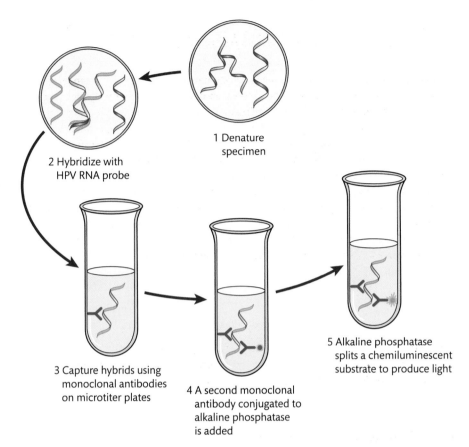

1 Denature specimen

2 Hybridize with HPV RNA probe

3 Capture hybrids using monoclonal antibodies on microtiter plates

4 A second monoclonal antibody conjugated to alkaline phosphatase is added

5 Alkaline phosphatase splits a chemiluminescent substrate to produce light

FIGURE 13.12
Diagrammatic illustration of HC2 signal amplification reaction.

BOX 13.6 Methodology of hybrid capture 2

DNA is extracted from a cervical sample, **denatured** and **hybridized** to an RNA probe that is complementary to the target DNA, forming an RNA/DNA hybrid, which is then captured onto a solid phase coated with antibodies specific for RNA: DNA hybrids. The latter are labelled for detection with multiple antibodies conjugated (bound or linked) to alkaline phosphatase. The addition of a chemiluminescent substrate results in emission of light (relative light units) in proportion to the number of copies of the target present in the sample. The assay is semi-quantitative as it identifies multiple HPV types rather than specific genotypes. The Hybrid capture 2® assay is highly sensitive, reproducible, and specific. The test can easily be performed on a standard piece of equipment (Figure 13.13) in a virology or cytology setting by scientists who are trained in RNA handling and clean techniques to prevent cross-contamination between samples.

Denaturation
The process of physically separating DNA into two separate strands of DNA using heat.

Polymerase chain reaction does have some limitations; these include risk of contamination leading to false positive results. However, most of the disadvantages have been minimized by the setting up of 'clean' areas within laboratories and the introduction of automated closed systems. False positive results can be avoided if scientists have adequate training and a thorough

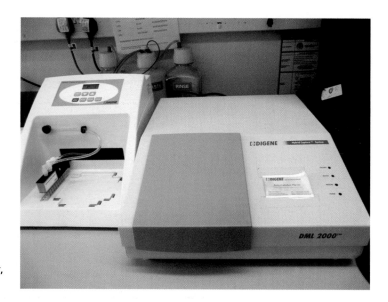

FIGURE 13.13

Digene HC2 system.
Reproduced with kind permission from Immunology and Infection Department, GSTS Pathology.

Viral load

This is measured in RNA copies per ml of blood and acts as an indicator of the amount of virus present in a sample.

understanding of the **viral load** associated with specific viral diseases. This ensures accurate interpretation of results and ability to spot anomalies when they occur.

SELF-CHECK 13.7

Why has PCR revolutionized molecular testing?

BOX 13.7 *Methodology of polymerase chain reaction*

(Figure 13.14a)

Thermal cycler

Equipment which amplifies segments of DNA via the polymerase chain reaction (PCR) process enabling millions of copy DNA sequences to be synthesized.

Real time PCR is carried out on microplates that are inserted into a **thermal cycler** (Figure 13.14b), which heats and cools the tubes to achieve the temperature required for each of the three temperature stages contained within a cycle.

Denaturation or melting of DNA by disrupting hydrogen bonds occurs at 94–98°C for 20–30 seconds and produces single strands of DNA.

Annealing is the process whereby the primer binds to the DNA strand; this occurs at 50–65°C degrees for 20–40 seconds.

Taqman probes

These are probes which are specially designed to increase the specificity of real-time PCR assays relying on Taq polymerases to cleave to complementary target sequence. Use of a fluorescent signal allows quantification of the product.

Elongation is the part of the process at which the DNA polymerase synthesizes a new and complementary strand of DNA by adding deoxynucleotide triphosphates. The time taken to complete elongation will depend on the DNA polymerase being used and the length of the DNA fragment to be amplified. Under optimum conditions the amount of DNA target is doubled.

Final elongation is performed at 70–74°C for 5–15 minutes to ensure remaining single-stranded DNA is fully extended.

The more popular method of quantitative analysis is obtained using **Taqman probes**.

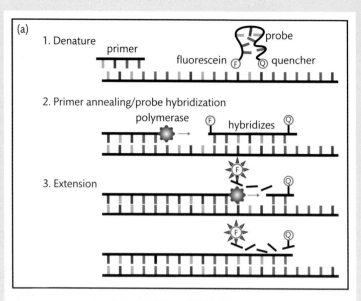

FIGURE 13.14

(a) Polymerase chain reaction showing denaturation, annealing, elongation, or extension using Taqman probe method. (b) Thermal cycler: a compact piece of laboratory equipment.

(b) reproduced with kind permission from DNA genetics laboratory, GSTS Pathology.

PCR has superseded many of the techniques previously used in molecular methodology. It can be used to detect **loss of heterozygosity** (LOH) and **microsatellite instability** (MSI), which are described below.

Loss of heterozygosity

Loss of heterozygosity (LOH) can occur when there is deletion of a large or small region of chromosome and with it a loss of all important cancer fighting tumour suppressor genes. LOH can be detected using PCR or FISH. In tumour transformation genetic alterations can occur,

Allele

One of two or more alternative forms of a gene.

leading to a significant shift in the normal balanced ratio of two **alleles**. In cytological samples normal cells can often outnumber tumour cells, causing erroneous results. **Laser microdissection** can selectively dissect sparse tumour cells for analysis. The big advantage of LOH is that it can analyse small DNA sequences for genetic imbalance.

BOX 13.8 *Laser microdissection*

Laser microdissection (LMD) is a useful tool to retrieve a small number of cells from a sparse sample. LMD can be used on paraffin blocks, ethanol fixed direct spreads, and LBC preparations.

Cells are identified and electronically marked on a computer screen, and are dissected using a heated laser beam through an inverted microscope (Figure 13.15).

A specially designed film allows fusion and capture of the cells into an Eppendorf tube for further molecular studies, including PCR and DNA sequencing analysis.

The skill and morphological knowledge of the operator is important to select and dissect the cells of interest, avoiding necrotic or fibrous material, for further mutational and genetic analysis.

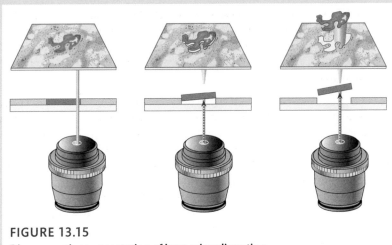

FIGURE 13.15
Diagrammatic representation of laser microdissection.
Reproduced with kind permission from Carl Zeiss Ltd.

Microsatellite instability

Microsatellite instability (MSI) is caused by defects in DNA mismatch repair proteins. Sequences are of variable length; a single microsatellite **locus** will exhibit different sequence lengths between individuals and between homologous chromosomes. These sequence length differences may result in two different alleles at a single locus in any individual, known as heterozygosity (if both alleles happen to be the same length this is called homozygosity). These repeat sequences are common in the human genome and can be inherited as in hereditary non-polyposis colorectal cancer (HNPCC). Microsatellite instability is known to be a key factor in gastric colorectal, endometrial, and ovarian cancers.

SELF-CHECK 13.8
Explain the benefits of laser microdissection.

In situ hybridization techniques

Fluorescent in situ *hybridization*

Fluorescent *in situ* hybridization (FISH) is primarily a cytogenetic technique used to detect the presence of DNA sequences on chromosomes, and is increasingly being used as an adjuvant method to provide additional diagnostic and prognostic information. Fluorescent probes are used to bind to chromosomes. FISH analysis performed on cytological or histological samples is carried out during **interphase** as opposed to the **metaphase** stage. This is because of the constraints around the processing of these samples, making metaphase analysis unsuitable.

There are many commercially available probes and by using different probes labelled with different fluorochromes it is possible to detect **translocations**.

Fluorescent *in situ* hybridization can also be used on unstained preparations, which is a major advantage.

Currently, the FDA have approved two systems: the UroVysion multitarget FISH testing for **aneusomy** detection (9p21 deletion) in urine samples (Figure 13.16a) and the scoring of HER2 gene amplification (chromosome 17q21) in breast cancers (Figure 13.16b), which can also be demonstrated using ICC, as discussed in Section 13.3. FISH is an appropriate line of investigation when cytomorphology is equivocal.

With use of computerized software it is possible to mark cells of interest on a conventional preparation electronically and carry out FISH analysis post-hybridization to gain further information on **chromosomal aberrations**. This also enables a retrospective review of cases which have been subjected to FISH analysis, allowing scientists to build up a library of interesting cases for review and educational purposes. Automated systems have been developed to count FISH signals, using three-dimensional analysis to capture signals and store images. Such technology has helped bring much needed standardization, validation, and quality control procedures for provision of reliable results. Scoring of multi-target FISH assays which use multiple probes is enhanced by also taking into consideration parameters such as nuclear size, irregular nuclear membrane, and inconsistent staining patterns.

Metaphase
A stage of cell division when chromosomes align along the equator of the dividing cell.

Translocation
A type of chromosomal aberration in which part of a chromosome is linked at a breakpoint to another chromosome. These can be detected with PCR and studied using FISH.

Aneusomy
Abnormal number of chromosomes.

Chromosomal aberration
A modification to the normal chromosome complement due to deletion, duplication, or rearrangement of genetic material.

BOX 13.9 Benefits of FISH

- Permits quantification of signal.
- Standardized process.
- Decreased subjectivity.
- Produces data and educational documentation.
- Allows for increased throughput of tests.

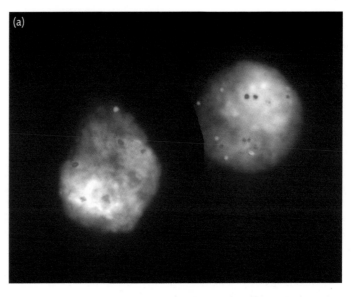

(a)

(b)

HER2 gene amplification and protein overexpression

Chromosome 17

HER2 gene
(normal copy number)

HER2 mRNA
(normal amount)

HER2 protein
(normal amount)

HER2 gene
(amplified copy number)

HER2 mRNA
(increased amount)

HER2 protein
overexpresssion

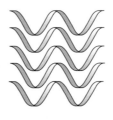

FIGURE 13.16
(a) UroVysion multitarget FISH. The
cell on the left shows normal diploid
complement, two coloured signals
per probe for all four loci in the probe
panel; cell on the right is aneuploid for
all loci, that is, chromosome gain.
(b) Diagrammatic illustration of HER2.
*(a) reproduced with kind permission
from Dr M. Neat, Cytogenetics
Department, GSTS Pathology.*

BOX 13.10 *Disadvantages of FISH*

- Relatively expensive.
- Interaction intensive.
- Little or no gain time in turnaround of results.
- Poor performance when small signals present as in 9p21.
- Not always definitive in equivocal or difficult cases.

Benefits of relocalization software

- Facilitates review of cases and allows more rapid reporting.
- Sparse atypical cells of interest can be relocated with ease.
- Diagnostic precision.
- Ability to link morphological findings to molecular genetics.

SELF-CHECK 13.9

What are the advantages of FISH over immunocytochemistry?

Chromagenic in situ hybidization

Chromagenic in situ hybridization or CISH, as it is commonly referred to, works in a similar way to FISH in that it visualizes the number of gene copies present in the nucleus. CISH differs in that it utilizes a chromagenic (coloured) substrate rather than a flurochrome. This removes the need for a fluorescent microscope and a specialized set-up within the laboratory. CISH is more easily integrated into a routine cytology setting as it is not dissimilar to ICC and can be interpreted with use of a counterstain and a light microscope. Studies have shown that CISH is as sensitive and specific as FISH, and considerably cheaper. However, multi-targeted FISH assays allow up to four different coloured probes to be analysed, whereas CISH allows only two.

Silver in situ hybridization (SISH)

This technique is similar to CISH but uses a silver reaction to demonstrate the number of gene copies within the nucleus (see Table 13.1).

SELF-CHECK 13.10

What are the limitations of FISH analysis?

Comparative genomic hybridization

Comparative genomic hybridization (CGH) is a more sophisticated cytogenetic hybridization technique which can detect gains or losses in chromosomal material within the tumour cell genome.

TABLE 13.1 In situ hybridization methods versus immunocytochemistry.

	FISH	CISH/SISH	ICC
Test	Specialist settings	Specialist settings	All laboratories
Microscopy	Fluorescent	Light	Light
Training	Specialist	Standard	Standard
Morphology	Not well preserved	Preserved	Preserved
Sample (tumour) size	Small sample	Large sample area	Large sample area
Interpretation	15–20 min	1–2 min	1–2 min
Cost	Expensive	Moderate expense	Inexpensive
Technology	Automated	Semi-automated	Automated

BOX 13.11 *Comparative genomic hybridization*

DNA material is extracted from a tumour sample and a known control and labelled with different flurochromes. Chromosomal gain or gene amplification will show an increase in fluorescence. Likewise, chromosomal loss will manifest less fluorescence. The main disadvantages of CGH are that it is labour intensive and not easy to automate. Array CGH (aCGH) has superseded conventional CGH as a high throughput technique which hybridizes sample DNA against arrayed probes. Scientists are excited about the potential for aCGH to produce a map of DNA sequence copy numbers to identify losses and gains of genetic material involving smaller regions of the genome. CGH may prove useful in the detection of primary tumours in effusion cytology and help understand the relationship between a primary tumour and its metastatic deposits.

Invader technology

Invader technology is a new concept, best demonstrated in HPV testing which uses invader chemistry to determine the presence of hrHPV types (Figure 13.17). This works at the molecular level to detect individual base pair changes associated with hrHPV strains. It uses signal amplification and a fluorescence detection methodology.

BOX 13.12 *Invader methodology*

An invader reaction uses a primary probe and an invader probe. The former hybridizes with target DNA, the invader probe overlaps and is recognized by the enzyme cleavase which cuts the primary probe and releases a **flap**.

This flap then participates in a second reaction and binds to a flurochrome containing an oligonucleotide called a quencher. This arrangement is called a FRET (flurochrome resonance energy transfer) cassette. The combination of flaps and FRET cassettes creates a fluorescence-based signal amplification. The signal can be measured by a fluorometer to confirm the presence or absence of hrHPV (Figure 13.18).

Flap
The latter region of the primary probe.

Tissue microarray technology

Tissue DNA microarray technology is used to examine gene expression in tissue extracts through RNA analysis. Paraffin blocks including cytological cell block preparations are a rich source of DNA and RNA, containing valuable genetic information on a diverse range of tumours.

Cross reference
Chapter 2.

Key Points

DNA microarrays can be used to analyse in parallel the expression of multiple genes in a single reaction.

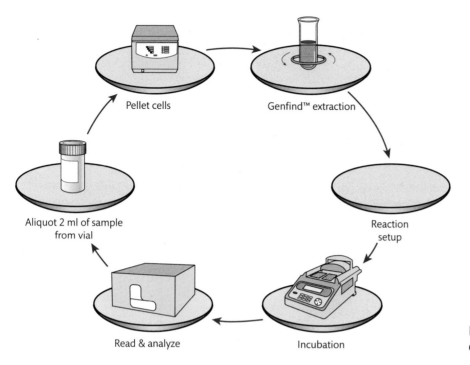

FIGURE 13.17
Cervista Workflow.

Pellet cells

Genfind™ extraction

Aliquot 2 ml of sample
from vial

Reaction
setup

Read & analyze

Incubation

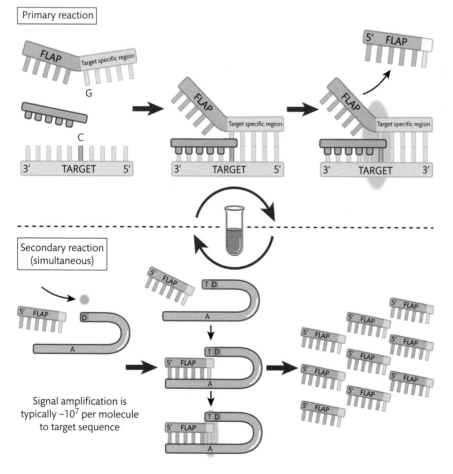

Primary reaction

FLAP   Target specific region

G

C

3'   TARGET   5'

FLAP

Target specific region

3'   TARGET   5'

5'   FLAP

FLAP

Target specific region

3'   TARGET   3'

Secondary reaction
(simultaneous)

5' FLAP   D

A

5' FLAP   T D

A

5' FLAP   T D

A

5' FLAP

A

5' FLAP   T D

A

Signal amplification is
typically ~10^7 per molecule
to target sequence

5' FLAP
5' FLAP
5' FLAP
5' FLAP
5' FLAP
5' FLAP
5' FLAP
5' FLAP
5' FLAP
5' FLAP

FIGURE 13.18
Cervista invader chemistry.

BOX 13.13 *Methodology of tissue microarrays*

Tissue is transferred from a conventional paraffin block so that tissues from multiple blocks or patients can be analysed on the same glass slide.

Multiple core needle biopsies are taken from carefully pre-selected areas (usually tumour) in existing embedded blocks.

Miniature core biopsies are then re-embedded in an arrayed master block (Figure 13.19). Most arrays will contain somewhere in the region of one hundred miniature core biopsies.

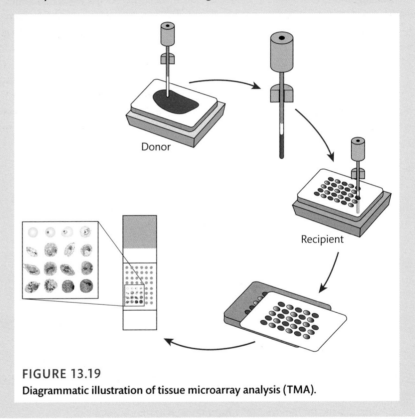

Donor

Recipient

FIGURE 13.19
Diagrammatic illustration of tissue microarray analysis (TMA).

Tissue microarray technology and cancers of unknown origin

Cross reference

Chapters 9 and 10.

Patients with a cancer of unknown primary have a very poor prognosis. Use of ICC panels to type a tumour has been addressed in Chapters 9 and 10. A primary cancer can spread to numerous sites throughout the body, including the liver, lungs, bones brain, lymph nodes, and serosal cavities. To treat cancer it is important to establish the primary (site of origin) to guide the clinician to the best form of therapy for that particular form of cancer. Therapies are wide-ranging and include surgery, chemotherapy, radiotherapy, hormonal, stem cell transplants, or a combination. Inability to detect the primary tumour could mean a worse prognosis for the patient. Molecular testing can provide additional information that will lead to better diagnosis and treatment therapies, which will ultimately provide a better standard of care for cancer patients. **Tissue microarrays** will enable multiple parallel testing.

SELF-CHECK 13.11

What is meant by a tumour of unknown origin?

In the same way as tissue microarrays are produced so also are genetic microarrays or **gene chips**. Sometimes this is referred to as 'lab in a chip' as the potential from such technology is likely to shift the way in which testing will be done in generations to come. Gene chip technology can rapidly screen thousands of genes to detect mutations. It is possible to compare messenger RNA from normal cells with cancer cells. By comparing which genes are active (expressed) or switched off it is possible to gather prognostic information which will help inform the choice of treatment for different cancers. Messenger RNA is the key to gene expression analysis and can determine not only how many and which genes are active within a cell but also the copy number, which is an indicator of the strength of gene expression. One of the areas in which gene chip technology is expected to be adopted is in HPV genotyping. Several commercial companies have already produced gene chips to type high-risk HPV. In-house production of microarray chips is laborious, and there are now many commercial companies providing chips containing 33,000 genes, which is thought to represent the entire human genome.

SELF-CHECK 13.12

What are the advantages of tissue microarray analysis?

BOX 13.14 *Advantages of tissue microarrays*

- Economical use of tumour material.
- Standardized preparations for research and other purposes.
- Can use wide range of demonstration techniques including ICC, FISH, CISH, and laser microdissection.
- Eliminates need for hundreds of conventional paraffin cut sections.
- Fewer reagents required for assay, which can be a significant cost benefit.
- Original diagnostic block is preserved.

13.7 Application of molecular techniques in cytopathology

In the previous section we have concentrated on the technical aspects of molecular testing. In this final section we summarize how these techniques can be applied in cytological practice. It is important for the reader to appreciate that although such techniques are of increasing benefit they will not replace conventional morphological diagnosis but complement it. Used selectively they can take diagnosis to the next level, especially in equivocal or diagnostically challenging cases, providing chromosomal and genetic insight which may give the predictor information needed to monitor and treat cancers. In this section we will focus on the most common solid tumours and use case studies to demonstrate the usefulness of molecular testing.

Urine cytology

Bladder cancer is the second most common urological malignancy. It is a chronic illness and requires careful surveillance and follow-up post-treatment. Recurrence can occur in up to 70% of urothelial tumours, with 5% progressing to invasive cancer. Diagnosing recurrent bladder cancer is one of the most challenging areas in urological and cytological diagnosis and is prone to many pitfalls. Urine samples can be tested for NMP22, a nuclear matrix protein, and MCM5, which are greatly increased in urothelial cells from a bladder cancer compared to normal levels found in normal cells.

Cross reference

Chapter 8.

Cystoscopy and cytological follow-up is key to monitor patients for recurrence or progression of previously treated bladder cancer. In some cases the cytology and cystoscopic examination can be equivocal. In these scenarios additional testing can help reach a definitive diagnosis. Promoter **hypermethylation** of tumour suppressor genes commonly occurs in bladder cancers and will affect the base sequence. These changes can be detected using PCR based methods.

Multi-targeted FISH analysis is an important tool in the follow-up of patients with a previously treated bladder cancer. One major advantage of FISH is that chromosomal genes can be analysed in the same cytological samples as were used for the conventional morphological diagnosis. The FDA approved UroVysion assay uses hybridization probes to detect irregularities on chromosomes 3, 7, and 17 and deletion of LSI (locus specific indicator) 9p21. Use of FISH significantly increases the sensitivity and specificity in both high and low-grade bladder lesions (18).

BOX 13.15 *UroVysion results*

A positive UroVysion (abnormal) result is defined as more than or equal to four cells containing gain (three copies or more) of two or more chromosomes of 3, 7, or 17 and more than or equal to 12 cells with a deletion of LSI p16 (9p21).

BOX 13.16 *Benefits of FISH in urine cytology*

- Detects gains in chromosomes 3, 7, 17, and deletion of LSI 9p21.
- Improves sensitivity and specificity compared to routine cytology.
- Resolves equivocal cytology cases, including those taken post-treatment.
- Predicts risk of recurrence and progression of bladder cancers.
- Streamlines patients of low risk and reduces number of cystoscopies.

Key Points

Umbrella cells are often seen in urine samples and have a high degree of polyploidy which will show abnormal FISH results. The experienced scientist and pathologist will be able to combine morphological and molecular knowledge to interpret the result as normal. This demonstrates why molecular pathology can complement but not replace morphology at least in the near future.

BCG therapy
Immunotherapy using Bacillus Calmette-Guerin, the same bacterium used for TB vaccination.

Cross reference
Chapter 8.

CASE STUDY 13.1

A 65-year-old male ex-smoker who worked in the textile industry presented to his GP with haematuria. He was referred for a urological opinion at his local hospital. Prior to cystoscopy he was asked to produce a voided urine sample which was sent for cytological analysis. The cystoscopy showed suspicious lesions which were biopsied for histological diagnosis. Cytological examination of the urine sample showed abnormal cells suspicious of high-grade urothelial carcinoma (Figure 13.20). Histology confirmed a grade 3 urothelial carcinoma, which had not invaded the muscle and was confined to the innermost layer of the bladder lining. The patient was discussed at MDT and a decision was taken to perform a transurethral resection followed by **BCG therapy**. The patient was followed up post-surgery with regular cystoscopy and cytology and remained disease free for two years. In year three, at a routine follow-up appointment the cytology showed some atypical cells and cystocopy findings showed a red patch in the bladder. The cytology sample was sent for FISH analysis using UroVysion. The results confirmed deletion of locus 9p21 and gain of chromosomes 3, 7, and 17. The case was reviewed at MDT and the patient was offered repeat cystoscopy and possible biopsy. Histological examination of biopsies showed a grade 3 urothelial carcinoma invading into deep muscle (stage 2b). The patient was treated with a radical cystectomy.

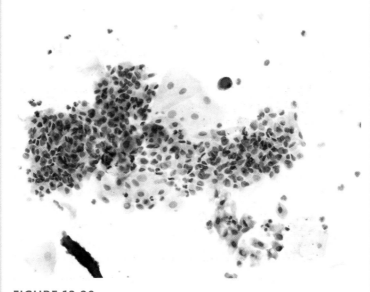

FIGURE 13.20
Pap stained follow-up urine sample from a patient with previous bladder cancer. Atypical cells present graded C3.

In what scenario may it be useful to carry out FISH analysis on patients with a previous bladder cancer?

The prostatic cancer assay PCA3 is a novel gene-based diagnostic test which can detect over-expression of prostate cancer gene messenger RNA in a simple urine sample, using transcription mediated amplification or TMA technology. The latter is outside the scope of this chapter; for further information consult the further reading list at the end of this chapter.

Respiratory cytology

Locus

A specific location on a chromosome. Loci is the plural of locus.

Cross reference

Chapter 10.

A multi-target FISH assay containing **locus** specific probes to chromosomes 6, 5p15, 7p12 (EGFR), and 8q24 (c-myc gene) is widely used to confirm tumour type and greatly improves sensitivity of the diagnosis. FISH is extremely sensitive and can be used on conventional ethanol fixed samples obtained at bronchoscopy. FISH can help reach a final diagnosis in equivocally difficult cases and has been shown to be successful in detecting early stage and peripheral tumours. Negative FISH is more in keeping with a reactive process than a true neoplasia.

Epidermal growth factor receptor

Epithelial growth factor is a family of proteins that when bound to the counterpart receptor (EGFR) on the cell surface and in the presence of **tyrosine kinase (TK)** are involved in cell growth, proliferation, differentiation, **apoptosis**, invasion, and metastasis. Mutation of EGFR gene causes an over-expression of EGFR in a significant number of non-small cell lung cancers and is responsible for promoting tumour growth.

Key Points

Tumours with EGFR mutations are more common in Asian women, non-smokers, and patients with adenocarcinoma.

Mutational analysis

This encompasses a wide range of methods, including PCR based techniques, DNA sequencing, and FISH. The method chosen will depend on the type of mutation.

Cross reference

Chapter 10, Section 10.2.

Samples obtained during bronchoscopy and **EBUS** procedures are assessed by a senior biomedical scientist for adequacy and material is taken back to the laboratory for further investigation. **Mutational analysis** can be carried out to detect EGFR mutation and over-expression, enabling clinicians to select patients for appropriate treatment (19).

EGFR inhibitor drugs such as **IRESSA (gefitinib)** and **TARCEVA (erlotinib)** work in a similar way by suppressing tyrosine kinase activity, thereby targeting EGFR. These drugs are currently being used to treat non-small cell lung tumours (NSCLC) which have become locally advanced or undergone metastatic spread to other parts of the body. Patients treated with TK inhibitors have shown a better response and progression-free survival (20). The only downside is that through time most NSCLCs become resistant to TK therapy. NICE has approved the use of IRESSA for patients with non-small cell carcinoma and who are EGFR positive.

Key Points

Tyrosine kinase TK is an enzyme which if constantly switched on is linked to many cancers. Inhibiting TK activity can be effective against cancer.

K-Ras

K-Ras protein is a member of the Ras protein family and is encoded by the K-Ras gene. K-Ras mutations (point mutation) have been detected in lung cancers, but to a lesser extent than EGFR. It is considered a poor prognostic factor and is increasingly being used as part of the work on lung and colorectal cancers prior to selecting patients for treatment.

Key Points

DNA sequencing analysis of PCR products is the gold standard for detection of point mutations such as K-Ras.

Key Points

Presence of K-Ras mutation renders EGFR inhibitors ineffective as a therapy.

EGFR and K-Ras mutations suggest there are two molecular pathways involved in lung adeno-carcinomas: one involving EGFR mutation and the other K-Ras mutation, and that tobacco may not be the major pathogenic factor in these particular tumours. In the USA all patients diagnosed with non-small cell cancer are routinely tested for EGFR and K-Ras mutations.

SELF-CHECK 13.14

Why is EGFR positivity important in patients with non-small cell carcinoma?

Gynaecological cytology

There are several methods which can be used to test for hrHPV or used to genotype HPV. HPV testing using FDA approved HC2 is currently the method of choice, although detection of messenger RNA by PCR, HPV proofer technology, and HPV genotyping using microarrays is also possible.

Expression of p16 has already been discussed in Section 13.3. Over-expression of p16 and Ki67 due to hrHPV E7 activity acts as a marker of proliferation and likely disease progression.

Key Points

High-risk HPV is the major cause of cervical cancer, though few infections progress to malignant disease. Progression to malignancy requires the over-expression of the E6 and E7 genes in the integrated HPV genome. It follows that the E6 and E7 mRNA transcripts could be useful markers of disease progression. HPV proofer test detects E6/E7 mRNA as opposed to DNA.

In the normal cervix hypermethylation is absent, making it a suitable marker for cervical cancer. DNA hypermethylation of certain genes is linked to severity of the lesion, 80% in invasive cancer and 5% in low-grade lesions.

CASE STUDY 13.2

A 73-year-old male presented with shortness of breath (SOB) and cough. A CT scan was performed, which showed a mass in the left main bronchus. There was also a pleural effusion. A lymph node adjacent to the carina was enlarged and there were suspicions of early metastatic deposit. The CT scan was consistent with a primary bronchial carcinoma. The patient underwent a bronchoscopy procedure. At bronchoscopy no lesion could be seen, but bronchial brushings were taken from areas which showed narrowing. On-site assessment of sample by a senior biomedical scientist confirmed adequacy. Bronchial washings were also taken.

The material was processed in the laboratory, and showed clusters (Figure 13.21a) and single atypical cells consistent with a non-small cell carcinoma (Figure 13.21b). A cell block was prepared from the bronchial washing and sent for immunocytochemistry. The ICC confirmed the cytological diagnosis. At MDM the treatment options were discussed, including oncology referral for palliative chemotherapy and radiotherapy. The cell block was sent for EGFR mutational analysis at a specialist referral centre. The sample was EGFR positive (Figure 13.21b) and the patient was considered suitable for treatment with the oral anti-cancer drug IRESSA (gefitinib) which acts to inhibit EGFR through tyrosine kinase.

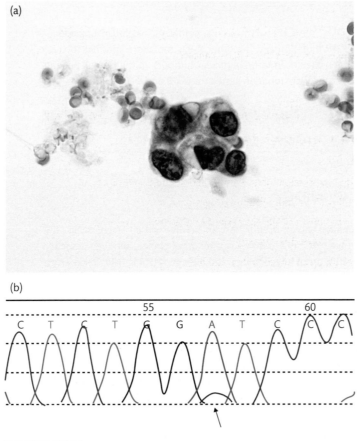

FIGURE 13.21

(a) Bronchial washings taken at bronchoscopy showing clusters of single atypical cells consistent with a non-small cell carcinoma. (b) Arrow shows EGFR mutation.

(b) reproduced with kind permission from Dr K. Tobal, Molecular Oncology Department, GSTS Pathology.

Microsatellite instability analysis to detect loss of heterozygosity and array CGH permit the study of gene alterations, copy numbers and over-expression. Loss of heterozygosity of 18q has been associated with decreased survival. Comparative genomic hybridization analysis has shown that more than 85% of invasive cervical carcinomas carry genetic imbalances which lead to an increase in copy numbers of chromosome arm 3. Band 3q26 contains the gene for the RNA component of telomerase (TERC). Additonal copies of TERC indicate 3q26 amplification, which is detected in high-grade lesions and also in low-grade where it is regarded as a marker of progression to high-grade disease and possible cancer (21).

Key Points

Telomerase is an enzyme which is responsible for adding DNA sequences to the DNA strand within the telomere region of the chromosome. Telomerase activation is seen in the majority of cancers and is thought to play a key role in cancer development. Hence its potential use in targeted drug therapy has been realized and clinical trials are underway to develop drugs which will switch off telomerase activity.

SELF-CHECK 13.15

What is the significance of the telomerase gene TERC in cancer?

BOX 13.17 *Benefits of molecular testing in cervical cytology*

- Detects hrHPV E6 and E7 expression by HC2 and PCR detection methods.
- HPV genotyping can inform the immunization programme.
- Detects increase copy numbers of TERC (3q21) and MYC(8q24).
- Resolves equivocal cases (borderline/ASCUS).
- Predicts risk of progression from low to high-grade disease.

Effusion cytology

From Chapter 9 the reader should have gained an understanding of the difficulties of diagnosing malignancy in effusion cytology and the effective use of characterizing tumours with the use of immunocytochemistry panels. The reader is referred back to Chapter 9, Section 9.5 for further reading on immunocytochemical markers in fluid cytology, before continuing with this section.

Cross reference
Chapter 9, Section 9.5.

The two main clinical scenarios are:

- Identification of metastatic cells which can be sparse in a cytological preparation.
- Differentiating reactive mesothelial cells from malignant mesothelioma.

Molecular methodology provides yet another set of tools to increase the accuracy of diagnosis. These novel techniques are highly unlikely to replace conventional morphology and ICC analysis in the foreseeable future. A combination of clinical, morphological, and molecular analysis, rather than reliance on one methodology, is likely to prove the most useful.

CASE STUDY 13.3

A 28-year-old woman attended her GP for a routine cervical test as part of the NHSCSP. A sample was taken with a Cervex® brush and placed in an LBC vial and sent to the laboratory for processing. The test showed minor atypical changes in squamous cells (Figure 13.22). A borderline report was issued, with a repeat in six months advised. The second sample showed similar changes and a second repeat test was advised. The third sample also showed the same atypical cells; the laboratory on this occasion referred the woman for a colposcopy examination. The colposcopy examination was normal and no abnormality was identified. A repeat cervical test and a punch biopsy was taken from a slightly red area of the epithelium. The cervical test and biopsy were all reported normal. The case was placed for review at the next colposcopy review meeting. On review it was decided to repeat the cytology and perform an HC2 test. Cytology was negative and HC2 test was negative for high-risk HPV types. The woman was discharged and advised to return to normal recall. At the time of writing, all cervical tests had been negative.

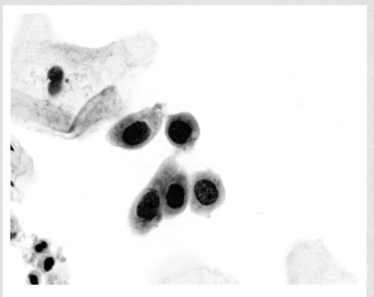

FIGURE 13.22
A Pap stained cervical LBC sample showing borderline changes, dyskaryosis in metaplastic cells.

The use of flow cytometry has been discussed already in Section 13.5. It is mentioned here to remind the reader that one of the most difficult areas of effusion cytology is differentiating reactive lymphocytes from a lymphoma or indeed other small cell tumours. Flow cytometry analysis is a useful adjuvant to conventional stains and complements ICC. A flow cytometer may not be readily available in all settings; however, it is possible to place samples in foetal calf serum and transport them to a facility which can carry out FCA. Establishing T cell clonality is an area where FCA can excel over ICC. FCA not only assists with phenotyping but also with sub-typing of lymphomas. FCA has proven useful in predicting disease progression in chronic lymphocytic leukaemia (CLL).

The molecular techniques discussed in Section 13.6 can be applied to effusion samples from the pleural, peritoneal, and pericardial cavities. Comparative genomic hybridization/aCGH, in particular, may yield the information needed to determine the site of origin of the primary tumour. As more genetic information is obtained on different tumour types, each in turn can be assigned a genetic signature which will allow their differentiation in subsequent samples. Array comparative genomic hybridization (aCGH) also will help our understanding of primary tumours and their metastatic deposits, as demonstrated. Telomerase is a useful marker and will play a key role in determining reactive versus benign versus malignant. Telomerase activation occurs in a large number of tumours and although it needs to be combined with morphological assessment the telomeric repeat amplification protocol (TRAP) assay has been used to some degree of success in effusion cytology (22).

CASE STUDY 13.4

A 28-year-old male attended his GP due to a painless neck swelling, cough, recurrent chest infection, and night sweats. He was referred for further investigation. Initial blood tests pointed towards a haematological abnormality. A CT scan showed a mediastinal mass with a distinct right pleural and left pericardial effusion. Fluid was drained and sent for cytology. Cytospin preparations stained with Pap (Figure 13.23a) and MGG (Figure 13.23b) showed a mature and immature lymphoid population which was in keeping with the haematology profile of acute lymphoblastic leukemia (ALL). An FNA of the neck node showed similar morphology to pleural fluid. The patient underwent a mediastinal and pleural biopsy which was sent for histological examination. Histological sections and immunocytochemistry profile

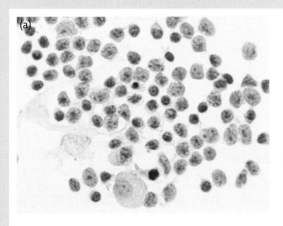
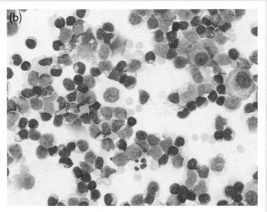

FIGURE 13.23
(a) A Pap stained preparation showing a population of mature and immature lymphoid cells with occasional mesothelial cell in background. (b) MGG stained preparation showing a population of mature and immature lymphoid cells with occasional mesothelial and red blood cells in background.
Reproduced with kind permission from Dr M. Moonin, Cytology Department, St Thomas' Hospital, London.

supported a diagnosis of T cell acute lymphoblastic lymphoma involving lung, mediastinal fat and pleural soft tissues. The tumour cells expressed CD3, CD4, CD5, CD7, CD8, CD19, CD38, CD56, CD79a, CD99, HLA-Dr and TdT.

The tumour cells were negative for CD13, CD25, CD34, CD61, CD68, CD117, glycophorin A, MPO, and TTF1, confirming that the tumour was not of lung origin. Cytology

washing samples were sent for flow cytometry and the results can be seen in Figure 13.24. In addition, and to establish the extent and severity of the disease, the patient had a bone marrow trephine performed. There was no evidence of bone marrow involvement by T cell acute lymphoblastic leukaemia or lymphoma. The patient was put on the UK–ALL chemotherapy regimen and currently is in remission.

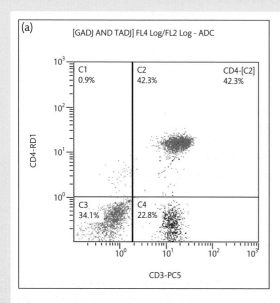 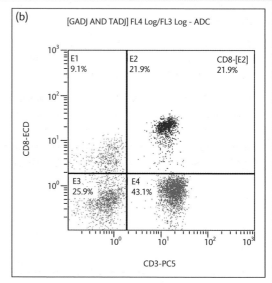

FIGURE 13.24

(a) Flow plot of T cells. C1-shows CD4 +ve cells only, C2-shows CD3/CD4 dual positivity, C3-negative cells, no expression, C4-CD3 +ve only. (b) Flow plot of T cells. E1–CD8 +ve, E2–CD8 + CD3 dual positivity, E3–CD3 only, E4–CD3 only.

Reproduced with kind permission from N. Kirkam, Specialist Haematology Laboratory, GSTS Pathology.

Mesothelioma

Differentiating mesothelial from carcinoma cells usually involves the use of an ICC panel. One of the main diagnostic dilemmas is the differentiation of reactive from malignant mesothelial cells. A conclusive diagnosis of mesothelioma may lead to compensation for the patient or their family even after death. Mesotheliomas tend to show recurrent abnormalities of chromosomes 6 and 7, and deletion of CDKN2A gene on chromosome 9p21. The latter gene encodes for p16 (tumour suppressor gene).

Cross reference

Chapter 9.

Fine needle aspiration cytology

Fine needle aspiration (FNA) is a well-established procedure in cytology with head and neck, lymph node, and breast pathology, along with lung and gastrointestinal samples (the latter collected during endoscopic procedures) being the most common. If performed by a skilled operator it can match histology, and in difficult clinical scenarios or in the case of a very ill patient it can remove the need for a biopsy or a difficult endoscopic procedure. FNA is a relatively easy method of aspirating superficial lumps and, with the aid of imaging, diagnostic samples can be obtained from deep-seated lesions. It is well documented and evidenced that having a trained biomedical scientist present during the procedure facilitates standardization of the process, immediate assessment of adequacy, and preliminary interpretation of the cells to determine the need for molecular testing.

Cross reference
Chapter 11.

Haematological disorders and sarcomas are two areas where FNA cytology and complementary molecular techniques have proved valuable. Sub-typing of lymphomas with the use of flow cytometry and immunocytochemistry has greatly contributed to diagnostic accuracy. The detection of chromosomal translocations with use of FISH probes and real time PCR can help in the diagnosis of lymphomas and sarcomas.

Diagnosis of thyroid cancer can also benefit from molecular testing on FNA samples (23). There are several mutations which, if detected, can contribute to the final diagnosis, namely BRAF, RAS, RET/PTC, and PAX8/PPAR gamma mutations.

One-stop breast FNA clinics still operate in many hospital settings despite competition from histological core biopsies. Cytological samples can be made from core tissue and stained with Diff-Quik™ and an on-site assessment of adequacy made. Positive cytology and biopsy will require testing for HER2 status to decide if a woman will respond to Herceptin treatment. Analysis of the TOP2A gene in breast cancer using ICC or FISH may prove helpful in establishing

CASE STUDY 13.5

A 37-year-old presented to her GP with a right breast lump. An urgent referral followed to a specialist one-stop breast clinic where the woman had a mammogram and an ultrasound, and an FNA was performed. The mammogram showed an area suspicious of tumour. Fine needle aspiration cytology confirmed malignancy and was reported as C5 (Figure 13.25a). A core biopsy was taken and cytological imprints made during the procedure showed cells similar to the previous FNA sample. Histology and immunocytochemistry confirmed a primary breast malignancy. Further histological sections were cut and stained for HER2. Results were equivocal and scored as 2 (Figure 13.25b); additional FISH analysis was requested which was strongly positive, as seen on the right of Figure 13.25c. The patient was discussed at MDM and a decision was made to commence the patient on Herceptin treatment. The woman continued to respond well to treatment three years post-diagnosis.

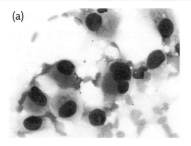
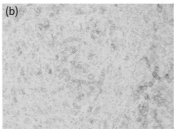
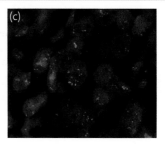

FIGURE 13.25
(a) MGG stained cytology imprint taken from a breast core biopsy showing malignant cells. (b) HER2 immunostained slide (2+). (c) FISH detection of HER2 gene showing HER2 negativity on left and HER2 positivity on right.
Reproduced with kind permission from Prof Sarah Pinder, Department of Histopathology, St Thomas' Hospital, London.

which genes are showing altered activity, by an increase in the number of gene copies. These changes can be indicative of an increased risk of tumour recurrence or decreased survival chances.

Pancreas and biliary tract

Another diagnostically challenging area is that of the pancreas. Very subtle changes can be interpreted as reactive when in fact they are from an adenocarcinoma or vice versa. Combined morphological and molecular assessment will help reach the correct diagnosis in more indeterminate cases. This is another example whereby morphological and molecular assessment, if carried out independently of each other, could result in a false positive diagnosis. A combined approach yields a more accurate diagnosis for the patient. Analysis for the K-Ras mutation may prove useful in material taken during EUS guided FNA to establish pancreatic cancer from pancreatitis.

In **cholangiocarcinoma** (primary tumour of the biliary tract) deletion of 9p21 is common and can be detected using multi-target FISH (chromosomes 3, 7, 17, and 9p21). Promoter hypermethylation is another feature of cholangicarcinoma, as is loss of heterozygosity (LOH). Loss of heterozygosity analysis of microdissected cells using microsatellite markers is also finding its way into the diagnostic repertoire. In the biliary tract subtle changes can occur which will trigger an atypical report. The cytopathologist can request additional FISH analysis which will help support a diagnosis consistent with a reactive process, or confirm an abnormality.

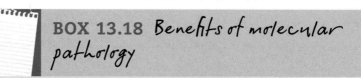

BOX 13.18 *Benefits of molecular pathology*

- **Aids in equivocal diagnosis.**
- **Gives prognostic and predictor information of tumours.**
- **Informs the use of targeted therapies and therapy.**
- **Enables service development and improvement for patients (personalized medicine).**
- **Increases skill sets for cytology staff.**

 SUMMARY

- Automated screening technology may in time replace or complement the manual screening process.

- High-risk HPV testing can be used to manage women with low-grade disease and as a test of cure.

- The cervical cancer vaccine is expected to reduce the incidence of cervical cancer.

■ Immunocytochemical and molecular testing will increase and optimize diagnostics and targeted therapies. Molecular testing will move away from single tests on multiple samples to multiple parallel testing on a single sample (tissue macroarrays).

■ Well trained scientists will be needed to develop, perform, and interpret molecular results. For molecular science to improve patient care and outcomes the biomedical scientist must be an integral part of the clinical team.

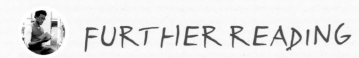

FURTHER READING

● Bruns D, Ashwood E, and Burtis C (2007) *Fundamentals of Molecular Diagnostics*. Saunders Elsevier, St Louis, Missouri.

● Cancer Research UK: http://www.cancerhelp.org.uk (accessed 2010).

● Dabbs DJ (2010) *Diagnostic Immunohistochemistry Theranostic and Genomic Applications*, third edition. Churchill Livingstone/Elsevier, New York.

● Feasibility Sub Group to Cervical Cytology Review Group. Cervical Cytology ThinPrep Imager (TIS) Feasability Study. A report from Scotland: http://www.pathologyscotland.org/cervical-cytology-review.htm (accessed 2010).

● Hunt J (2008) Molecular pathology in anatomic pathology practice. A review of basic principles. *Arch Pathol Lab Med* 132, 248–60.

● Institute of Cancer Research: http://publications.icr.ac.uk (accessed 2010).

● Pfefier JD (2006) *Molecular Genetic Testing in Surgical Pathology*. Lippincott, Williams and Wilkins, Philadelphia.

DISCUSSION QUESTIONS

1. Do automated screening systems have a role to play in a modern cytology laboratory? If so explain.

2. What will be the impact on the NHS Cervical Screening Programme of vaccinating young girls?

3. What benefit can molecular testing bring to cytological practice?

Answers to the self-check questions, and tips for responding to the discussion questions, are provided in the book's Online Resource Centre.

 Visit www.oxfordtextbooks.co.uk/orc/shambayati/

Glossary

Aberrant S phase induction occurs when there is a premature or delayed entry into S phase of the cell cycle from G1/S checkpoint where DNA is checked for errors prior to replication.

Absolute alcohol is water-free alcohol.

Acetic acid a weak organic acid commonly used during colposcopy to demonstrate abnormal areas of epithelium.

Acetowhite epithelium a white patch that appears on abnormal areas of the cervix when a dilute solution of acetic acid is applied.

Acinar is a term used to describe round clusters of cells.

Acquired immune deficiency syndrome (AIDS) final stage of disease caused by HIV, which gradually causes suppression of the immune system. The individual will become susceptible to infections and developing tumours.

Acrosome a membrane-bound compartment found in the head of a sperm. It contains the enzymes that allow penetration of the outer surface of the egg.

Actinomyces-like organisms bacteria found in cervical samples from women fitted with an intrauterine contraceptive device (IUCD). They may also be seen in the absence of an IUCD. The most common species is *Actinomyces israelii*.

Adenocarcinoma cancer of glandular epithelium.

Adjuvant additional or supplementary test.

Age-standardized rate is a statistical method of comparing cancer rates and mortality between different geographical populations. This calculation removes the age bias by adjusting the rate to take into account how many people of different ages are in the population being looked at.

Albumin a protein that is soluble in water. Used in cytology to enhance the adhesion of cells to a glass slide.

Alkaline phosphatase an enzyme commonly used as an antibody tag in immunocytochemistry techniques.

Allele alternative form of a gene.

Alveolar macrophage type of macrophage present in the pulmonary alveolus. Its function is the removal of foreign matter.

Amosite also known as brown asbestos, used widely in fire protection.

Amplification an increase in the number of copies of a gene in a cell.

Ampulla a sac-like area where sperm is stored.

Anaerobic coccobacilli round to rod-shaped bacteria that only grow in the absence of oxygen.

Anaplastic carcinoma of thyroid type of thyroid cancer which is very aggressive in nature.

Anaplastic cells display significant pleomorphism, hyperchromasia with high nuclear to cytoplasmic ratio. Anaplastic cells do not show any differentiation.

Anemometer a device used to measure wind speed.

Aneuploid a decrease or increase in the number of chromosomes.

Aneusomy abnormal number of chromosomes.

Angiogenesis the development of new blood vessels.

Angiogenic growth factors proteins that stimulate angiogenesis. Examples include vascular endothelial growth factor (VEGF) and platelet derived growth factor (PDGF).

Angiography imaging of the blood vessels by X-ray, usually after administration of radio-opaque dyes.

Anisonucleosis variation in nuclear size.

Anoxia poor oxygen supply.

Anterior pelvic clearance an extensive operation for treatment of bladder cancer in women, where bladder, urethra, ovaries, uterus, and the upper part of the vagina are removed. Internal lymph nodes that lie within the pelvis are also usually removed during this operation.

Anthracyclines drugs used in cancer chemotherapy.

Antibody a molecule that is produced by the immune system and binds to an antigen.

Antibody a substance produced by plasma cells in response to bacteria, viruses and other foreign particles.

Anticoagulant medicine that reduces the clotting of blood. Anticoagulants are given to patients who are at risk of thrombotic (blood clot) disorders.

Antigen a molecule that is recognized by the immune system and stimulates the production of an antibody.

Antihuman IgG antibodies are antibodies that will bind to all human-derived IgG type antibodies.

Apoptosis programmed cell death.

Apoptotic bodies the remains of cells following apoptosis.

Arias-Stella reaction endocervical glands lined by a single layer of cells with abundant clear cytoplasm and large hyperchromatic nuclei. Often associated with oral contraceptive use.

Aromatic amines organic molecules which contain one or more benzene rings.

Artefact any undesired or deliberate alteration in a cell preparation introduced by a specimen processing technique.

Asbestos a naturally occurring fibrous mineral which was used for its unique insulating qualities. Its use has been banned in many countries due to its associations with lung cancer.

Ascites accumulation of fluid in peritoneal cavity.

Atrophy/atrophic reduction in size of a tissue or organ. Epithelial atrophy refers to the thinning of epithelial tissue.

Atypia refers to deviation from normal, or not 'typical' of what is normally accepted. Its use in cytology refers to changes that are short of malignancy but raise some concern.

Atypical carcinoid these neuroendocrine tumours are clinically more aggressive than 'typical' carcinoids and metastasize more frequently.

Atypical metaplasia is recognized as dissociated squamous cells which show increase in nuclear/cytoplasmic ratio and sometimes keratinization.

Audit trail a chronological sequence of audit records containing evidence pertaining to the execution of a process.

Autolysis cellular breakdown caused by the release of damaging intracellular enzymes.

Avidin a large molecular weight glycoprotein with a high binding affinity for biotin. It can be readily combined with chromogenic tags such as horseradish peroxidase and, just like biotinylated antibodies, is also provided in commercially available ICC kits.

Avidin-Biotin complex (ABC) an immunocytochemistry technique that makes use of avidin (an egg white protein) and biotin (a vitamin) to create a link between an enzyme tag and a secondary antibody reagent.

Axillary clearance this surgical technique is used in the treatment of breast cancer by removing axillary lymph nodes. The aim of this surgery is to eradicate local disease and to determine prognosis to guide adjuvant therapy.

Azoospermic a condition where no spermatozoa can be found in the semen.

B and T cell clonality when a diagnosis of lymphoma is made it is important to establish T and B clonality or cell line which is dominant in the tumour. This information can help with treatment selection and management of the disease.

Bacillus Calmette-Guerin vaccine (BCG) is a treatment for carcinoma *in situ* and high-grade bladder carcinoma. BCG, which is an inactivated form of the bacterium *Mycobacterium tuberculosis*, is mixed in with a saline solution and instilled directly into the bladder via a catheter.

Bacteria (singular: bacterium) unicellulular microorganisms, typically only a few micrometres in length.

Bacterial vaginosis a vaginal or cervical condition characterized by a thin, milky vaginal discharge and a foul, fishy odour. The condition is caused by an imbalance of vaginal flora and a predominance of anaerobic bacteria, particularly *Gardnerella vaginalis*.

Basal cells in the respiratory tract are cuboidal cells that have the potential to differentiate into other cell types. Basal cells are the most immature form of squamous cell, found along the basement membrane, virtually never seen in cervical samples.

Base pairs in molecular biology, two nucleotides on opposite complementary DNA or RNA strands that are connected via hydrogen bonds are called a base pair—adenine, guanine, cytosine, and thymine.

Basement membrane a thin proteinaceous sheet separating epithelium from the underlying connective tissue.

BCG therapy immunotherapy using Bacille Calmette–Guerin, the same bacterium used for TB vaccination.

BD cytorich™ blue preservative an ethanol based preservative solution suitable for collection and transport of cytological material.

Benign non-cancerous mild, non-invasive, harmless.

Benign tumour a non-cancerous growth.

Ber EP4 this antibody reacts to all epithelial cells except the superficial layer of squamous epithelia cells, hepatocytes and parietal cells (acid secreting cells of stomach). It is useful in serous effusion cytology as it does not label mesothelial cells and is rarely reactive with mesotheliomas.

Bilharzia common name given to schistosomiasis, after the German physician Theodor Bilharz, who first described the cause of urinary schistosomiasis in 1851.

Binucleate having two nuclei.

Biological therapy is a form of therapy that uses the body's own immune system to attack cancer cells, or uses medication to interfere with receptors in the body that are involved in cell division.

Biomarker provides information at the genetic level by determining changes in DNA sequence which are associated with a specific disease or predisposition to a disease.

Biopsy a piece of tissue taken for diagnostic and/or treatment purposes.

Biotin a low molecular weight vitamin that can be easily conjugated to primary antibodies as a component of commercially available ICC kits.

Bladder washings a sampling method which involves washing the bladder. Saline is introduced to the bladder via a catheter and removed for microscopical examination.

Blue blobs degenerative parabasal cells in which the cytoplasm of the cell has taken up haematoxylin.

Blueing the process of converting the colour of nuclei from red to dark blue in the Papanicolaou staining technique.

Borderline nuclear changes cytological changes of uncertain significance.

Bronchial brushing a sample obtained from the respiratory tract using a specially designed brush.

Bronchial carcinoids are rare tumours arising from neuroendocrine cells. Two types of bronchial carcinoids are recognized: the first and most common is referred to as typical carcinoid. This is a low-grade tumour, capable of local invasion, but it rarely metastasizes. The second type is referred to as atypical carcinoid, which has the potential to metastasize.

Bronchial intraepithelial neoplasia a pre-invasive stage. It is thought that intraepithelial neoplasia of bronchial epithelium and alveolar epithelium precedes the development of lung carcinoma.

Bronchial neuroendocrine tumours are a group of tumours arising from the neurendocrine cells in the bronchus. Their behaviour varies considerably from relatively benign, as in typical carcinoid, to very aggressive small cell carcinoma.

Bronchoalveolar cell carcinoma is a non-small cell lung cancer and is a subtype adenocarcinoma. It arises from terminal air sacs and spreads diffusely through the lung.

Bronchoscopy is a technique of visualizing the bronchial epithelium using a miniature camera fixed to the end of a flexible tube known as a bronchoscope.

Brushing (of cells) removal of cells from a body surface using a specially designed brush.

Bubble gum cells contain large vacuoles and displaced nuclei, and are thought to be an inflammatory response to an intrauterine contraceptive device.

Buffy layer (buffy coat) the layer of cells overlying the packed red blood cells in a centrifuged body fluid.

Burkitt's lymphoma is a rare type of cancer of the lymphatic system.

Cachexia is severe weight loss and muscle mass associated with disease seen in cancers, AIDS, and other chronic progressive diseases.

Calretinin is a calcium binding protein. Antibodies to this protein are reactive with cells of mesothelial origin. This antibody stains both the nucleus and the cytoplasm of the cell.

Cancer a disease characterized by the uncontrolled growth of cells which invade the surrounding tissues and spread to distant parts of the body.

Candida is a yeast like fungus that is normally present as normal flora of the mouth, intestinal tract, and vagina amongst others, but has potential to cause infection (called candidiasis or thrush).

Capsid the protein coat of a virus.

Carbowax® the proprietary name for a family of polyethylene glycols (PEG) used as additives in cytological fixatives. The PEG coats the cells and acts as a barrier to help prevent cell damage.

Carcinoembryonic antigen (CEA) is glycoprotein expressed by foetal cells, and only expressed in small amounts by normal adult epithelial cells. Its expression is greatly increased in many adenocarcinomas, including those of colonic, pancreatic, and lung origin. It is useful in effusion cytology as it is not expressed by mesothelial cells or mesothelioma.

Carcinogen a substance or agent that causes cancer.

Carcinogenesis a process by which normal cells undergo change to become cancer cells.

Carcinoma a malignant tumour of epithelial cells. Epithelia are the covering of the surfaces of the body.

Carcinoma *in situ* an alternative term for CIN3.

Carcinoma *in situ* (early stage urothelial carcinoma) has a flat growth pattern that spreads along the surface of bladder. These tumours are histologically graded as high grade and the majority will eventually become invasive.

Cartwheel chromatin is the description given to the chromatin pattern of plasma cells, which resembles the spokes of a cartwheel. Plasma cell nuclei can also resemble the face of a clock, hence the alternative term 'clockface' chromatin.

CD cluster of differentiation commonly abbreviated to CD is a nomenclature used to identify and type cell surface molecules on white blood cells, e.g. CD4.

CD117 see c-kit.

CD15 is found on neutrophils. It is expressed in Hodgkin's disease, some T cell lymphomas, and leukaemias.

CD30 is expressed in activated T and B cells and also in Hodgkin's disease.

CD34 is expressed by haematopoietic stem cells and gastrointestinal stromal tumours (GISTs).

CD45 is a protein present on surface of lymphoid cells. CD45 is also known as leukocyte common antigen.

CD56 antibody to CD56 reacts with many normal and neoplastic tissues. It is useful in identification of pulmonary small cell carcinoma.

CD68 a glycoprotein expressed by macrophages.

Cell block a way of processing a cytology sample as if it were a solid tissue specimen. Cells are immobilized in a solid medium which is then processed histologically.

Cell culture medium a specially formulated liquid or gel used to support the growth of cells.

Cell engulfment is a term to describe when one cell appears to be engulfing another cell by wrapping around it.

Cell enrichment the process by which the proportion of neoplastic cells in a specimen is increased, usually by depleting the sample of unwanted cells.

Centrifugation is the use of centrifugal force (g-force) to isolate cells or other particles from the medium in which they are suspended. The suspension is rotated at a speed which forces cells away from the axis of rotation.

Cercaria small free-living larval stage of trematodes.

Cervarix® a bivalent vaccine produced by GSK.

Cervical broom a device used to take cell samples from the cervix.

Cervical glandular intraepithelial neoplasia a tumour of the glandular cells of the uterine cervix that is confined within the epithelial layer. There is no invasion of the underlying stroma.

Cervical glandular intraepithelial neoplasia an abnormal growth of tissue in the cervix that is restricted to glandular epithelium. There is no invasion of the underlying stroma.

Cervicitis inflammation of the cervix.

Cervix the lower part of the uterus where it joins the top of the vagina.

Charcot-Leyden crystals are orangeophilic needle-shaped structures derived from degenerating eosinophils. They are seen in patients with allergic disorders, such as asthma.

Chemotherapy the treatment of malignant disease by means of cytotoxic drugs.

Chlamydia trachomatis an obligate pathogen that causes several human diseases. In women it can cause cervicitis and pelvic inflammatory disease.

Cholangiocarcinoma adenocarcinoma of bile duct.

Chromatid either of the two strands formed when a chromosome duplicates itself during the early stages of cell division. The chromatids are initially joined together by a structure known as a centromere but separate later, one going to each daughter cell.

Chromagenic *in situ* hybridization (CISH) uses a coloured substrate rather than a fluorchrome.

Chromatin margination aggregation of chromatin at the nuclear border, forming a thickened rim.

Chromatin the proteins and nucleic acids that make up the substance of the nucleus.

Chromatin distribution the variation in the spatial density of chromatin granules over the area of the nucleus.

Chromatin pattern refers to the degree of coarsening of the chromatin granules as seen under the microscope.

Chromogen a substance capable of conversion into a visible pigment or dye.

Chromogranin a glycoprotein located on neurosecretory granules.

Chromosomal aberration modification to the normal chromosome complement due to deletion, duplication, or rearrangement of genetic material.

Chromosomes structures found in the cell nucleus that contain the genetic material: human cells contain 23 pairs of chromosomes making a total of 46.

Chronic lymphocytic leukaemia (CLL) is a type of slow-developing leukaemia affecting the lymphocytes.

Chronic sialadenitis chronic inflammation of salivary gland.

Chrysotile also known as white asbestos; its curly, flexible fabric could be woven into fire resistant fabrics.

Chylous a fluid containing lymphocytes and small fat globules that appears milky in colour.

Cilia fine hair-like structures found on the surface of some types of cells. Cilia may be motile or non-motile.

CIN1 the earliest recognizable stage of CIN in which neoplastic basal cells are confined to the lowest third of the epithelium.

CIN2 the stage of CIN in which neoplastic basal cells are confined to the lowest two-thirds of the epithelium.

CIN3 the most severe stage of CIN in which neoplastic basal cells span more than two-thirds of the epithelium, and may involve the full thickness.

Cisplatin a platinum-based chemotherapy drug useful for treatment of many cancers, including bladder cancer, testicular cancer, and small cell anaplastic carcinoma.

Cytokeratin 5/6 (CK5/6) is an antibody to cytokeratins 5 and 6. This antibody stains mesothelium and squamous carcinoma, but is usually negative in adencocarcinoma.

Cytokeratin 20 (CK20) is expressed by many epithelial cells and carcinomas arising from those epithelia, including gastrointestinal epithelium and colorectal carcinoma.

Cytokeratin 7 (CK7) is present in many epithelial tissues including lung, breast, and ovary, amongst others.

c-kit a cytokine receptor expressed on the surface of hematopoietic stem cells as well as other cell types. Also called CD117, this is a cell membrane-spanning signalling molecule whose expression can be demonstrated by immunocytchemistry.

Clara cells produce various proteins which protect the bronchiolar epithelium. They also have the potential to differentiate into ciliated cells.

Clearing agent a substance used to raise the refractive index of a stained preparation so that it becomes optically clear.

Clearing the process of rendering stained preparations optically clear by immersion in a clearing agent.

Clinical governance the system through which health organizations are accountable for continuously improving the quality of their services and safeguarding high standards of care.

Clinical Pathology Accreditation Ltd (CPA) is a regulatory body which oversees quality in the UK, conducting peer review visits every three years to registered laboratories.

Clinician person with overall responsibility for the care of a patient.

Clockface chromatin see cartwheel chromatin.

Clonal expansion an increase in the number of genetically identical cells arising through cell division.

Clone (as in cancer development) a population of genetically identical cells.

Clue cells squamous epithelial cells whose surface is completely covered with the microorganisms that cause bacterial vaginosis.

Cofactors contributing factors.

Cohesiveness (of cells) the tendency of cells to stick together.

'Cold' coagulation a form of treatment for cervical intraepithelial neoplasia in which a hot probe is used to burn away the abnormal cells.

Collection fluid a fluid into which cytological samples are placed following collection. The fluid is usually a fixative but can also be a cell culture medium.

Colloid nodules non-cancerous (benign) overgrowths of normal thyroid tissue.

Colloid oncotic pressure is a type of osmotic pressure due to proteins present in the blood.

Colovesical fistula a connection between the colon and bladder.

Colposcope a special microscope designed for examining the uterine cervix and vagina.

Colposcopic assessment a close examination of the cervix using a magnifying instrument called a colposcope.

Colposcopy a method of viewing the epithelial surfaces of the vagina and cervix using a magnifying instrument called a colposcope.

Columnar epithelial cells long narrow cells lining the endocervical canal.

Computed tomography a non-invasive imaging technique that uses a large series of X-rays to generate a three-dimensional image of the inside of the body.

Computerized tomography computer imagery formed from multiple X-rays.

Cone excision involves the surgical excision of a cone-shaped portion of the ecto- and endocervix, including the removal of the entire squamocolumnar junction (SC junction) of the cervix.

Conjugation (in immunocytochemistry) the chemical binding of an antibody with a visible marker or tag.

Connective tissue diseases are a group of diseases that primarily affect the connective tissues in the body. Many connective tissue diseases feature autoimmune activity, where the body's own immune system is involved in an inflammatory response against its own tissues. Systemic lupus erthematosus and rheumatoid disease are two examples of connective tissue disease.

Connective tissue supports and binds other kinds of tissue.

Contact inhibition the natural arrest of cell division when cells come into contact with one another.

Containment level a numerical system (1–4) describing the standard to which a laboratory is designed and built for the purposes of containing hazardous microorganisms.

Contrast-enhanced chest CT during the CT scan, a radio-contrast agent or 'contrast media' is administered intravenously in order to provide higher image quality.

Conventionally spread cells placed directly on a slide and manually spread to form a thin layer.

Corpora amylacea are round, glass-like substances that are often seen in male urine samples. They originate from the prostate gland. They have no diagnostic significance. Corpora amylacea are also rarely seen in respiratory samples.

Corpus luteum the structure that develops from an ovarian follicle following ovulation. The corpus luteum produces oestrogen and progesterone, which are needed to maintain the endometrium.

Corticosteroids a family of drugs closely related to naturally produced cortisols, which are produced by the adrenal gland. They are useful as they have an anti-inflammatory effect.

Counterstain a stain with a contrasting colour to the primary stain. Counterstaining in cytology is achieved by staining the cytoplasm of cells a different colour to the nuclei.

Coverage the proportion of the target population that is screened.

Coverslip a thin sheet of glass that is placed over slide preparations to protect cells and facilitate microscopy.

Creatinine is waste product of creatine phosphate metabolism in muscle tissue. Measurement of serum creatinine level is used to monitor renal function.

Creola bodies are clusters of ciliated epithelial cells sometimes seen in sputum samples of patients with asthma.

Crocidolite is also known as blue asbestos. This type of asbestos is the strongest type and highly resistant to acids.

Cryotherapy a form of treatment for cervical intraepithelial neoplasia in which a cold probe is used to freeze away the abnormal cells.

Crystal deposition diseases a group of disorders characterized by the precipitation of certain types of crystals in the joints and connective tissue.

Cuboidal cells cube-shaped cells.

Curschmann's spirals are strands of mucus that are formed in the lumen of small bronchi.

Cyanophilia/cyanophilic an affinity for green/blue dyes.

Cyclophosphamide this chemotherapeutic agent is used in combination with other drugs in treatment of many cancers, including lymphoma and leukaemia. It is also used in treatment of some autoimmune disorders.

Cystoprostatectomy this surgical procedure involves removal of the bladder, prostate, lower end of ureters, and sometimes the urethra.

Cystoscopy this is an endoscopic procedure that is used to examine the bladder.

Cytocentrifugation a technique used to centrifuge small volumes of fluid (generally up to 0.5 ml) directly onto a glass slide.

Cytogenetics the examination of cell samples for the purpose of detecting and interpreting chromosome abnormalities.

Cytokeratin 7 present in many carcinomas, including the majority of lung, breast, ovarian, and endometrial adenocarcinomas.

Cytokeratin 20 present in many carcinomas, including gastrointestinal and transitional cell carcinomas.

Cytokines these protein chemical messengers are produced by the cells of the immune system and tumour tissue and exert various systemic effects in the body.

Cytological surveillance the monitoring of women with low-grade cytological abnormalities by repeat cervical cytology.

Cytolysis cellular breakdown.

CytoLyt® A methanol based preservative solution suitable for collection and transport of cytological material.

Cytomegalovirus (CMV) cytomegalovirus is a member of the herpes virus family that is acquired by most people during their lives, but rarely causes disease unless the person is immunocompromised.

Cytopathic effect the effect of a disease on the morphology of a cell.

Cytopathic characteristic pathological changes in a cell.

Cytopathology the study of disease based on the morphology of individual cells.

Cytoplasmic bridges thin strands of cytoplasm linking the spermatids.

Cytoplasm the substance of a cell that lies between the nucleus and the cell membrane.

Cytotoxic drug a drug that aims to kill neoplastic cells selectively.

Cytotoxic any agent that is toxic to cells and causes cell death.

Decoy cells a term given to polyomavirus infected cells. These cells exhibit nuclear abnormality which is similar to cancer and can act as a 'decoy' and confuse the cytologist.

Degeneration the deterioration and ultimate death of cells and tissues.

Dehydration the removal of water from a stained preparation by immersion in absolute alcohol.

Denaturation the process of physically separating DNA into two separate strands using heat.

Density gradient a method used for separating bloodstained serous fluids.

Density gradient fluid a solution of natural or synthetic compounds forming a fluid of known specific gravity.

Desmin an antibody to muscle tissue useful in serous effusion cytology as it reacts with benign mesothelial cells.

Diagnostic cytology the discipline involved in examining cells from the symptomatic patient, in an effort to diagnose the reason for illness.

3,3′-Diaminobenzidene (DAB) an organic compound used in immunocytochemistry techniques as a chromogenic substrate for horseradish peroxidase.

Diff-Quick™ is a commercial brand of Romanowsky stain. Its high concentration of dyes allows rapid staining of cells. It is a May Grünwald Giemsa stain for rapid on-site assessment and diagnosis.

Differential cell count technique of counting cell types in a sample and expressing these as percentages of the total number of cells in the sample.

Differential diagnosis the processes of distinguishing between different diseases.

Differentiation (in Papanicolaou staining) the process of removing excess haematoxylin from nuclei in order to accentuate the chromatin pattern.

Differentiation the process of cell specialization. Differentiated cells have well defined morphology and a specific physiological function.

Differentiator a substance used for differentiation in a staining procedure. In the Papanicolaou technique this substance is acidified alcohol.

Diploid possessing two copies of each gene.

Direct referral a system where the clinical referral of a patient is initiated by the clinic.

Disobliteration therapy opening up of closed channels; in the context of the respiratory tract it refers to opening of closed airways due to disease.

Disposal contractor a company authorized to dispose of waste safely and legally.

Dissection the appropriate cutting and sampling of biopsies.

Dithiothreitol (DTT) a chemical compound $(C_4H_{10}O_2S_2)$ that can be used as a mucolytic agent.

Diverticular disease is a condition where the inner, lining layer of the colon bulges out through the outer muscular layer. These out-punchings are called diverticula.

DNA is the nucleic acid that contains the hereditary information used in the development and functioning of all living organisms. Its main role is to store genetic information.

DNA sequencing learning the order of the four bases (C, G, A, and T) which make up DNA.

DNA translation the first step in the cellular process that produces proteins using the information contained in DNA.

Ductal carcinoma (of breast) is the most common type of breast cancer arising from the ducts.

Dyskaryosis a morphological abnormality of cell nuclei.

Dyskaryotic having an abnormal nucleus.

Dyskeratosis abnormal keratin formation, of which there are several forms, including hyperkeratosis and parakeratosis.

Dysplasia an alternative term for cervical intraepithelial neoplasia.

Dyspnoea shortness of breath (SOB).

Ectocervix the portion of the cervix projecting into the vagina.

Ectopy also called **ectropion**, this term describes a reddened granular appearance of the cervix. Also known as the transformation zone. It is in this zone that transformation of endocervical columnar epithelium to metaplastic squamous epithelium takes place. This is a normal physiological process.

Ectropion see ectopy.

Effusions fluid accumulating in the serous cavities.

Electrocautery the stoppage of bleeding using an electrically heated metal instrument.

Electrodiathermy a form of treatment for cervical intraepithelial neoplasia in which the area of abnormal cells is cut away using a loop of wire and an electric current.

Embryonal tissue is tissue derived from the embryo.

Empyema the collection of pus and debris in the pleural space.

End-stage renal disease is the complete failure of the kidneys to remove waste and produce urine. Dialysis or kidney transplantation is the only treatment.

Endobronchial ultrasound guided transbronchial needle aspiration (EBUS-TBNA) in this procedure a bronchoscope is inserted via the mouth into the lungs. An ultrasound probe which is located at the tip of the bronchoscope provides images and guides the bronchoscopist to target suspicious areas for fine needle aspiration.

Endocervical adenocarcinoma cancer of the glandular epithelium in the endocervix.

Endocervical canal the narrow passageway between the uterus and vagina between the internal and external os. Also called the endocervix.

Endocervix see endocervical canal.

Endometrial polyp an overgrowth of endometrium, usually benign.

Endometritis inflammation of the endometrium.

Endometrium the epithelial lining of the uterus.

Endoscope a lighted tube which is inserted deep into body cavities in order to view the internal surfaces.

Endoscope a lighted tube that is passed into the body and through which the observer can visualize the surfaces within. The word endoscope originates from the Greek: *endon* = within and *skopos* = target.

Endoscopic ultrasound (EUS) is a procedure that combines endoscopy with ultrasound. An ultrasound probe at the tip of the endoscope allows high quality ultrasound images of organs inside the body.

Entamoeba gingivalis a harmless amoeba normally found in the oral cavity.

Enterovesical fistula normally there is no connection between the urinary system and alimentary canal. An enterovesical fistula is an abnormal connection between the bladder and alimentary system.

Enzyme a protein that increases the rate of chemical reactions.

Eosin azure (EA) a mixture of eosin and light green used as a counterstain in the Papanicolaou technique.

Eosin a red dye used as a cytoplasmic stain in the Papanicolaou staining technique.

Eosinophil a type of white blood cell with eosinophilic cytoplasmic granules. Eosinophils are an important component of the immune system.

Eosinophilia/eosinophilic in the context of this chapter eosinophilia refers to an affinity for the dye eosin ('loves eosin').

Eosinophilic pneumonia encompasses a group of diseases that occur when eosinophils gather in the lung and can cause damage to the alveoli.

Epidemiology the study of the cause, frequency, and distribution of diseases in populations.

Episome a piece of viral genetic material that exists independently of the host cell DNA.

Epithelial casts these casts form from the epithelial cells lining the tubules. They are indicative of disease.

Epithelial membrane antigen (EMA) is a glycoprotein that is present in a wide variety of epithelial cells. It is useful in fluid cytology as antibodies to this protein are immunoreactive with most adenocarcinomas and mesotheliomas, but unreactive in benign mesothelial cells.

Epithelial mesothelioma is a histological subtype of malignant mesothelioma, where the cells have an epithelial appearance. It is the most common type of mesothelioma.

Epithelioid histiocytes also macrophages, look like epithelial cells and are seen in granulomatous inflammation.

Epithelium tissue that lines the body cavities and the surfaces of organs.

Equational division also called second meiotic division, where sister chromatids finally split, creating a total of 4 haploid cells with 23 chromosomes.

Equivocal used in cytopathology reporting to indicate uncertainty.

Ergonomic hazard can be defined as any object, system, or environment with potential to cause human harm through poor design.

Ergonomics the science of designing tasks, equipment, and workplaces for the purposes of human comfort, efficiency, and safety.

Erionite a naturally occurring fibrous mineral which is carcinogenic and has been linked to causing mesothelioma.

Erythrocyte also called red blood cells, these are the cells responsible for carrying oxygen and carbon dioxide to and from body tissues.

Ethanol a colourless flammable liquid used as a cytological fixative and dehydrating agent.

Etoposide a chemotherapy drug used in treatment of many cancers, including lung, stomach, and non-Hodgkin's lymphoma.

Eukaryotic cells contain a membrane-bound nucleus and membrane-bound structures called organelles.

Eversion the movement of the squamocolumnar junction so that endocervical columnar epithelium comes to lie on the vaginal portion of the cervix. Eversion results from the change in shape of the cervix in response to increasing levels of ovarian hormones.

Ewing's sarcoma a rare type of malignant bone tumour.

Excision margin this tumour was completely excised and this reduces the likelihood of local recurrences.

Exfoliation (of cells) the shedding of loosely held cells from their parent tissue.

Exfoliative cytology the study of cells shed spontaneously from the body surfaces.

Exogenous hormone a hormone originating from outside the body that is administered for a medical reason.

Exon the coding part of a gene.

Expectorate to cough up material from the respiratory tract.

Extensive stage disease small cell carcinoma is described as extensive stage when it has metastasized to outside the chest.

External genitalia in human females the external genitalia refers to the vulva and clitoris.

External os the opening of the ectocervix.

External Quality Assurance (EQA) the monitoring and evaluation of a service by an external body.

Extracutaneous occurring outside the skin.

Exudate a protein-rich fluid that is released from the blood vessels into the tissues as a result of an inflammatory response, or neoplastic process.

Faecaluria mixing together of urine with faeces which is usually due to presence of a fistula

Fail-safe a system designed to prevent a failure of patient care due to inaction of one or more of the agencies involved in screening.

Fallopian tubes two fine tubes leading from the ovaries to the uterus.

False negative a test result that suggests the absence of a condition when the condition is actually present.

False positive a positive test result in the absence of the disease being tested for.

Feathering the presence of elongated slender nuclei projecting perpendicularly from the edge of a sheet of dyskaryotic endocervical cells, giving the appearance of a bird's feather.

Ferruginous or asbestos bodies asbestos fibres engulfed by macrophages and coated with iron.

Fertilization the union of a sperm and an ovum leading to the development of an embryo.

Fibrillary stroma fibre-like stroma.

Fibrin protein involved in clotting of blood.

Fibroid a benign tumour arising in the uterus.

Fiducial markings on the ThinPrep® slide ensure accurate localization of cells on each slide; the system will alert the user if markings are out of alignment.

Fields of view (FOV) areas, not necessarily abnormal, deemed of interest and selected by the imager for manual review.

Filariform infective stage in the life cycle of a nematode.

Fimbriae finger-like projections at the distal end of each of the Fallopian tubes.

Fine needle aspiration (FNA) cytology the study of cells aspirated from a tissue using a narrow gauge needle and syringe.

Finger clubbing is the change in the shape of fingertips and nails that is seen in some patients with lung cancer and heart disease. The fingertips appear larger and nails curve more than usual. It is thought to be due to fluid accumulation in the fingertips.

Fistula an abnormal opening that joins two organs or epithelial surfaces together.

Five-year survival rate the percentage of patients that are alive five years after the diagnosis of their disease.

Fixation the process of stabilizing the morphology and chemical structure of cells, rendering them impervious to microbial attack and facilitating subsequent demonstration techniques.

Fixatives are often liquid agents used for the above purpose.

Flagella (singular: flagellum) long hair-like motile projections that extend from some unicellular organisms.

Flagellum a whip-like extension of certain cells, e.g. sperm, that functions as an organ of locomotion.

Flap refers to the latter region of the primary probe.

Fluorescent *in situ* hybridization (FISH) is used to detect presence or absence of DNA sequences on chromosomes.

Fluorodeoxyglucose (FDG) is a radioactive compound used in PET scanning. It is similar in structure to glucose and taken up by tissues and cells that use high levels of glucose, such as brain, kidneys, and tumour cells. The distribution of high levels of FDG is captured by the scanner and analysed.

Fluorophore a fluorescent dye.

Foetus the unborn offspring from the eighth week after fertilization until birth.

Follicle stimulating hormone a gonadotropin released from the anterior pituitary which stimulates the maturation of one or more follicles within the ovary.

Follicular adenoma is a common benign thyroid tumour.

Follicular carcinoma of thyroid a malignancy of thyroid follicular cells.

Follicular phase the first half of the menstrual cycle. Refers to the development of a follicle in the ovary.

Food and Drug Administration (FDA) United States Food and Drug Administration, which regulates and supervises the food and drug industry on behalf of the public.

Formalin an aqueous solution of formaldehyde used as a tissue fixative in histology.

Fraying see feathering.

Fume extraction hood a device used for handling noxious chemicals that draws in air and expels it from a building via suitable ducting. It is designed to limit the user's exposure to chemical fumes.

Fumigation a method of decontamination that completely fills an area with a gaseous disinfectant.

Fungi (singular: fungus) a single-celled or multicellular microorganism that may or may not cause infection.

Gametogenesis egg development.

Gardasil® a quadravalent vaccine produced by MERCK.

Gardnerella vaginalis a bacterial opportunistic pathogen that normally resides in the vagina.

Gastrointestinal this refers to oesophagus, stomach, and duodenum.

Gel electrophoresis a technique used for the separation of deoxyribonucleic acid (DNA), ribonucleic acid (RNA), or protein molecules using an electric field applied to a gel matrix.

Gene chips are grids composed of DNA which are transferred to a glass slide.

Genetic marker an identifiable DNA sequence that can be used to indicate a specific disease or an increased risk of the disease.

Genome a full set of chromosomes carried by an organism.

Germ cell tumour a neoplasm arising from germ cells present in the ovary or testis.

Gestation the period of pregnancy.

Glacial acetic acid water-free (i.e. pure) acetic acid. Used in cytology preparation for the destruction of red blood cells.

Glandular cells are specialized for the secretion of products of metabolism, e.g. mucin.

Glomerular disease a term given to diseases that affect the filtration units in the kidneys, the glomeruli.

Glycogen an intracellular source of stored energy derived from glucose.

Glycoprotein a substance that consists of proteins combined with carbohydrates. They are mainly protective in function but may also play a role in cell-cell interactions.

Goblet cells glandular cells whose sole function is the production and secretion of mucin.

Goitre enlargement of normal thyroid gland.

Golgi apparatus a network of stacked membranous vesicles present in most living cells. It is involved in the production of cell secretions.

Gonadotropin a hormone released from the anterior pituitary, which stimulates ovarian activity.

Gonadotropin-releasing hormone a hormone released from the hypothalamus which stimulates the secretion of gonadotropins from the anterior pituitary.

Graafian follicle a mature ovarian follicle that ruptures during ovulation to release the ovum.

Granular casts these are either formed from depositions of proteins or are broken down cellular casts.

Granulomatous inflammation this inflammatory response is characterized by accumulation of macrophages. Macrophages that are epithelia-like in appearance are called epithelioid. Macrophages may also fuse together to form giant cells.

Granulosa cells surround the developing oocyte in an ovarian follicle and persist within the corpus luteum following ovulation. Granulosa cells produce the sex hormones oestrogen and progesterone in varying quantities, depending on the stage of the menstrual cycle.

Grocott's methenamine silver method uses a solution of methenamine silver that reduces to a black precipitate on contact with *Pneumocystis jiroveci* and other pathogenic fungi.

Ground glass the term used to describe the pale bland chromatin pattern in cells infected with herpes simplex virus.

Ground glass nuclei in cytology describes nuclear features that appear due to margination of chromatin which gives the appearance of 'ground glass'. Often used to describe nuclear features due to herpes virus infection.

Haematein the oxidized and active form of haematoxylin.

Haematemesis vomiting of blood, the source is generally from the upper GI tract.

Haematopoietic stem cells generate all blood cell types.

Haematopoietic system the organs and tissues responsible for the production of blood cells, principally bone marrow, spleen, tonsils, and lymph nodes.

Haematoxylin a natural dye extracted from the logwood tree, *Haematoxylin campechianum*, is used to stain cell nuclei.

Haematoxylin and eosin (H&E) the most widely used histological staining technique that uses haematoxylin to stain nuclei and eosin to stain cytoplasm and other structures.

Haematoxyphilic an affinity for haematoxylin.

Haematuria the presence of red blood cells in urine. There are many causes of haematuria, including trauma, infections, stones, benign enlargement of prostate gland, and malignancy. It is called macroscopic haematuria when blood can be seen with a naked eye, and microscopic when only a small amount is present and requires microscopy or urine test for detection.

Haemocytometer a device which was originally designed for counting blood cells, and can also be used for counting other types of cells.

Haemolysis the breakdown of red blood cells.

Haemopoietic blood cell types including myeloid (monocytes and macrophages, neutrophils, basophils, eosinophils, erythrocytes, megakaryocytes/platelets, dendritic cells) and lymphoid lineages (T cells, B cells, NK cells).

Haemoptysis coughing up blood.

Haemostasis the stoppage of bleeding by natural (i.e. blood clotting) or artificial (e.g. electrocautery) means.

Hazard a situation with the potential to cause harm to a person's life, to their health, to property, or to the environment.

Helical or spiral CT is a three-dimensional CT scanner that can produce high resolution images with lower X-ray dose to the patient.

Heparin one of the commonly used anticoagulants.

HER2 (also called Her2/neu, c-erb-2 or erb-2) is a protein made by the HER2/neu gene. Receptors on the protein can bind to cancer cells causing them to divide and grow.

Herceptin, otherwise known as trastuzumab, is a drug which can be used to treat HER2 positive breast cancers.

Herpes simplex virus a virus that infects the skin and nervous system, causing watery blisters.

Heterogeneity (as in tumour development) the presence of different types of cells within a tumour. Sub-populations of cells may have a different genetic make-up, growth rate, and susceptibility to specific drug therapies.

High efficiency particulate air (HEPA) filter a type of filter used in microbiological safety cabinets for trapping particles down to the size of 0.3 μm, which will include the vast majority of microorganisms found in clinical specimens. Filtration of laboratory air through an HEPA filter renders the air safe for evacuation to the environment.

High-grade intraepithelial lesion moderate dyskaryosis or more severe squamous abnormality.

High-grade squamous dyskaryosis is the name given to moderately to severely abnormal appearing cells on a Pap smear.

High-grade urothelial carcinoma cells show obvious cytological abnormality. High-grade lesions have a higher tendency to become invasive.

High negative predictive value is a marker of accuracy, i.e. that the tests called negative by the imager have been correctly classified.

Hilum the area where the bronchus, blood vessels, nerves, and lymphatics enter or leave the lung.

Histiocyte see macrophage.

Histological section a thin slice of tissue that has been placed on a glass slide and stained ready for microscopic examination.

Histology the science that deals with the examination of tissues under a microscope.

Histopathology the study of disease at the tissue level.

HMB45 is an antibody used for marking malignant melanoma. It is a monoclonal antibody that reacts with melanomas and occasional carcinomas.

Hodgkin cells are large atypical single nucleated cells seen in Hodgkin's lymphoma.

Hodgkin's disease a type of lymphoma (cancer arising from lymphocytes). It is characterized by the presence of Reed-Sternberg and Hodgkin cells.

Hodgkin's lymphoma is a type of lymphoma typified by presence of Reed-Sternberg cells.

Holding category a category of cytological report in which a decision regarding the precise nature of the cellular changes is deferred until follow-up tests are performed.

Homeostatic control regulation of the internal bodily environment to maintain a constant, stable condition.

Honeycomb sheets the term used to describe the appearance of sheets of normal endocervical cells, which can resemble the wax structure inside a beehive.

Hormone replacement therapy treatment with oestrogen and progesterone to artificially boost the levels of these hormones during the menopause.

Hormone a chemical released from cells in one part of the body which exerts effects on cells in another part of the body. Hormones can be thought of as chemical messengers.

Horseradish peroxidase an enzyme commonly used as an antibody tag in immunocytochemistry techniques.

HPV proofer test detects E6/E7 messenger RNA as opposed to DNA.

HPV testing a test for the presence of HPV in clinical specimens.

HPV vaccine a preparation containing harmless HPV proteins that provides immunity to infection by certain types of HPV.

Human chorionic gonadotropin a hormone secreted by trophoblastic cells in pregnant women.

Human epidermal growth factor 2 see HER2.

Human Genome Project (HGP) completed in 2001, is one of the greatest scientific achievements to date. The human genome contained within a set of chromosomes is estimated to contain 25–30,000 genes.

Human immunodeficiency virus (HIV) a rotavirus causative agent in AIDS.

Human papillomavirus a papillomavirus capable of infecting humans and that causes warts and neoplastic conditions. Over one hundred types have been identified. High-risk types are implicated in the development of malignant conditions in epithelial tissues.

Hyaline casts are formed by deposition of glycoprotein in the tubules. They can be seen in concentrated urine of normal individuals.

Hybridized conjugated or bound.

Hybridization the process of combining single-stranded nucleic acids from different sources to form double-stranded molecules. A common technique used in molecular biology.

Hydrogen peroxide the substrate for the enzyme horseradish peroxidase in immunocytochemistry techniques.

Hydrostatic pressure in blood vessels is due to the height of the blood above, which exerts a force on the blood lower down the body.

Hyperchromasia more darkly stained than normal.

Hyperchromatic crowded cell groups darkly stained groups of cells in which there is considerable overlap of cells and nuclei, a common cytological presentation of high-grade dyskaryosis.

Hyperkeratosis a form of excessive cellular keratinization in which densely orangeophilic cytoplasm renders cell nuclei invisible.

Hypermethylation the addition of a methyl group to DNA can prevent gene expression. Tumour suppressor genes may be inactivated by hypermethylation.

Hyperplasia or hyperplastic is the physiological response to stimuli. Hyperplasia may lead to increase in cell size and number.

Hypoalbuminemia a condition where levels of albumin in the blood serum are low.

Hypochromasia staining less intensely than normal.

Hypothalamus a small but complex portion of the brain, which, among other functions, regulates the secretion of hormones from the pituitary.

Hysterectomy an operation to remove the uterus and cervix that can also involve the removal of adjoining structures, such as the fallopian tubes and ovaries.

Iatrogenesis damage or disease caused unintentionally by medical treatment or any other form of medical intervention.

Iatrogenic changes alterations seen in cervical samples as a result of iatrogenesis.

IgG antibodies Y-shaped protein molecules secreted by B lymphocytes as a primary immune defence. Each arm of the Y has one of a pair of identical binding sites that can combine with the complementary sites of foreign antigens.

Ileal conduit a reservoir for the collection of urine, constructed from a piece of ileum after removal of the bladder.

Imaging (in medicine) the process used to create images of parts of the body to aid a diagnosis.

Imatinib a biological therapy used in treatment of gastrointestinal stromal tumours (GISTs).

Immature squamous metaplasia the early stage of squamous metaplasia in which the constituent cells have not yet fully differentiated into mature squamous epithelial cells.

Immortalization (in tumour development) the potential for limitless cell division.

Immune response a specific bodily defence mechanism that protects the body against pathogens and other foreign particles.

Immunocompromised describes the state of a person's immune system that is less able or has lost the ability to fight infections. Examples of immunocompromised people include those with AIDS and those receiving chemotherapy or drugs to suppress the immune system after organ transplantation.

Immunocytochemistry a demonstration technique that uses specific antibody-antigen reactions to detect cell constituents.

Immunodeficient lacking immunity or unable to fight infections.

Immunohistochemical a technique involving the application of antibodies raised to specific antigens found within a particular cell type.

Immunoreaction the specific reaction between an antibody and its corresponding antigen.

Immunosuppressive therapy drugs given to patients who have had an organ or bone marrow transplant to suppress their immune system to prevent rejection.

Immunotherapy a form of treatment that appears to work by initiating a local immune reaction against a tumour. Intravesical therapy for treatment of bladder cancers with BCG is an example of immunotherapy.

Impalpable unable to be felt by the clinician on a manual examination of the patient.

Improved Neubauer haemocytometer a design of haemocytometer.

In vitro is an artificial environment outside a living organism.

Inadequate unsuitable for reporting.

Inappropriate re-epithelialization the replacement of damaged epithelium by a type of epithelium that is not normally found at that site. Tubo-endometrioid metaplasia is an example in the cervix.

Incidence the number of new cases of a disease arising within a specified time period.

Infection invasion of the body by pathogenic microorganisms.

Inflammation the protective response of the body to injury.

Inflammatory changes the term used to describe the many cellular changes seen in cervical samples when they are taken during an inflammatory process.

Inflammatory exudate a mass of polymorphs that accumulates in a tissue in response to inflammation.

Informed patient consent permission obtained from a patient for a procedure or test to be carried out, based upon a clear understanding of the facts and possible implications of the procedure.

Inspissated mucus refers to thickened mucus.

Intermediate cells a partially mature squamous cell.

Intermediate filaments are a type of cell structural protein. There are numerous types of intermediate filaments, some of which are tissue specific. Antibodies to intermediate filament protein can be used to identify various tissues.

Internal os the opening of the endocervix into the uterine cavity.

Internal sexual organs also referred to as the internal genitalia; in human females these are the vagina, cervix, uterus, fallopian tubes, and ovaries.

Interphase the stage in the development of a cell following mitosis or meiosis, during which the nucleus is not dividing. In cells preparing for a further division the DNA in the nucleus is duplicated during interphase.

Interstitial cells of Cajal the interstitial cells of Cajal are thought to be gut 'pace makers' and are involved in peristalsis.

Interstitial lung is the term used to describe various types of inflammatory diseases that affect the tissue around the air sacs. These diseases generally cause progressive scarring of the lung tissue, which eventually interferes with oxygen absorption.

Interstitial nephritis is inflammation of the kidney that involves the tissues surrounding the renal tubules.

Interstitium is the supporting tissue surrounding the cells. In the respiratory tract it refers to the surrounding tissue around the alveolar spaces.

Intraepithelial confined within the epithelial layer.

Intranuclear within the nucleus of a cell.

Intranuclear grooves is used to describe folds or apparent condensation of chromatin within the nuclei.

Intranuclear viral inclusions aggregates of viral proteins within the nucleus of infected cells.

Intrauterine contraceptive device an object placed within the uterus to prevent fertilization and to inhibit implantation of a fertilized egg into the endometrium.

Intravenous urogram (IVU) also known as intravenous pyelogram is a radiological procedure for visualizing the kidneys.

Intravesical therapy involves placing medication directly into the bladder to treat bladder cancer.

Intron the non-coding part of a gene.

Invader technology is known as a third-wave technology. Cleavase recognizes and cleaves a three-dimensional **invader** structure formed by hybridization of two overlapping oligonucleotides to the target sequence (the cleavage of one of the oligonucleotides releases a flap that initiates a secondary cleavage reaction with a fluorescence resonance energy transfer (FRET) label.

Invasion (of tumours) the process by which malignant cells enter the bloodstream or lymphatics and spread throughout the body.

Invasive squamous cell carcinoma cancer of squamous epithelium.

IRESSA (Gefitinib)/TARCEVA (erlotinib) EGFR inhibitor drugs which suppress tyrosine kinase activity and are found to be effective against cancer.

Irradiation the use of radiation to treat cancer.

Ischaemic necrosis localized death of tissue caused by a lack of blood supply.

IUCD cells cells with high NCR and prominent nucleoli, often bi- or multinucleated, typically seen in the presence of an intrauterine contraceptive device.

Ivermectin a broad spectrum anti-parasitic drug effective against *Strongyloides stercoralis*.

Karyolysis complete dissolution of the nucleus.

Karyopyknosis condensation of chromatin.

Karyorrhexis nuclear fragmentation.

Karyotype the number, form, size, and arrangement of chromosomes in the nucleus of a cell.

Keratin a tough insoluble protein found in the epidermis, hair, nails, and other tissues, where it serves a protective function.

Keratinizing stratified squamous epithelium.

Ki67 (MIB1) is a cell proliferation marker. Ki67 positivity in tumour cells can act as a prognostic marker and help predict the likelihood of survival and recurrence in certain tumour types. It is a useful marker of cell proliferation.

Knife cone the removal of a cone-shaped portion of the ecto- and endocervix using a scalpel.

Knowledge and skills framework an agenda that defines and describes the knowledge and skills that staff need to apply in their work to deliver quality services.

Köhler illumination a type of microscope 'set up' in which the image of the light source is focused on the focal plane of the sub-stage condenser.

Koilocyte a squamous cell with one or more enlarged hyperchromatic nuclei and deep perinuclear clearing, caused by infection with human papillomavirus.

Koilocytosis is the term used to describe the presence of koilocytes in a specimen.

Labia majora two lips of skin on each side of the vaginal orifice outside the labia minora.

Labia minora two thin folds of skin between the vaginal orifice and the labia majora.

Labour the process of giving birth.

Lactating pattern the pattern of parabasal cells and lower level intermediate cells seen in post-partum cervical samples taken from women who breast-feed.

Lactic acid an organic acid produced by anaerobic respiration. *Lactobacilli* are the most likely source of lactic acid in the vagina.

Lactobacillus a group of non-pathogenic rod-shaped bacteria residing in the vagina.

Large loop excision of the transformation zone (LLETZ) the process of cutting away the whole of the transformation zone with a thin wire loop that has been heated using an electric current.

Large-volume centrifugation is centrifugation of 'large' volumes of fluid, generally up to 25 ml.

Laser cone the removal of a cone-shaped portion of the ecto- and endocervix using a laser beam.

Laser microdissection a method of selecting sparse cells of interest from a cytological preparation or a paraffin block using a laser.

Laser therapy/treatment a form of treatment for cervical intraepithelial neoplasia in which a high-energy beam of laser light is used to burn away the abnormal cells, by boiling intracellular water thus bursting the cells.

Latent phase the stage of a virus's life cycle during which there is no virus production and the viral genome remains dormant.

LE cells are seen in effusions due to SLE. The LE cell is a macrophage or neutrophil which has ingested nuclear matter from another cell.

Length mutations a collective name for insertions, deletions, duplications, and repeats of nucleic acids in DNA sequence.

Leucocytes white blood cells.

Leucophagocytosis infiltration of the cytoplasm of epithelial cells by white blood cells.

Leukaemia a group of malignant diseases characterized by the proliferation of immature leucocytes in the bone marrow and their accumulation in the circulation.

Leukoplakia a raised white patch of epithelium.

Levothyroxine a synthetic form of thyroxine.

Light green a blue/green dye used as a cytoplasmic stain in the Papanicolaou staining technique.

Limited stage disease small cell lung cancer is described as limited stage disease, if the tumour is confined to chest at the time of the diagnosis.

Lipofuscin a brown substance found in various cells, associated with breakdown of fats and proteins.

Liquid based cytology any cytology processing technique that relies on the immediate transfer of cells into a liquid preservative when the specimen is collected.

Lithiasis the formation of calculi.

Lithotomy position a common position for medical examinations. The patient lies on their back with knees bent and spread apart through the use of stirrups.

Liver cirrhosis scarring of the liver, which will eventually lead to loss of liver function. There are many causes for cirrhosis, including alcoholism and infection with hepatitis B and C virus.

LLETZ cone the removal of a cone-shaped portion of the ecto- and endocervix using an electrified hot wire loop.

Lobectomy is the removal of the entire lobe of a lung.

Locus a specific location on a chromosome.

Loss of heterozygosity this phenomenon can occur in cancer when both forms of a tumour suppressor gene are damaged (by mutation).

Low-grade and high-grade malignancy cells in a low-grade malignancy appear similar to normal cells under the microscope. Low-grade tumours usually grow slowly and are less likely to spread. In contrast, cells in a high-grade malignancy appear abnormal and are more likely to spread quickly.

Low-grade and high-grade tumours this refers to how the cells appear under the microscope. Histopathologists use various cytological and architectural criteria to determine the grade of a tumour. In a low-grade tumour the cells look like normal cells and are 'well differentiated'. Low-grade tumours generally grow slowly. Cells in a high-grade tumour appear very abnormal and 'poorly differentiated'.

Low-grade cytological abnormality borderline changes and/or mild dyskaryosis.

Low-grade squamous intraepithelial lesion an alternative term for mild dyskaryosis.

Low-grade urothelial carcinomas have cells that look similar to normal cells. These lesions are less likely to progress and spread.

Lugol's iodine a solution of iodine and potassium iodide in water applied to the cervix during colposcopy. Normal tissue stains brown due to its high glycogen content, while neoplastic tissue does not stain.

Lung fibrosis formation of excess connective tissue in the lung. This causes a decrease in the oxygen diffusion capacity.

Luteal phase the second half of the menstrual cycle; this refers to the development of an active corpus luteum in the ovary.

Luteinizing hormone a gonadotropin released from the anterior pituitary, which stimulates ovulation to occur.

Lymph node metastasis the spread of cancer cells from their primary site to one or more regional lymph nodes.

Lymph node stations is the term used to describe anatomical position of lymph nodes for lung cancer staging. These are numbered from 1 to 14.

Lymphadenopathy disease of the lymph nodes, but is commonly used to describe enlarged lymph nodes.

Lymphoblast an immature lymphocyte.

Lymphocyte a type of white blood cell with an important role in the immune system.

Lymphoma a solid tumour of lymphoid tissue.

Macrophage also known as a histiocyte, the macrophage is a type of white blood cell whose role is to engulf and digest cellular debris and foreign particles such as bacteria. They are longer lived than the polymorphs and have additional roles as part of the immune system.

Malignant (as in tumour growth) the capacity to invade tissue and spread to other parts of the body.

Malignant transformation the process by which a normal cell becomes a cancer cell.

Malignant tumour cancer.

Mammography X-ray image of breast structure.

Maturation is the process of cell development and specialization.

Mature metaplastic cells fully differentiated cells derived from the end-stage of squamous metaplasia. These cells are morphologically indistinguishable from mature squamous epithelial cells.

May-Grünwald-Giemsa a type of Romanowsky stain.

Mediastinum the area between the lungs that contains the heart, aorta, oesophagus, trachea, and lymph nodes.

Medullary carcinoma of thyroid type of thyroid cancer that arises from parafollicular cells.

Meiosis a specialized form of cell division in which there are two successive divisions (meiosis I and II) without any chromosome replication in between.

Melaena passing black faeces. This is commonly due to bleeding in the upper GI tract, usually in the stomach or duodenum.

Melan A an antibody used for diagnosis of malignant melanoma.

Melanoma a malignant tumour originating in the melanocytes of the skin and other tissues.

Membrane filtration the forcing of a cell suspension through a thin porous membrane. The membrane traps cells and particles that are too large to pass through the pores.

Menarche a woman's very first menstrual cycle, usually occurring between 9 and 17 years of age.

Menopause the cessation of ovarian function at around the fifth decade of life.

Menstrual cycle the periodic discharge of blood and endometrial debris through the vagina.

Menstrual debris a mixture of blood, inflammatory exudates, and endometrial cells, which can be seen in cervical samples taken during the menstrual phase of the menstrual cycle.

Menstrual phase the period of the menstrual cycle during which menstrual bleeding occurs.

Mesothelial cells the cells forming the mesothelium.

Mesothelium membrane covering the body cavities: the pleural, peritoneal, pericardial, tunica vaginalis testis in men and tunica serosa uteri in females are lined with mesothelium.

Messenger RNA a specific type of RNA which carries genetic information from the DNA in the cell nucleus into the cytoplasm.

Metaphase a stage of cell division when chromosomes align along the equator of the dividing cell.

Metaplasia transformation of one type of mature epithelium to another, usually as a response to some stimulus.

Metastasis the process by which malignant cells spread to distant parts of the body.

Metastatic malignant disease of unknown primary origin.

Microbiological safety cabinet (exhaust protective cabinet) a ventilated enclosure providing protection, to the user and the environment, from airborne microorganisms that may be generated when handling clinical specimens.

Microglandular hyperplasia histological proliferation of endocervix to form numerous small glands. Often associated with oral contraceptive use.

Microinvasive carcinoma the preclinical stage of invasive carcinoma.

Microorganism a microscopic organism.

Microsatellites or simple sequence repeats (SSRs) repeated sequences of DNA, which if they undergo mutation, can become shorter or longer in length, becoming unstable.

Microtomy the cutting of thin slices of tissue for microscopy.

Microvilli microscopic hair-like protrusions that are present on the cell surface of many cells.

Mild dyskaryosis mild cytological changes in squamous epithelial cells that are predictive of CIN1 or more severe abnormality.

Minichromosome maintenance chromosome protein complex which plays a key role at the onset of DNA replication and during the elongation process. They are useful markers of disease.

Minimally invasive technique medical or surgical procedures that cause least possible trauma to the patient.

Miracidia one of larval stages of trematodes.

Mitochondria a structure in the cell cytoplasm in which food molecules are broken down in the presence of oxygen and converted to energy in the form of ATP.

Mitomycin C is a chemotherapeutic agent is used in the treatment of many cancers, including breast, rectal, and bladder cancer. It is used intravesically in treatment of bladder cancer.

Mitosis the process of cell division in somatic cells (i.e. body cells, as opposed to germ cells).

Mitotic figure a nucleus undergoing chromosomal division during mitosis to produce two identical daughter nuclei.

Mitotically active cells undergoing cell division.

Mixed or biphasic mesothelioma this subtype of mesothelioma has features of both epithelial and sarcomatoid mesothelioma.

Moderate dyskaryosis cytological changes in squamous epithelial cells that are predictive of CIN2 or more severe abnormality.

Moderately differentiated carcinoma a carcinoma composed of cells that bear a moderate resemblance to their normal counterparts.

Molecular biology the study of genes, chromosomes, and the molecules from which they are made—the nucleic acids (DNA and RNA).

Molecular cytology this is the science of visualizing genomic events within cells.

Molecular pathology the study of disease at the molecular level.

Molecular techniques these techniques investigate the cell at the submicroscopic level. They are used to identify genetic mutations and chromosome abnormalities. They are helpful in refining the diagnosis in particular tumour types and evaluating the probable effectiveness of various treatment options. They are also used to detect genetic mutations which may predispose a person to a particular disease or condition later in life.

Monolayer one cell thick.

Monoclonal (as in tumour development) a group of genetically identical cells derived from the same ancestral cell by repeated division.

Monoclonal antibody to cytokeratins 5 and 6 (CK 5/6) is useful in identification of squamous carcinoma and cells of mesothelial origin.

Mordant a substance that assists staining by forming a link between a dye molecule and a cell component.

Morphology the study of physical shape and size. Cell morphology is the basis of cytology.

Mortality the number of deaths caused by a disease within a specified time period.

Mortality the number of deaths from a disease in a defined population over a specified period of time.

Mosaicism an abnormal pattern of small blood vessels on the cervix resembling a tiled floor or 'chickenwire'.

Moulded nuclei the squeezing together of epithelial cell nuclei leaving no gaps between them. This is a characteristic feature of herpes simplex virus infection.

Mounting medium (mountant) a substance used for adhering a coverslip to a glass slide preparation.

Mounting of the coverslip the action of applying a thin glass plate over the cells on a slide using a clear liquid medium as cement. When dried this protects the cells for storage and enhances the microscopical image.

Mucin a family of high molecular weight glycoproteins found in mucus.

Mucoid resembling mucus.

Mucolysis the breakdown of mucus by chemical or mechanical means.

Mucus a viscous bodily secretion containing mucin.

Multidisciplinary cancer network team a group of medical professionals who are specialists in cancer diagnosis and treatment. The team follow nationally agreed guidelines to ensure they deliver the best care for cancer patients.

Multidisciplinary team a group of healthcare professionals from a variety of disciplines who work together to provide an integrated approach to patient care.

Multinucleated cells containing many nuclei.

Muscle invasive urothelial carcinoma muscle invasive tumours are those that at presentation have invaded muscularis propria. These tumours are high grade and behave aggressively.

Muscularis propria this is the muscle layer in the bladder that contracts during urination.

Mutation a change in the DNA sequence of a gene.

Mutational analysis involves a wide range of methods, including PCR based techniques, DNA sequencing, and FISH. The method chosen will depend on the type of mutation.

Myoepithelial cells these smooth muscle-like cells are found in the basement membrane of glandular tissue and are involved in secretory activity of the gland.

Myometrium the muscle layer of the wall of the uterus.

National Health Application Infrastructure Services a national database holding information about the health history and demographics of individuals.

National Health Service Cervical Screening Programme (NHSCSP) cervical screening programme in the UK, launched in 1988, which aims to reduce mortality from cervical cancer.

Natural history the course that the disease takes if it is left without treatment or any other form of medical intervention.

Navicular cells glycogenated intermediate squamous cells.

Near monolayer a term used in LBC and refers to a cellular preparation which comprises one to two layers of cells, compared to the conventional smear which was composed of many cell layers, making it more difficult for the cytologist to screen and for automated systems to scan.

Necrosis localized tissue death in response to disease or injury.

Necrotic debris remains of dead cells.

Needle track seeding the growth of tumour cells along the site of a biopsy that can occasionally occur after core biopsy and rarely after FNA.

Negative and positive feedback control the means by which bodily functions are regulated. Also known as homeostatic control.

Negative predictive value the measure of the accuracy of a negative test result.

Negatives the individuals that do not have the target condition.

Nematode elongated unsegmented worms belonging to the phylum Nematoda.

Neoplasia the abnormal growth of new tissue.

Neoplasm an abnormal growth of new tissue. Also known as a tumour.

Neoplastic new growth. The term is often used to describe a malignant growth.

Nephropathy disease of the kidney.

Nervous tissue consisting of neurons (nerve cells) and supportive cells, specialized for responding to stimuli and transmitting impulses from one part of the body to another.

Neuroendocrine cells antibodies to this glycoprotein are useful in identification of tumour of neuroendocrine origin.

Neuroendocrine cells in respiratory tract receive neural input and can release various messenger molecules into the blood. In the lung they may act as receptors for oxygen levels.

Neuron specific enolase (NSE) is expressed by many tumours, including those of neuroendocrine origin.

Neutrophil see polymorphonuclear neutrophil leucocyte.

Non-diploid state occurs when the normal arrangement of 23 matching chromosome pairs has been subjected to a chromosomal alteration.

Non-Hodgkin's lymphoma (NHL) a diverse group of lymphomas compromising the majority of lymphoid malignancies.

Non-invasive technique procedures that do not involve cutting or entering a body cavity.

Non-invasive urothelial papillary carcinomas are tumours that do not invade underlying tissue and have a papillary (nipple-like) growth pattern that projects into the bladder.

Non-keratinizing stratified squamous epithelium multi-layered epithelium that does not normally accumulate the protein keratin.

Non-small cell lung cancer (NSCLC) a broad division of lung cancers encompassing all epithelial cancers that are not small cell carcinoma.

Normal flora microorganisms that normally reside in the body and have no known harmful effects.

Normal recall In the UK screening programme a woman is called every three years for a cervical test between the age of 25 and 50 and every five years from 50 to 64 years of age. This range is referred to as normal recall.

Normochromasia normal staining intensity.

Normochromatic cells with normal staining colour.

Northern blot a hybridization technique for detecting specific RNA sequences.

Nuclear immunoreactivity with p63 is seen in squamous cell carcinomas and in urothelial carcinomas. Adenocarcinomas generally do not express p63.

Nucleic acid molecules carrying genetic information within cells. There are two types of nucleic acid: deoxyribonucleic acid (DNA) and ribonucleic acid (RNA).

Nucleocytoplasmic ratio (N/C ratio) the ratio of the size of the nucleus to the size of the cell. In cytology the N/C ratio is judged visually rather than measured specifically, and is an important indicator of the type and behaviour of a cell.

Nucleoli (singular: nucleolus) small dense structures composed largely of ribonucleic acid within the nucleus of cells.

Nucleus the central controlling body within a cell containing the genetic information for maintaining life.

Obligate pathogen a microorganism that will always cause an infection if it is present in a tissue, regardless of the immune status of the host.

Occult malignancy a metastatic malignancy of unknown primary site.

Oesophagogastroduodenoscopy (OGD) an endoscopic procedure to examine the oesophagus, stomach, and duodenum.

Oestrogen and progesterone receptor status (ER and PR) some breast cancers have receptors for oestrogen and progesterone. Cells that have these receptors are called receptor positive. Tumour cells that are receptor positive respond to hormone therapies which aim to block these receptors and reduce the growth of these tumours.

Oestrogen receptors (ER) are expressed by some breast carcinomas. Oestrogen receptors are required for oestrogen stimulated growth and proliferation of breast cancers. Immunocytochemical expression of ER in malignant cells in serous effusion is useful in identification of metastatic breast carcinoma.

Oestrogen a sex hormone released from the ovary in response to pituitary gonadotropins.

Oestrous cycle a hormone-dependent cycle in female mammals involving ovulation and periods of heat.

Oncogene a potentially cancer-inducing gene.

Oncologist a physician who specializes in the study, diagnosis, and treatment of cancer patients.

One-stop breast clinic here the patient can expect a consultation with a specialist, any necessary imaging tests, and, if needed, a core biopsy or FNA all on the same day.

Oocyte an immature egg.

Opportunistic pathogen a microorganism that does not normally cause an infection but will do so if the immune system of the host is compromised.

Oral contraceptive a hormone taken orally for contraceptive purposes.

Orange G (OG) a synthetic orange dye used as a counterstain in the Papanicolaou technique.

Orangeophilia/orangeophilic an affinity for the dye Orange G.

Ovary the reproductive organ of the human female.

Ovulation the cyclical (roughly once every 28 days) release of a mature egg (ovum) from the ovary.

Ovulatory phase the phase of the menstrual cycle during which a surge of luteinizing hormone stimulates the rupture of the Graafian follicle and release of a mature egg from the ovary.

Ovum mature egg.

p16INK4A an intracellular protein that accumulates during cell proliferation. It is advocated as a marker for neoplasia.

p16 is a tumour suppressor gene located on chromosome 9p21. It plays an important role in regulating the normal cell cycle and differentiates persistent hrHPV from transient hrHPV. Over-expression of p16 is seen in cells where hrHPV oncogenes have initiated.

p63 nuclear immunoreactivity with p63 is seen in squamous cell carcinomas and in urothelial carcinomas. Adenocarcinomas generally do not express p63.

Pale dyskaryosis a form of dyskaryosis in which the cell nuclei are unusually pale staining and in which the surrounding non-dyskaryotic cells display a normal staining reaction.

Palisaded strips the term used to describe the side-by-side arrangement of normal endocervical columnar cells, resembling a picket fence.

Palliation minimization of the pain and suffering from a disease.

Palliatively treating the patient to reduce the severity of disease, where curing is not possible.

Palliative care is any treatment that serves to reduce the severity of the disease symptoms.

Palpable able to be felt by the clinician on manual examination.

Pancreatitis is inflammation of the pancreas.

Pancreatobiliary system this refers to pancreas, bile duct, and gall bladder.

Papanicolaou stain a versatile staining technique developed by George Papanicolaou. The most commonly used staining method in cytology.

Papillaroid nipple-like growth or projection. This term is often used to describe cellular architecture.

Papillary carcinoma most common type of thyroid cancer with a peak onset between 30 to 50 years. This cancer has a very good prognosis.

Parabasal cell an immature, squamous cell.

Paracentesis the procedure for removing the fluid from the peritoneal cavity.

Parakeratosis a form of abnormal keratinization in which pyknotic nuclei are visible within deeply orangeophilic cells.

Parietal layer the outer layer of mesothelium covering the wall of the cavity.

Parotidectomy removal of the parotid gland.

Pastette a disposable plastic pipette.

Pathological long bone fracture a pathological fracture arises when a disease process weakens the bone significantly so that fractures occur during normal activities such as walking.

Many disease processes, including infections, generalized bone diseases such as osteoporosis, and metastatic tumours can cause pathological fractures. Malignant tumours that metastasize to the bone include carcinomas of the lung, breast, prostate, thyroid, and kidney, amongst others.

Pelvic inflammatory disease inflammation of the uterus, fallopian tubes and/or ovaries.

Pericardiocentesis procedure for removing fluid from the pericardium.

Pericardium mesothelium enveloping the heart.

Periodic acid Schiff technique is a method for demonstrating carbohydrate. The periodic acid oxidizes glucose residues, forming aldehydes that react with Schiff reagent to give a magenta colour. The technique can be used for demonstrating pathogenic fungi and a variety of other constituents of cell samples.

Peripleural mass around the pleural.

Peristalsis rhythmic contraction in the GI tract.

Peritoneal mesothelium mesothelial layer covering the abdominal organs.

Peritoneal washing a procedure whereby physiological saline is introduced into the peritoneal cavity and then removed by suction. The aspirated fluid is examined for tumour cells.

Personal protective equipment (PPE) laboratory clothing designed to protect the wearer from chemical or microbiological hazards. Examples include gloves, goggles, and laboratory coat.

Petri dish a clear shallow dish normally used to grow bacteria, but used in cytology for the examination and sampling of mucoid specimens such as sputum.

Phagocytes from Greek word *phagein* = to eat and *cyte* = cell. These cells ingest foreign matter.

Phenacetin this pain killer, which has similar uses to paracetamol, was initially banned from general use in 1968 as it was linked to bladder and kidney cancer. The ban was later revoked but its use is highly restricted.

Phenotype (of a tumour) the observable physical or biochemical characteristics of a tumour.

Physical intervention any form of intervention resulting in a physical trauma.

Physiological saline (also known as isotonic saline) a sterile solution of 0.9 g l^{+1} sodium chloride used in medical applications such as irrigation of body cavities and intravenous infusion (also known as isotonic saline).

Pituitary gland a pea-sized endocrine gland that lies beneath the hypothalamus at the base of the brain. The pituitary secretes many different hormones, including follicle stimulating hormone and luteinizing hormone.

Placenta the organ that connects the foetus to the uterine wall to enable the uptake of nutrients, elimination of waste, and exchange of gases via the mother's blood supply.

Plasma cell a type of white blood cell whose specific function is the synthesis and secretion of antibodies.

Pleomorphic adenoma commonest salivary gland tumour. This benign tumour consists of variable numbers of epithelial and myeoepithelial cells that sit in a background of myxoid material (mucoid).

Pleomorphism variability in size and shape. This term is usually used when describing malignant tumours.

Pleural mesothelium mesothelial layer covering the pleural surfaces.

Pleural plaques pleural thickening seen on CT on the inner surface of the ribcage and diaphragm which indicate previous exposure to asbestos.

Pleurectomy/decortication surgical removal of the pleura for treatment of mesothelioma.

Ploidy the number of sets of chromosomes within a cell.

Pneumaturia passage of gas from the urethra during urination. This occurs if there is an enterovesical fistula.

Pneumocystis carinii see *Pneumocystis jiroveci*.

Pneumocystis jiroveci, previously called *Pneumocystis carinii*, is a yeast-like fungus, the causative agent in pneumonia.

Pneumocyte type of cell lining the alveoli in the lungs.

Pneumonectomy is the surgical procedure to remove an entire lung.

Pneumonia is infection of the lungs, usually due to infective agents such as bacteria, viruses, or fungi.

Pneumothorax is presence of air in the pleural space. This leads to collapse of the lung, causing pain, discomfort, and breathing difficulties.

Polygonal angular in shape (i.e. shaped like a polygon).

Poly-L-lysine a polypeptide used to coat glass slides in order to enhance the adhesion of cells, thus reducing the chances of cell loss during processing.

Polymerase chain reaction (PCR) a molecular biology technique whereby multiple copies of a segment of DNA are produced from a single initial strand.

Polymorphonuclear neutrophil leucocyte a type of white blood cell with neutrally-staining cytoplasmic granules. Its function is to eat, or phagocytose, foreign particles and damaged cells.

Polymorphs, also known as neutrophil cells, a short-lived phagocytic white blood cell having a lobulate nucleus and neutrophil granules in the cytoplasm. Produced in large numbers to combat bacterial infections.

Polymorph see polymorphonuclear neutrophil leucocyte.

Polyomavirus human polymomaviruses are a genus of viruses belong to the family of *Polyomaviridae*.

Polyomavirus is an oncogenic DNA virus belonging to the family *Papovaviridae*. There are five types of polyomavirus that have been found in humans. Of these, the BK and JC viruses infect the urinary tract.

Point mutations occur when one base is substituted for another.

Poorly differentiated carcinoma a carcinoma composed of cells that bear a poor resemblance to their normal counterparts.

Positive predictive value the measure of the accuracy of a positive test result.

Positives the individuals that have the target condition.

Positron emission tomography (PET) scanner (a machine similar to a CT scanner in appearance) detects positrons (subatomic particles) emitted by a radionuclide in the organ or tissue being examined. PET scanning measures metabolic activity; in this procedure a radionuclide such as fluorodeoxyglucose (FDG) is administered into a vein. After a short waiting period, this chemical accumulates in biologically active tissues. Most commonly these days, the PET scanner is combined with CT scanning, which allows for simultaneous visualization of metabolically active areas in relation to anatomical site. PET scanning is commonly used when staging the tumour.

Positron a subatomic particle with the same mass as an electron and an equal but positive charge.

Post-menopausal atrophic pattern the pattern of deep parabasal cells lying singly and in sheets seen in cervical samples taken after the menopause.

Post-partum after childbirth.

Praziquantel (Biltricide) is an anthelmintic effective against flatworms.

Pre-cancer the stage of an abnormal growth of tissue that precedes the likely development of a malignant tumour.

Precipitating agent a substance that achieves fixation by rendering cellular proteins insoluble.

Premalignant an abnormality with the potential to become malignant in the future.

PreservCyt® a methanol-based preservative solution suitable for collection and transport of cytological material.

Prevalence the proportion or percentage of the population that have the target condition or disease.

Primary antibody (in immunocytochemistry) an antibody that is specifically targeted toward the antigen to be detected in an immunocytochemical reaction.

Primary follicle an oocyte with its surrounding layer of protective cells, found within the ovary.

Primary oocyte immature egg.

Primary tumour a tumour growing at the anatomical site from which it arose.

Primordial follicle a biologically inactive structure within the ovary containing an immature oocyte surrounded by granulosa cells.

Prior notification list a list of individuals identified as requiring a screening test.

Productive phase the stage of a virus's life cycle during which the virus's genetic material takes over the host cell functions to produce new viruses.

Progenitor cells can differentiate into different cell types.

Progesterone a sex hormone released from the corpus luteum during the second half of the menstrual cycle.

Prognosis a forecast of the course of a disease and its probable outcome.

Proliferative phase the first half of the menstrual cycle, during which time the endometrial lining thickens in response to oestrogen secretion from the ovary.

Prostaglandins a group of potent hormone-like substances found in many bodily tissues especially semen. They influence a wide range of physiological functions, such as contraction of smooth muscle and modulation of inflammation.

Prostate gland a gland in males located at the base of the bladder. The prostate gland opens into the urethra and secretes a milky fluid that is a major component of semen.

Prostatic acid phosphatase (PAP) antibody to PAP is useful in identification of metastatic adenocarcinoma of the prostate.

Prostatic specific antigen (PSA) is a protein produced by the cells of the prostate gland. Antibodies to this protein are useful in identification of metastatic adenocarcinoma of the prostate.

Protozoa (singular: protozoan) single-celled organisms containing nuclei.

Psammoma body a rounded deposit of calcified protein occurring in some benign and malignant neoplasms. Most commonly seen in papillary cancers including thyroid and ovary.

Pseudohyphae false hyphae that form as a result of elongation of fungal spores.

Pseudostratified epithelium this describes the arrangement of cells. All the cells are in contact with basement membrane, but their nuclei are placed at different planes giving the impression of multilayered epithelium.

Pseudostratification a single layer of columnar cells that appears to be multilayered because the cell nuclei rest at different heights in adjacent cells. A characteristic feature of cervical glandular intraepithelial neoplasia.

Punch biopsy removal of a small piece of tissue using specially designed forceps for histological assessment.

Punctation the end-on appearance of abnormal blood vessels seen on the cervix during colposcopy, appearing as small red dots when viewed under magnification.

PUNLUMP this category of papillary lesions was defined by the World Health Organization (WHO) in 1994 to categorize a group of papillary lesions that are slow growing and unlikely to spread. These lesion were called papillary urothelial lesions of low malignant potential or PUNLUMP.

Purulent composed of pus.

Pyknosis condensation of chromatin in nucleus.

Pyknotic see karyopyknosis.

Quality Assurance Reference Centres (QARCs) are regional centres which oversee the quality of cervical screening and colposcopy services on behalf of the NHSCSP.

Quality assurance the systematic evaluation of services provided to ensure that standards of quality are being met.

Quality control methods for detecting defects.

Quality management a method of ensuring that the design, development, and implementation of a service are efficient, and effective.

Quality standards defined standards against which services are measured.

Quality a word used to define products and services.

Quality-adjusted life years disease-free years of life saved by a screening programme.

Quintiles five equal proportions of 20% each.

Radiation treatment the treatment of cancer using X-rays or gamma rays.

Radical cystectomy surgical removal of bladder and adjoining organs for treatment of bladder cancer. This procedure differs in different sexes: in men the prostate is also removed, while in women the ovaries, fallopian tubes, and uterus are often removed.

Radiologist a physician who specializes in interpreting images of the body.

Radiotherapy a form of treatment that involves the use of electromagnetic radiation (e.g. gamma rays) to destroy malignant tissue.

Rapi-Diff® a commercial brand of Romanowsky stain which is used for rapid staining of the cell.

Reactive changes (also called inflammatory changes) the term used to describe the cellular changes seen in cytological samples when they are taken during an inflammatory process.

Reactive hyplerplasia this is a benign reversible enlargement of a lymph node due to a stimulus such as bacterial or viral infection.

Reactive mesothelial cells this is the term used to describe mesothelial cells that are undergoing proliferation due to a stimulus.

Red atrophy an overall change in cytoplasmic staining from green to bright orange/red that occurs in atrophic cervical samples.

Red blood cell casts are indicative of glomerular damage.

Red blood cell see erythrocyte.

Reductional division this is also called the first meiotic division, where the chromosome number is reduced from diploid (46 chromosomes) to haploid (23 chromosomes).

Reed-Sternberg cells large bi- or multinucleated cells seen in Hodgkin's lymphoma.

Refractive index a numerical value describing the extent to which light is slowed down or bent as it travels through a medium. RI is calculated by dividing the velocity of light in a vacuum by the velocity of light in the medium. Air has an RI of 1.0, whereas glass typically has an RI in the range of 1.50–1.54 (i.e. light travels slower in glass than it does in air).

Regeneration the process of repair or replacement of lost or damaged cells.

Regressive staining the deliberate overstaining of a cell component followed by removal of the excess dye.

Renal calyces system of ducts that carries the urine from the kidneys to the ureters.

Renal failure a condition where the kidneys are not functioning adequately.

Renal pelvis the central area in the kidney that urine passes through before it is funnelled into the ureter.

Repair see regeneration.

Reserve cell hyperplasia the controlled proliferation of reserve cells as an initial step in the process of squamous metaplasia.

Reserve cells small undifferentiated epithelial cells lying on the basement membrane. They can be found beneath the columnar cells of the endocervix but are rarely identified in cervical samples. Reserve cells in the respiratory tract are also known as basal cells and have the potential to differentiate into other cell types.

Respiratory epithelium is a type of epithelium found in the respiratory tract. Its function is to moisten the airways and form the first line of defence against the pathogens and foreign matter.

Retrograde ejaculation is entry of semen into the bladder instead of going out through the normal ejaculation route via the urethra.

Retrograde ureteropyelography, sometimes called retrograde pyelogram, is a combination of cystoscopy and X-ray examination which allows visualization of the bladder, ureters, and renal pelvis. During a cystoscopy, contrast dye is introduced into the ureters via a catheter and X-ray images are taken before and after introduction of the dye to allow visualization of any abnormal areas.

Rhabditiform early larval stage of a nematode.

Rheumatoid disease is the name given to a group of chronic inflammatory diseases that affect many tissues, but largely the joints.

Rheumatoid pleuritis is a pleural effusion that occurs in some patients and may occur as a systemic manifestation of rheumatoid arthritis.

Risk the likelihood that harm from a particular hazard is realized.

Risk assessment a formal procedure for determining the level of risk from an identified hazard.

Risk control the practices, procedures, equipment, and training that are put in place to minimize or eliminate risk.

Risk rating number (RRN) a numerical measure of the degree of risk, calculated by multiplying the likelihood of an adverse event by its severity.

Romanowsky stains a group of stains originally designed for staining of blood films, but now adapted for staining cytological preparations.

Rosette a circular group of pseudostratified dyskaryotic endocervical cells.

Round cells cells, other than mature sperm cells, found in semen. Round cells could be immature sperm cells, inflammatory cells, or epithelial in origin.

Saccomanno technique the use of a blender to achieve the mechanical breakdown and homogenization of mucus.

Sampling error in terms of specimen collection for cytology, sampling error can be defined as the failure to collect cells that are representative of the site being sampled, or to collect them in sufficient quantity to make a reliable diagnosis.

Sarcoidosis an inflammatory condition that causes formation of granulomas in different organs, including lungs, lymph nodes, and skin.

Sarcoma malignancy arising from connective tissue cells.

Sarcomatoid mesothelioma a subtype of mesothelioma. The cells in sarcomatoid mesothelioma appear as elongated spindle-shaped cells.

Scalloped the term used to describe a rounded group of cells with a knobbly border.

Schistosoma haematobium a parasitic flatworm found in Africa and the Middle East, causing urinary schistosomiasis.

Schistosoma a fluke worm parasite that can cause disease in humans.

Scraping (of cells) removal of cells from a body surface using a specially designed scraping device.

Screening intervals the period of time between each screening test.

Screening tool a means of testing apparently healthy people for presence of a disease or disorder.

Secondary antibody (in immunocytochemistry) an antibody that binds to a primary antibody in an immunocytochemical reaction. Secondary antibodies are typically labelled with probes to make them visible.

Secondary tumour a tumour that develops as a result of metastasis from a primary tumour.

Secretory phase the second half of the menstrual cycle, during which time there is an increase in secretory activity of the endometrial glands.

Sections the thin slices of tissue for microscopy.

Selective neck dissection a surgical excision, commonly employed in the treatment of head and neck cancers, which removes lymph nodes on one side of the neck.

See and treat in the context of cervical screening, is the policy of treating women at the first colposcopy appointment if the clinician considers there to be a high probability of high-grade neoplasia.

Semenogelin a protein produced by seminal vesicles.

Seminal vesicle a pair of small glands that are situated behind the bladder. Seminal vesicles contribute fluid to semen.

Seminiferous tubules are found inside lobules in the testes within which spermatozoa are produced. There are about 750 per testicle.

Sensitivity the ability of a screening test to identify the positive individuals within the population.

Sensitivity the probability of getting a correct diagnosis of disease rather than a false negative.

Sensitivity of a screening test a measure of the ability of a screening test to identify the positives.

Septate hyphae a term used in mycology to describe partition division of fungi.

Septum see septate hyphae.

Serous fluid the lubricating fluid, produced by the mesothelium, which allows for the movement of organs in the body cavity.

Sertoli cells are found in the walls of the seminiferous tubules of the testis. They provide an anchoring point and source of nutrition for developing spermatids and also form the blood testes barrier. They are named after Italian physiologist Enrico Sertoli (1842–1910).

Settling chamber a cylindrical open-ended container fixed to a glass slide into which a cell suspension is placed. Cells settle onto the glass slide under their own weight.

Severe dyskaryosis cytological changes in squamous epithelial cells that are predictive of CIN3 or more severe abnormality.

Signal amplification the generation of several molecules of a visible substance at the site of detection of each single molecule of interest in a specimen.

Signal amplification amplifies the signal in the presence of target DNA or RNA without increasing the amount of nucleic acids.

Signet ring cell a cell with a large cytoplasmic vacuole which displaces the nucleus towards the edge of the cell, giving the appearance of a signet ring.

Simian virus 40A a DNA virus found in both monkeys and humans and has the potential to cause tumours in monkeys.

Small cell carcinoma an aggressive cancer usually associated with the lung.

Small cell dyskaryosis severe dyskaryosis in small squamous cells.

Small cell lung cancer a highly malignant tumour arising from neuroendocrine cells.

Small lymphocytic lymphoma this is B cell lymphoma which is similar to chronic lymphocytic leukaemia.

Smear a cytological preparation made by spreading a specimen (or a sample of it) directly onto a glass slide.

Sodium citrate trisodium citrate solution may be used as an anticoagulant when collecting serous effusions.

Somatic mutations are mutations in cells that don't make eggs or sperm.

Southern blot a hybridization technique for detecting specific DNA sequences.

Southern blotting transfer of DNA molecules from thick agarose gel by capillary or an electric field. Once on the membrane, the molecules are immobilized and can be detected at high sensitivity by hybridization or antibody labelling.

Specific gravity the density (or 'heaviness') of a substance compared to that of water. If the liquid floats on water it has a specific gravity less than 1.0. If it sinks then its specific gravity is greater than 1.0.

Specificity the probability of correctly identifying the absence of disease rather than getting a false positive.

Spermatogenesis the process by which spermatozoa are produced in the testes.

Spermatogonia a type of stem cell found in the wall of the seminiferous tubules that undergoes mitosis to produce types A and B spermatogonium cells.

Spermiogenesis is the process of transforming the spermatids into mature sperm cells.

Spider cells a common description for the morphology of immature squamous metaplastic cells. The term refers to the cytoplasmic projections resembling a spider's legs.

Sputum a viscous secretion of the lower respiratory tract.

Squamocolumnar junction the abrupt transition from stratified squamous epithelium to columnar epithelium in the endocervical canal.

Squamoid shaped like a squamous cell.

Squamous cell carcinoma cancer of squamous epithelium.

Squamous cells cells specialized for the protection of underlying tissues.

Squamous differentiation is the process whereby one type of epithelium takes the physical characteristics of squamous cells. This sometimes occurs in malignant tumours, including urothelial, breast, and endometrial carcinoma.

Squamous metaplasia–respiratory tract squamous metaplasia occurs as a response to injury. It is commonly seen in smokers. The cells appear as a sheet of polygonal shaped cells. They are similar in appearance to that which occurs in the cervix.

Squamous metaplasia a gradual replacement of columnar epithelium of the endocervix by stratified squamous epithelium in response to exposure of the columnar epithelium to the acid vaginal environment.

Stage (of cancer) the extent to which a cancer has spread.

Staging (in cancer) a numerical system (usually numbers I–IV) describing the extent to which a cancer has spread. Generally it takes into account the size of the tumour, depth of invasion, lymph node involvement, and presence of metastases.

Staining applying dyes or chemical compounds to colour various components of tissue or the cell, in order that they are more easily visualized under the microscope.

Standard operating procedures prescriptive instructions on how to perform a task safely and accurately.

Stellate star-shaped.

Stenting a stent is a tube, commonly made of wire, mesh, or plastics, placed in open structures in the body to keep them open. In lung cancer, a stent can sometimes be used to keep an airway open that is narrowed or blocked due to a tumour.

Stoma this is a surgically created opening between an organ and the outside of the body, constructed to allow the passage of urine or waste products.

Stratified epithelium contains multiple cell layers and is highly specialized for providing protection.

Stroma connective tissue that lies beneath the basement membrane of epithelia.

Strongyloides stercoralis a parasite belonging to the nematode family that can infect humans.

Stylet this is a thin wire inserted into the needle to provide extra support and rigidity to the needle.

Subepithelial lymphoid follicles a collection of lymphocytes beneath an epithelium.

Superficial and intermediate squamous cells are the top layer cells of stratified squamous epithelium. Intermediate cells form the layer below.

Superficial cells the mature surface cells of stratified squamous epithelium.

Supernatant the liquid that lies above sediment after a fluid is centrifuged.

Surfactant a lipoprotein secreted by alveolar cells. Its function is to reduce the surface tension of fluids in the lung.

Surgeon a physician who treats diseases and other conditions by operative methods.

Suspicious lesion a tissue mass that the clinician thinks could be neoplastic.

Synaptophysin a glycoprotein present in neuroendocrine cells. Antibodies to this protein are useful in identifying tumours of neuroendocrine origin.

Syncitium a sheet of cells containing many nuclei but without visible cell borders.

Systemic lupus erythematosus or SLE is an autoimmune disease of the connective tissues that can affect various organs, including the serous cavities.

Taqman probes Taq polymerase is a thermostable DNA polymerase which is commonly used in the polymerase chain reaction (PCR).

Target amplification increases the number of copies of a target DNA or RNA by *in vitro* synthesis of nucleic acids, PCR being the most popular way of amplifying.

Target population the group of individuals that the screening programme aims to benefit.

Targeted therapies target therapy is aimed at selectively killing cancer cells whilst preserving surrounding healthy tissue.

Telomerase an enzyme that adds DNA sequence repeats to the end of DNA strands in the telomere (end) regions of chromosomes.

Testes the two male reproductive glands located in the cavity of the scrotum. The testes are divided up into lobules which contain the seminiferous tubules.

Thermal cycler a piece of laboratory equipment which amplifies segments of DNA via the polymerase chain reaction (PCR) process enabling millions of copy DNA sequences to be synthesized.

Thiabendazole a medication used to treat fungal and parasitic infections. It is effective against *Strongyloides stercoralis*.

Thoracocentesis, or pleural tap, is the medical procedure for removal of effusion from the pleural space.

Thrombin a protein involved in the coagulation process (clotting) in blood.

Thrombomodulin is a membrane protein. Antibodies to this protein are immunoreactive with mesothelial cells and some adenocarcinomas.

Thyroid function tests (TFTs) a biochemical blood test which is carried out to measure thyroid activity. Initially the level of TSH is measured and if this is elevated or decreased, levels of T4 (thyroxine) and T3 are re-measured.

Thyroid lymphoma type of cancer of the lymphatic system that arises in the thyroid gland.

Thyroid nodule any abnormal growth in the thyroid gland that appears as a lump.

Thyroid transcription factor 1 (TTF1) nuclear immunoreactivity with TTF1 antibody is useful in identifying adenocarcinomas of pulmonary origin. Small cell anaplastic carcinoma of the lung also stains positive with this antibody. It is a nuclear transcription factor that is expressed in normal lung, in thyroid, and in the cancers arising from these organs.

Thyroiditis inflammation of the thyroid gland.

Thyroxine or T4 is a major thyroid hormone which has a wide metabolic activity.

Tight junctions are structures that are associated between two cells and form an impermeable barrier to fluids.

Tinctorial staining the application of dyes to cell preparations in order to produce differentially coloured cell constituents.

Tingible body macrophages contain the remnants of ingested lymphocytes.

Tissue banking is a means of collecting material for future diagnostic and research use. It is highly regulated under the Human Tissue Act (HTA) and requires formal consent from a patient before their tissue can be used in any research or clinical trial.

Tissue microarray miniature needle core biopsies are placed in a master array block to enable multiparallel genetic testing.

Top hat formation a term used to describe the morphology of groups of endometrial cells. Imagine looking at a gentleman's top hat from above and you would see a dark central circle and an outer paler circle.

TOP2A a gene located at chromosome 17q21. Patients with TOP2A amplified tumours benefit from anthracycline therapy.

Topoisomerase II alphaEnzyme is involved in the separation and reconfiguration of DNA strands during transcription.

Total thyroidectomy an operation which removes the whole of the thyroid gland.

Transbronchial through the wall of a bronchus.

Transcription a process which copies DNA genetic information into RNA language.

Transformation (in tumour development) the changes that a normal cell undergoes as it becomes a cancerous cell.

Transformation zone the area of the cervix where the physiological transformation from columnar epithelium to metaplastic squamous epithelium takes place.

Transitional cell carcinoma (TCC) transitional cell carcinoma or urothelial carcinoma is a type of malignancy that arises in the urinary tract.

Transitional epithelium was the term originally given to epithelium lining the urinary tract—so-called as it was originally thought to be a transition between squamous and glandular epithelium.

Transition mutation is a type of point mutation involving substation of one base pair for another by replacement of one purine by another purine, and of one pyrimidine by another pyrimidine. There is no change in purine-pyrimidine orientation.

Translation a process which occurs in the cytoplasm of cells and translates nucleic acids into proteins.

Translocation type of chromosomal aberration in which part of a chromosome is linked at a breakpoint to another chromosome.

Transport medium a specially formulated liquid used for transporting clinical specimens to the laboratory for examination.

Transthoracic fine needle aspiration aspiration through the chest wall or thoracic cavity.

Transudate fluid low in protein content specific gravity. Transudates are caused by an imbalance of hydrostatic and oncotic pressure and are associated with kidney, heart, or liver failure.

Transurethral resection of bladder tumour (TURBT) is a treatment for superficial bladder cancers. Under general anaesthesia the cystoscope is passed through the urethra and after visualization of the tumour a special electrically heated wire loop is used to remove the tumour and seal the blood vessels.

Transversion occurs when a purine (adenine or guanine) replaces a pyrimidine (cytosine or thymine).

Trematode flattened oval or worm-like parasitic animals.

Triage in the context of cervical screening triage is a process or test that helps to sort women with low-grade cytological abnormalities into groups, based on their risk of underlying high-grade neoplasia.

Trichomonas vaginalis a single celled motile protozoan that can cause infection in the urogenital tract. In males it commonly infects the urethra and in females it can infect the vagina.

Trimester one of three periods of approximately three months into which pregnancy is divided.

Triple approach in breast cancer diagnosis this approach or triple assessment in breast cancer diagnosis involves clinical examination, imaging (mammogram or ultrasound), and biopsy (FNA or core biopsy). When the three modalities agree a very high rate of accuracy is achieved.

Trophoblasts cells that surround and nourish the developing embryo and eventually develop into the placenta.

True negatives the negative individuals identified as negative by the screening test.

True positives the positive individuals identified as positive by the screening test.

Tuberculous pleuritis pleural infection by *Mycobacterium tuberculosis*.

Tubo-endometrioid metaplasia a form of inappropriate re-epithelialization following surgery to the cervix, in which epithelium resembling that of the Fallopian tube and/or endometrium replaces cervical epithelium.

Tubo-endometrioid metaplasia the presence in the cervix of glands resembling those normally found in the Fallopian tubes or endometrium. A form of inappropriate re-epithelialization following surgery to the cervix.

Tubular necrosis death of tubular cells that form the tubules in the kidney. It is one of the causes of renal failure.

Tumour an abnormal growth of new tissue. Also known as a neoplasm.

Tumour diathesis the mixture of necrotic cell debris, inflammatory exudates, and blood that can accompany an invasive tumour.

Tumour marker a substance in the body that is associated with the presence of cancer.

Tumour necrosis factor-alpha (TNFα) a protein which is known to induce cell death in tumour cells. It also has a pro-inflammatory effect in conditions such as rheumatoid disease and Crohn's disease.

Tumour suppressor gene a gene that acts to inhibit cell division, thus helping to protect against cancer.

Tunica serosa uteri mesothelium covering the internal reproductive organs in females.

Tunica vaginalis testis mesothelium covering the testis.

Typical carcinoid this is the least biologically active member of the bronchial neuroendocrine tumours. They rarely metastasize.

Tyrosine kinase (TK) an enzyme which is involved in the transfer of a phosphate group from ATP to protein. It plays a key role in cell regulation.

Ultrasound imaging a non-invasive test that uses high-frequency sound waves to create images of organs and other structures inside the body.

Umbrella cells these are large superficial cells found in urothelium. The superficial cells are referred to as umbrella cells as they cover the intermediate cells.

Undifferentiated cells completely lack specialization and have an ill-defined morphology.

Undigested food particles fragments of plant tissue and meat fibres are sometimes observed in sputum samples.

United Kingdom National External Quality Assessment Service (UKNEQAS) is an external body that oversees various quality assurance schemes.

Upper urinary tract brushings sampling of upper urinary tract (ureters and renal pelvis) by brushing the area of concern.

Uptake the participation in the screening programme of the invited individuals.

Ureter a small muscular tube that propels urine from the kidney to the bladder.

Urethra a small tube that carries the urine from the bladder to outside the body.

Urinary calculi more commonly known as urinary stones, they are solid structures formed from minerals present in urine.

Urinary casts are cylindrical structures that form in the tubules and collecting ducts of the kidney. The casts dislodge and pass into the urine. Urinary casts are seen in certain diseases.

Urinary catheter a short piece of thin plastic tubing that is inserted into the bladder to allow urine to drain freely into a bag.

Urostomy an opening on the surface of the abdomen to drain urine. A bag is attached to this opening to collect the urine.

Urothelial carcinoma malignancy that arises from the urothelial epithelium present in the urinary system.

Urothelium specialized epithelium lining the urinary tract.

Uterine cervix the neck of the uterus.

Uterus the major internal reproductive sex organ of females.

Vacuole an enclosed compartment within cells which may contain fluids or occasionally solid matter.

Vagina the hollow muscular tubular tract leading from the cervix to the exterior of the body in females.

Vesicular the term used to describe finely granular and even chromatin pattern.

Video-assisted thoracoscopic surgery (VATS) is a form of minimally invasive surgery, commonly known as 'keyhole' surgery.

Viral integration is the insertion of viral DNA into host cell DNA.

Viral load is measured in RNA copies per ml of blood and acts as an indicator of the amount of virus present in a sample. The more active the virus the higher the viral load.

Viral loads are used to monitor viral activity in patients receiving drug therapy.

Virus a microorganism that is much smaller than a bacterium and has no independent metabolic activity. Viruses can only replicate within a cells they infect.

Visceral layer the inner layer of mesothelium covering the organ.

Voided urine naturally passed urine.

Vortex a spinning flow of fluid.

Vulva part of the external genitalia of a woman.

Warthin's tumour is the second most common benign salivary gland tumour. It consists of transformed epithelial cells (oncoytes—containing numerous mitochondria) within a lymphoid tissue stroma.

Wart the common name for a growth of tissue caused by human papillomavirus.

Washing (of cells) removal of cells from a body surface or cavity by rinsing with physiological saline.

Wedge resection is the surgical removal of small wedge-shaped tissue that includes the tumour and small area of normal surrounding tissue.

Well-differentiated carcinoma a carcinoma composed of cells that bear a close resemblance to their normal counterparts.

Wet fixation is achieved by immersion of slide preparations into a fluid fixative or by spraying them with an aerosol fixative.

White blood cell a type of cell found in the circulating blood system and classified by the presence or absence of granules in the cytoplasm.

White blood cell casts are indicative of infection in the kidney.

Wilm's tumour gene product (WT-1) antibodies to WT1 protein, the product of WT1 gene, react with mesothelium and are unreactive with a majority of adenocarcinomas, with the exception of some ovarian carcinomas.

Work-related upper limb disorder (repetitive strain injury) a term used to describe a variety of musculoskeletal problems that may be related to repetitive work tasks. The disorder is characterized by numbness, pain, and weakening of muscles in the hand, wrist, arm, and shoulder.

Wreath formation a term used to describe the morphology of endometrial cell groups, which can often resemble a circular flower wreath.

Xylene an aromatic hydrocarbon used as a solvent and clearing agent in the preparation of cytological specimens.

Ziehl-Neelson (ZN) stain a biological stain used in identifying acid-fast bacilli such as *Mycobacterium tuberculosis*, the causative agent in TB.

References

Chapter 2

Ergonomic working standards for personnel engaged in the preparation, scanning, and reporting of cervical screening slides. NHSCSP publication number 17: http://www.cancerscreening.nhs.uk/cervical/publications/nhscsp17.html (accessed 2010).

HSE (2003) *Safe Working and the Prevention of Infection in Clinical Laboratories and Similar Facilities*. HSE Books, Sudbury.

Chapter 8

ARC GLOBOCAN (2002) Cancer Incidence, Mortality and Prevalence Worldwide (2002 estimates): http://www-dep.iarc.fr/ (accessed 2010).

ISD Online (2009) Cancer Incidence, Mortality and Survival data: http://www.isdscotland.org/isd/183.html (accessed 2010).

Northern Ireland Cancer Registry (2009) Cancer Incidence and Mortality: http://www.qub.ac.uk/research-centres/nicr/Data/OnlineStatistics/ (accessed 2010).

Office for National Statistics (2009) Cancer Statistics Registrations: Registrations of Cancer Diagnosed in 2006, England: http://www.statistics.gov.uk/statbase/Product.asp?vlnk=8843 (accessed 2010).

Welsh Cancer Intelligence and Surveillance Unit, Cancer Incidence in Wales: http://www.wales.nhs.uk/sites3/page.cfm?orgid=242&pid=18135 (accessed 2010).

Chapter 9

1. Light RW and Hamm H (1997) Malignant pleural effusion: would the real cause please stand up? *European Respiratory Journal* **10**, 1701-2.

2. Cooke WE (1924) Fibrosis of the lungs due to the inhalation of asbestos dust. *British Medical Journal* **2** (3317), 140-2, 147.

3. Health and Safety Executive: http://www.hse.gov.uk/statistics/causdis/mesothelioma/index.htm (accessed 2010).

4. Metintas M, Hillerdal G, and Metintas S (1999) Malignant mesothelioma due to environmental exposure to erionite: follow-up of a Turkish emigrant cohort. *European Respiratory Journal* **13**, 523-6.

5. Rivera Z, Strianese O, Bertino P, Yang H, Pass H, and Carbone M (2008) The relationship between Simian virus 40 and mesothelioma. *Current Opinion in Pulmonary Medicine* **14** (4), 316-21.

Chapter 10

1. Linder J (2000) Lung cancer cytology, something old something new. *Am J Clinical Pathology* **114**, 169-71.

2. Tyson EB (1957) The development of the bronchoscope. *J Med Soc N J* **54** (1), 26-30.

3. Wang KP, Marsh BR, Summer WR, Terry PB, Erozan YS, and Baker RR (1981) Transbronchial needle aspiration for diagnosis of lung cancer. *Chest* **80** (1), 48-50.

4. Ferlay J, Autier P, Boniol M, Heanue M, Colombet M, and Boyle P (2007) Estimates of the cancer incidence and mortality in Europe in 2006. *Ann Oncol* **18** (3), 581-92.

5. Office for National Statistics, Cancer Statistics registrations: registrations of cancer diagnosed in 2006, England: http://www.statistics.gov.uk/ (accessed 2010).

6. ISD Online. Cancer Incidence, Mortality, and Survival data: http://www.isdscotland.org/isd/CCC_FirstPage.jsp (accessed 2010).

7. Welsh Cancer Intelligence and Surveillance Unit, Cancer Incidence in Wales: http://www.wales.nhs.uk/sites3/home.cfm?OrgID=242 (accessed 2010).

8. Northern Ireland Cancer Registry. Cancer Incidence and Mortality: http://www.qub.ac.uk/research-centres/nicr/ (accessed 2010).

9. Quinn M, Cooper N, and Rowan S (2005) *Cancer Atlas of the United Kingdom and Ireland 1991-2000*. Office for National Statistics, Palgrave Macmillan, Basingstoke.

10. Office for National Statistics Mortality Statistics: Cause. England and Wales 2007 London: http://www.statistics.gov.uk/statbase/product.asp?vlnk=618 (accessed 2010).

11. Scottish Executive Health Department (2001) *Cancer Scenarios: an aid to Planning Cancer Services in Scotland in the Next Decade*. The Scottish Executive, Edinburgh.

12. Bach PB, Silvestri GA, Hanger M, and Jett JR (2007) Screening for lung cancer: ACCP evidence-based clinical practice guidelines, 2nd edition. *Chest* **132** (3 Suppl.), 69S-77S.

13. Lubin JH and Caporaso NE (2006) Cigarette smoking and lung cancer: modeling total exposure and intensity. *Cancer Epidemiol Biomarkers Prev* **15** (3), 517-23.

14. Taylor R, Najafi F, and Dobson A (2007) Meta-analysis of studies of passive smoking and lung cancer: effects of study type and continent. *Int J Epidemiol* **36** (5), 1048-59.

15. Hoffmann D and Hoffmann I (1997) The changing cigarette, 1950-1995. *J Toxicol Environ Health* **50**, 307-64.

16. Samet JM (1989) Radon and lung cancer. *Journal of the National Cancer Institute* **81** (10), 745–58.

17. Darby S, Whitley E, Silcocks P, *et al.* (2005) Radon in homes and risk of lung cancer: collaborative analysis of individual data from 13 European case-control studies. *BMJ* **330** (7485), 223.

18. Alberg AJ and Samet JM (2003) Epidemiology of lung cancer. *Chest* 2003 **123** (1 Suppl.), 21S–49S.

19. National Institute for Clinical Excellence (2005) Lung cancer. The diagnosis and treatment of lung cancer, in Clinical Guidance 24: http://www.nice.org.uk/CG24 (accessed 2010).

20. Rami-Porta R, Crowley JJ and Goldstraw P (2009) The revised TNM staging system for lung cancer. *Ann Thorac Cardiovasc Surg* **15** (1) 4–9.

Chapter 11

1. Kocjan G, Chandra A, Cross P, *et al.* (2009) BSCC Code of Practice—fine needle aspiration cytology. *Cytopathology* **20** (5), 283–96.

2. Ribeiro A, Vasquez-Sequeiros E, Wiersema LM, *et al.* (2001) EUS-guided fine-needle aspiration combined with flow-cytometry and immunocytochemistry in the diagnosis of lymphoma. *Gastrointest Endosc* **53**, 485–91.

3. Takahasi K, Yamao O, Okubo K, *et al.* (2005) Differential diagnosis of pancreatic cancer and focal chronic pancreatitis by using EUS-guided FNA. *Gastrointest Endosc* **61**, 76–9.

4. Non-operative Diagnosis Subgroup of the National Coordinating Group for Breast Screening Pathology (2001) *Guidelines for Non-operative Diagnostic Procedures and Reporting in Breast Cancer Screening.* NHSBSP Publication No. 50. Non-operative Diagnosis Subgroup of the National Coordinating Group for Breast Screening Pathology.

5. Perros P (ed.) (2007) *British Thyroid Association, Royal College of Physicians. Guidelines for the Management of Thyroid Cancer,* second edition. Report of the Thyroid Cancer Guidelines Update Group. Royal College of Physicians, London.

6. Redman R, Zalaznick H, Mazzaferri EL, and Massoll NA (2006) The impact of assessing specimen adequacy and number of needle passes for fine-needle aspiration biopsy of thyroid nodules. *Thyroid* **16** (1), 55–60.

7. O'Donnell ME, Salem A, Badger SA, *et al.* (2009) Fine needle aspiration at a regional head and neck clinic: a clinically beneficial and cost-effective service. *Cytopathology* **20** (2), 81–6.

8. Batsakis JG, Sneige N, and el-Naggar AK (1992) Fine-needle aspiration of salivary glands: its utility and tissue effects. *Ann Otol Rhinol Laryngol* **101** (2 Pt 1), 185–8.

9. Alsohaibani F, Girgis S, and Sandha GS (2009) Does on-site cytotechnology evaluation improve the accuracy of endoscopic ultrasound-guided fine-needle aspiration biopsy? *Can J Gastroenterol* **23** (1), 26–30.

Chapter 13

1. Parker EM, Foti JA, and Wilbur DC (2004) Focalpoint slide classification algorithms show robust performance in LBC slides. *Diag Cytopathol* **30**, 107–10.

2. Ayivi J, Cas F, Paugam M, Prat JJ, and Bergeron C (2010) Impact of the implementation of ThinPrep®, an automatic liquid based cytology system, replacing manual LBC technology in a leading French laboratory (Poster P2-010). Laboratoire Cerba, Cergy Pontoise, France. 17th International Congress of Cytology, Edinburgh, Scotland, 16–20 May 2010. *Acta Cytol* **54**, 481.

3. Halford JA, Batty T, Boost T, *et al.* (2009) Comparison of the sensitivity of conventional cytology and the ThinPrep® Imaging System for 1,083 biopsy confirmed high grade squamous lesions. Cytology Department QML Pathology Brisbane, Australia. *Diagnostic Cytopathology.*

4. Bolger N, Heffron C, Regan I, *et al.* (2006) Implementation and evaluation of a new automated interactive image analysis system. Dept of Cytopathology, Coombe Hospital, Dublin, Ireland. *Acta Cytol* **50** (5), 483–91.

5. Boost T (2009) A comparison of screening times between the ThinPrep® Imager and conventional cytology. Cytology Department Queensland Medical Laboratory Australia. *Diag Cytopathol* **37** (9), 661–4.

6. Davey E, d'Assuncao J, Irwig L, *et al.* (2007) Accuracy of reading LBC slides using the ThinPrep® Imager compared with conventional cytology: prospective study. School of Public Health, University of Sydney, Australia. *BMJ* **335** (7609), 31.

7. Passamonti B, Bulletti S, and Camilli M (2007) Evaluation of the FocalPoint GS system performance in a an Italian population based screening of cervical abnormalities. *Acta Cytologica* **51** (6), 865–71.

8. Kitchener HC, Almonte M, Thomson C, *et al.* (2009) HPV testing in combination with liquid-based cytology in primary cervical screening (ARTISTIC): a randomised controlled trial. *Lancet Oncology* **10** (7), 672–82.

9. Franco EL, Bosch FX, and Cusick J (2007) Preventing cervical cancer: unprecedented opportunities. *Outlook* **23** (1).

10. Bosch FX, Castellsagué X, and Sanjosé S de (2008) HPV and cervical cancer: screening or vaccination. *British Journal of Cancer* **98** (1), 15–21.

11. Littlejohn P (2006) Trastuzumab for early breast cancer: evolution or revolution? *Lancet Oncology* **7** (1), 22–23.

12. Bergeron C, Ordi J, Schmidt D, Trunk MJ, Keller T, and Ridder R (2010) p16 conjunctive p16 INK 4a testing significantly increases accuracy in diagnosing high grade cervical intraepithelial neoplasia. *American Journal of Clinical Pathology* **33** (3), 395–406.

13. Carozzi F, Confortini M, Dalla Palma P, *et al.* (2008) Use of p16-INK4a overexpression to increase the specificity of HPV testing: a nested substudy of the NTCC randomised controlled trial. *Lancet Oncology* **9**, 937–45.

14. Ikenberg H, Griesser H, Angeloni C, Bergeron C, Dachez R, and Puig-Tintore LM (2010) Ki67 diagnostic performance of p16/Ki67 dual stained cytology for the detection of cervical disease. Results from the Pan European 27,000 Women PALMS trial. 17th International Congress of Cytology, Edinburgh, Scotland, 16–20 May 2010. *Acta Cytol* **54**, 410.

15. Malinowski DP (2005) ProExC Molecular diagnostic assays for cervical neoplasia: emerging markers for the detection of high grade cervical disease. *Biotechniques* **38**, S17–S23.

16. Schmitt F, Longatto-Filho A, Valent A, and Vielh P (2008) Molecular techniques in cytopathology practice. *Journal Clinical Pathology* **61**, 258–67.

17. Nonogaki S, Wakamatsu A, Filho AL, *et al.* (2005) Molecular strategies for identifying human papilloma infection in routinely processed samples: focus on paraffin sections. *J Low Genit Tract Dis* **9** (4), 219–24.

18. Bubbendorf L, Grilli B, Sauter G, Mihatsch MJ, Gasser TC, and Dalquen P (2001) Multiprobe FISH for enhanced detection of bladder cancer in voided urine specimens and bladder washings. *American Journal of Clinical Pathology* **116**, 79–86.

19. Zielinski SL (2005) Study Examines Role of EGFR Gene Mutations in Lung Cancer Development. *J Natl Cancer Inst* **97**, 325.

20. Ciardiello F (2008) EGFR antagonists in cancer treatment. *New England Journal of Medicine* **358**, 1160–74.

21. Andersson S, Wallin K-L, Hellström A-C, *et al.* (2006) Frequent gain of the human telomerase gene TERC at 3q26 in cervical adenocarcinomas. *British Journal of Cancer* **95**, 331–8.

22. Reis-Filho JS and Schmitt FC (2005) Fluorescent in situ hybridisation, comparative genomic hybridisation and other molecular techniques in the analysis of effusions. *Diagnostic Cytopathology* **33**, 294–9.

23. Nififorova MN (2009) Molecular diagnostics and predictors in thyroid cancer. *Thyroid* **19** (12), 1351–61.

Index